CAMBRIDGE EARTH SCIENCE SERIES

Editors:
A.H. Cook, W.B. Harland, N.F. Hughes,
J.G. Sclater

A history of Persian earthquakes

A history of Persian earthquakes

N.N. AMBRASEYS & C.P. MELVILLE

Department of Civil Engineering, Imperial College, London

CAMBRIDGE UNIVERSITY PRESS

Cambridge

London New York New Rochelle

Melbourne Sydney

CAMBRIDGE UNIVERSITY PRESS
Cambridge, New York, Melbourne, Madrid, Cape Town, Singapore, São Paulo

Cambridge University Press
The Edinburgh Building, Cambridge CB2 2RU, UK

Published in the United States of America by Cambridge University Press, New York

www.cambridge.org
Information on this title: www.cambridge.org/9780521241120

First published 1982
This digitally printed first paperback version 2005

A catalogue record for this publication is available from the British Library

Library of Congress Catalogue Card Number: 81-15540

ISBN-13 978-0-521-24112-0 hardback
ISBN-10 0-521-24112-X hardback

ISBN-13 978-0-521-02187-6 paperback
ISBN-10 0-521-02187-1 paperback

Contents

Preface

Our choice of title for this book deliberately echoes Davison's *A History of British Earthquakes*, which was published by the Cambridge University Press early in the 1920s. This was one of the first in recent times to make a systematic study of seismic activity in a particular country, which we have tried to emulate for Iran. By calling our study a history we wish also to emphasise the importance of time in the unfolding of geologic processes, and of investigating the past when attempting to understand the present.

It was not clear in advance just how much the study of historical events could lead to a better understanding of the generic cause of earthquakes, the processes of continental deformation, and of earthquake risk. The benefit of being able to refer to observations over a period more than ten times longer than the eighty years that have elapsed of this century, however, was obvious. A striking illustration of the likely value of such historical data, and one that gave a germinal impulse to retrieving them, came simply from comparing two maps of world seismicity. The first (figure 1) was compiled in the mid-1800s by a painstaking solitary scientist, and the second (figure 2) was compiled in the mid-1900s, the result of a multi-million dollar effort by a group of seismologists. Both the similarities and differences between these maps show that the former was anticipatory of later 'discoveries'. Depicted solely on the basis of pre-instrumental, historical data, one can see on the 1857 map almost all the plate boundaries we know today. The data used to construct this map are as crude as the hypothesis or theory of plate tectonics that makes one look for such boundaries. However, on the same map one can also clearly see seismically active regions, such as the Jordan Rift and Eastern Anatolian Thrust zones, as well as Eastern China, these being shown as almost totally inactive on the twentieth-century map.

There is more to be seen in Mallet's map than appears to be there at first sight. In particular, it shows the results of interdisciplinary research that can come to fruition not through the agency of a national or international committee for planning or financing research in global tectonics and seismicity (which would probably cause the project to flounder by setting up unimaginative constraints such as a time limit), but by the efforts of dedicated individuals such as Mallet and a few like him, in the days when one had time and was able to read and write in languages other than Fortran.

The need to test observations of short-term seismicity against longer-term trends identified from historical studies has long been recognised and partially fulfilled by previous investigators. If it is easier to criticise rather than praise their efforts, it should be said that they laboured under a disadvantage in that much essential original material was not readily accessible to them. To avoid this drawback, the present study draws on several specialist disciplines to investigate most factors contributing to an identification, assessment and analysis of earthquake data throughout Persian history. The method of interpretation of historical earthquakes in terms of modern concepts evolved gradually in the process of this work, and what is presented here may serve as a starting point for the development of a method of multi-disciplinary study. The fact that we happened to concentrate on Persia, which is not the easiest region available for investigation, is immaterial. Much that emerged should prove relevant to other parts of the world as well.

The term 'Persia' is used to underline the scope of our

intentions, for although modern Iran forms our central interest, its present political boundaries were not fixed until the mid-nineteenth century. It is desirable to look beyond these boundaries, not only because it is unrealistic to confine an investigation of geologic processes to such artificial limits, but also because Persia itself once extended far beyond them. The field of Persian history at various times in the past has stretched northwest into Armenia and Georgia (now in Turkey and the USSR), east into Afghanistan and northeast up to the Oxus and the lands beyond Bukhara and Samarqand. These bordering regions thus come within the wider sphere of Iran's historical past and information in varying quantities is available in Arabic, Persian and other oriental sources.

For our present purposes the area thus generally defined as Persia is delineated by the 24th and 40th degrees of north latitude and the 44th and 66th degrees of east longitude (see figure 3, which also shows the physical topography of the region). Selection of this area permits discussion of any events affecting though not originating in Iranian territory, which contribute to the seismicity of the country. It is also determined by the fact that further extension to the west, for which a wealth of data is available, would more than double the volume of an already over-large book; while to the east, the genuine lack of data concerning earthquakes is considered to be significant and therefore needs to be demonstrated.

Although scattered indications of earthquake effects go

Figure 1. Map of the world showing the distribution of earthquakes, prepared by Robert Mallet on the basis of historical data and presented to the British Association in 1857 (28th Report of the British Association, 1858). Note the similarity of this map to the map of short-term, modern seismicity prepared on the basis of modern instrumental data shown in figure 2.

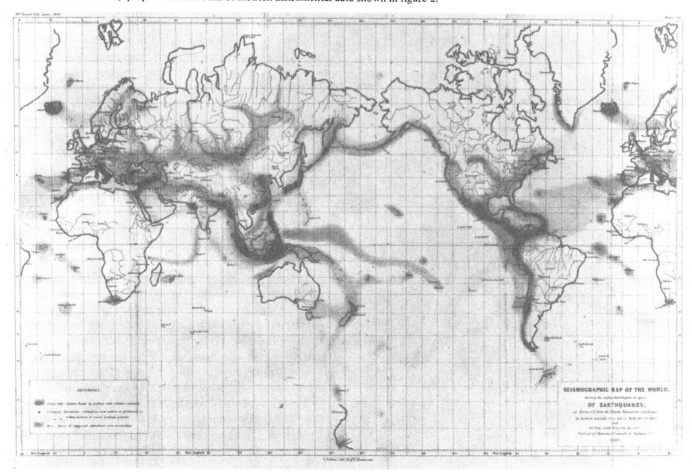

back as far as the third millennium B.C., adequate documentary coverage of individual events does not begin until the advent of the Islamic period in the seventh century A.D. Chapter 1 discusses the various sources that have been found to contain information, and relates their record of events to the prevailing historical circumstances. We can thus form some idea of the completeness of our data, and appreciate how various factors have influenced the distribution of earthquakes recorded prior to the twentieth century.

Non-instrumental, descriptive data (macroseismic data) are retrieved from a variety of documentary materials and also from direct observations made in the field. Chapter 2 deals with the type of information contributed by field studies, which have involved collecting oral or literary data about local earthquakes and also first-hand investigation of regional earthquake effects. This information in turn provides a practical context within which early and modern events should be discussed on a uniform basis. It also permits creation of yardsticks against which they can be classified.

The central chapter presents a description of the largest and most interesting earthquakes that have been identified up to 1979, utilising all the macroseismic data available to us. These accounts illustrate many of the concepts previously outlined. They also form the basis for later theoretical analysis, large magnitude earthquakes being not only the most important from the human point of view, but also the most informative to the earth scientist or earthquake engineer.

Instrumental recordings of earthquakes began around the

beginning of the present century, and have subsequently come to yield an accurate supply of precise technical data. Chapter 4 describes the early development of the seismograph network around Persia and the problems caused by deficiencies in the quality of its resources. Early epicentral locations and subsequent attempts at relocation are shown to remain generally inferior compared with those based on macroseismic data. In view of the non-homogeneity of existing magnitudes for earthquakes in Persia before the early 1960s, magnitudes had to be re-calculated uniformly for the whole period. The problem of estimating magnitude values for events for which instrumental data are lacking or inadequate is then examined, and an approach is made to assessing magnitude as a function of other parameters. As these can generally be determined from macroseismic data, it is possible to assign magnitudes to historical events.

In the process of acquiring and classifying this information, a considerable number of new data have emerged. Many hundreds of epicentral locations and magnitudes have been re-calculated or assessed, which, together with other information on damage and Intensities, we intend to publish separately.

The range and type of data assembled allow one to proceed confidently towards an analysis of the seismicity of Persia. This is done largely in terms of the characteristics of earthquake occurrence in certain broadly defined zones that can be identified as coherent units. No attempt is made, however, to use the data for mapping seismic hazard in Iran or for the

Figure 2. Worldwide distribution of instrumentally determined locations of earthquakes for 1961–7. Compare with the figure 1 map, prepared more than a century earlier. (Barazangi & Dorman 1969.)

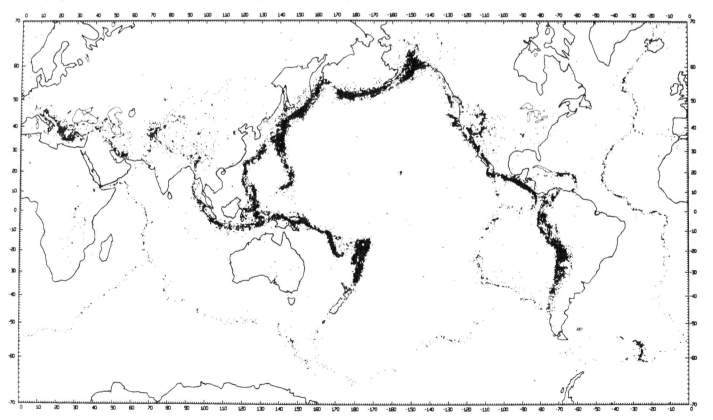

preparation of regional vulnerability analysis. This is considered to be the task and responsibility of the relevant Iranian authorities.

One disadvantage of an inter-disciplinary study lies in the demands it makes on its audience. While pursuing a specific goal, it nevertheless draws, sometimes perfunctorily, on the resources of several different fields of learning, and in so doing may satisfy none of them completely, thus falling between however many stools are represented. Most seismologists are not linguists as well; most historians are not also engineers. Examination of earthquakes within their historical context means that many of the data and deductions presented in this work should be of interest to the orientalist, historical geographer or sociologist, as well as to the earth scientist and engineer. The emphasis throughout the book swings from one side to the other. Nevertheless, we hope that an overall balance has been maintained, and we have tried to present the investi-

gation in such a way that contributing facts and ideas are consistently accessible and intelligible to whoever cares to follow them through, given of course a certain initial interest and subsequent concentration. With this in mind, the text has been kept as free as possible from technical jargon, and much that would have over-burdened or distracted attention from the flow of the exposition has found a place in the footnotes, for the length and number of which we make no apology. Even though we have been unable to resolve all the questions to which we hoped this inter-disciplinary study would provide a solution, we nevertheless feel that the curiosity and enthusiasm with which the work was started has produced some rewarding results.

N.N. Ambraseys
C.P. Melville

London
1981

Figure 3. Physiographical map of Iran (*Camb. Hist. Iran*: i (1968), p. 1).

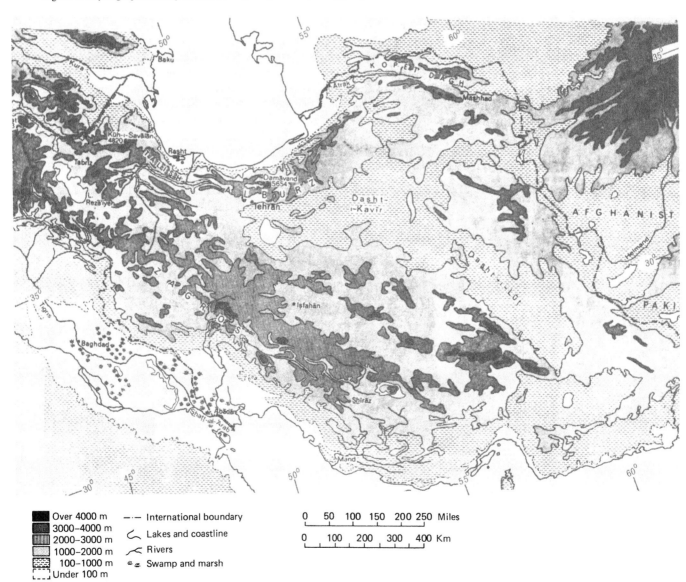

■ Over 4000 m	—·— International boundary
▨ 3000–4000 m	⌒ Lakes and coastline
▥ 2000–3000 m	⌇ Rivers
▢ 1000–2000 m	⁂ Swamp and marsh
▨ 100–1000 m	
⌐⌐ Under 100 m	

0 50 100 150 200 250 Miles

0   100   200   300   400 Km

Acknowledgements

This work has grown over a period of almost two decades and has owed much to the contribution of individual personalities at various crucial stages. It would be impossible to mention here the names of all the people and institutions that have provided information, facilities, or financial assistance over the years. We hope that anyone who has not been mentioned will not suppose that their help has not been appreciated.

Ali Akbar Moinfar has done so much to make this work possible during difficult times that he must be mentioned before any other. The late Professor Sedrak Abdalian and Dr Michael Fournier d'Albe were among those whose help in the early stages of this work contributed more than they realised. In the field, invaluable help was provided by John Tchalenko, and by old Iranian students of Imperial College. Also, we would like to thank Jean Vogt, Jean Aubin, Chahriyar Adle and Anne Kroell in Paris for kindly supplying information and also Norman Falcon and Anthony Hughes in London.

For unpublished Arabic manuscripts thanks are due to Abdel Moneim Omar in Cairo, and to Kevork Bardakjian for his great help with Armenian texts. We would like to thank the Soviet Geophysical Committee which, through our late colleague Evgeni Savarenski, provided us with unpublished consular documents. Professors V. Menage and D.J. Wiseman of the London School of Oriental Studies helped with Ottoman Turkish and Sumerian texts, and Pierre Stahl with reliable seismological data during the first stages of the establishment of a seismological service in Iran. Professor Markus Båth, Dr Vit Kárník, Professor David Vere-Jones and Cina Lomnitz read parts of the text and offered very useful advice. We would like to thank Sandy Morton for always being prepared to have his brain picked and for reading through a portion of the typescript.

Financial support for our work throughout this period was provided by UNESCO, by the National Environment

Research Council, by the Plan and Budget Organisation in Tehran, and by Imperial College of Science and Technology in London.

We are most grateful for the facilities afforded by many libraries, such as the Royal Geographical Society's library in London, the Millikan Library in Pasadena, the Lenin Library in Moscow, and the Public Record Office in Kew.

We also wish to thank Clarice Bates for allowing us to use photographs from the album of her father, the late A.D. Hovhannissian of Tabriz.

Lastly, we would like to thank Cambridge University Press, and in particular Alan Cook, for asking us to write this book, and Corinne Gibbons for her help in preparing the manuscript.

Definitions

Aftershocks are secondary shocks following the main earthquake. An *epicentre* is the point on the Earth's surface verti-
cally above the focus, the location in space where the first
motion occurs. *Focal depth* is the vertical distance between
the focus and the Earth's surface in an earthquake or aftershock. The *focus* is the point within the Earth which marks
the origin of an earthquake (hypocentre). *Intensity* is defined
by a numerical index describing the effects of an earthquake at
a particular location on man, on man-made structures and on
the ground itself. The number is rated on the basis of an earthquake Intensity scale, that in common use being the Modified
Mercalli Scale of 1931 with grades indicated by Roman
numerals from I to XII (Richter 1958: 135, Newmark and
Rosenblueth 1971: 217 and Ambraseys 1973). *Isoseismals* are
the lines drawn as boundaries between regions of successive
Intensity ratings. The adjective *macroseismic* denotes information or data acquired without the help of instruments. The
magnitude of an earthquake is a measure of the energy
released by the shock. Finally, the *meizoseismal* or *epicentral
region* of an earthquake is defined as the region of intense
shaking in the near-field of the event, usually within the isoseismal of highest Intensity.

Transliteration

In the course of our research we have had to read books written in a number of languages with a non-Latin script. Transliteration from these languages is normally designed to reproduce the symbols of the original, but this can lead to a text that is daunting to look at and difficult to read. The need for a system of transliteration arises primarily in the case of spelling place names and, to a lesser extent, other proper names.

As a fundamental principle, place names throughout the book are spelt as they are written in the script of the country in which they are located; this generally involves a transliteration. Because of the fluctuating demarcation of boundaries over the long period covered, and because places formerly in Armenia or now in Turkey and Russia were once under Persian administration, it is in practice necessary to refer to places as they were known in the historical context in which they are cited. Their equivalents are given as identified, along with other modern names conforming to their current indigenous spelling. Particularly in the northwest of our area, some names are given a standard spelling, chosen arbitrarily for its familiarity, such as Erivan (for Yerevan) and Tiflis (for Tbilisi).

E.G. Browne once observed that it is both easier and more philosophical to transliterate on a fixed and definite principle than to decide in each case whether a given spelling has or has not been sanctioned by usage. A rigid system for the transliteration of Arabic and Persian has only been followed, however, in the bibliography, where names of authors and titles of books have been given in accordance with the system used in the *Cambridge History of Iran*. Persian variant forms of Arabic consonants are used only if they occur in a place name or personal name. In the body of the text, this system has been greatly relaxed, notably by the omission of macrons and diacritics. This in turn makes a number of consonants indistinguishable from one another, but this is not likely to bother a specialist. Persian variant forms generally approximate

Persian pronunciation, thus Faizabad (for Arabic Faiḍābād), although the tendency has been to retain the Arabic 'th' throughout (pronounced 's' in Persian). Place names in Iran are spelt consistently on the basis of such a transliteration from the Persian. On the other hand, common Islamic terms such as Ramaḍān or Qāḍī retain their basic Arabic spelling, thus Ramadan, Qadi. The absence of macrons makes long and short vowels also indistinguishable, the vowels 'o' and 'e' generally being avoided, though the latter is invariably used before the Persian 'silent h', which is given as 'eh'. This usage has spilled over into words where the h is not silent, such as Tehran and *deh* (village), the latter being correctly spelt only in the word *dihistan* (village district). Other minor inconsistencies, firmly supported by modern usage, have escaped the pedantic axe, such as the spellings 'Herat', 'Mosul' and 'Melli'. Similarly, it is obviously out of the question to transliterate correctly the names of Iranian authors who themselves spell their names in English at variance with our own system. This last observation applies to all bibliographical references using Latin characters in the original, these merely being reproduced as they stand.

Transliteration from Turkish presents additional problems, thanks to the change from Arabic to Latinised script instituted by Atatürk in 1928. Modern Turkey impinges only marginally on our area, and in deference to the wider geographical and historical scope of the book, Ottoman Turkish, used for the bulk of the period covered, is transliterated effectively as though it were Persian. The system used in the *Redhouse Dictionary* (Istambul 1968), which has both Arabic and Latinised characters, is thus not followed where the Arabic or Persian form of names and titles is more familiar. In the absence of diacritics, ç is written 'ch' to distinguish it from c (Persian 'j') and 'sh' and 'kh' are preferred to ş and h, when reproducing both Ottoman and modern Turkish script; thus we have Ja'far for Cafer, Chelebi for Çelebi, shaikh for şeyh, pasha for paşa, *tarikh* for *tarih*, etc. Place names in Turkey may thus be modified slightly from spellings found on modern maps, which are used for reference in conjunction with H. Kiepert's map, *'Provinces asiatiques de l'Empire Ottoman'*, published in 1884, on a scale of 1:1 500 000.

Kiepert's map also provides a reference for place names in Armenia and Georgia. In addition, Armenian place names found in contemporary historical accounts are spelt as in Hübschmann, *Die altarmenischen Ortsnamen* (Strasburg 1904), with an accompanying map. Hübschmann's standard academic transliteration, which uses various Latin characters artificially, is unintelligible to a general reader and names have therefore been transcribed phonetically according to the scheme in Gulbekian, 'A phonetic transcription from Armenian to English', *Ararat*: II (New York, Summer 1961). The same transcription has also been used for personal names.

Russian place names reported in Latin characters are adopted directly from the original text. Names of people and titles of books are similarly reproduced if given originally in Latin script; transliteration from the Cyrillic script follows the system in L.I. Callaham, *Russian—English Technical and Chemical Dictionary* (New York 1961).

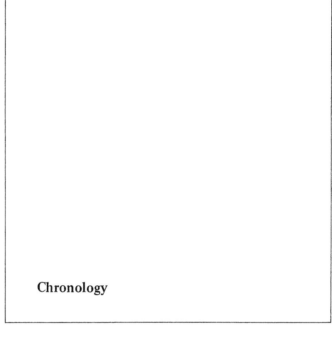

Chronology

Several calendars have been used to date the earthquakes recorded in Persia, notably the Seleucid (Sel.), Armenian (Ar.), Muslim *hijra* (H.) and Persian *shamsi* (Sh.) calendars. Details of these, and others less frequently employed, may be found in V. Grumel, *Traité d'études Byzantines, I: La chronologie* (Paris 1958), which has been used to convert dates to the Christian era.

The Persian solar year begins in March, conversions involving the addition of 621 years to the *shamsi* date for the first nine months of the year and 622 for the last three: thus Farvardin 1350 falls in 1971, but Isfand 1350 in 1972.

The Muslim *hijra* calendar is lunar and began on 16 July, A.D. 622. Conversions are calculated from the tables of H.G. Cattenoz, *Tables de concordance des Ères Chrétienne et Hégirienne* (Rabat 1961). Where a conversion is given, the Muslim year comes first, e.g. 704/1304; the Christian year is the one that forms the longest part of the Muslim year in case of overlap, unless it can be established that the other is more accurate.

Throughout the book, dates are given A.D. unless otherwise indicated, as by one of the suffixes noted above. However, Persian *shamsi* and Muslim *hijra* years are not suffixed when the month is also given. For this reason, the months of the two calendars are set out below, for easy reference and to encourage familiarity with their sequence:

Persian solar months	*Muslim lunar months*
1. Farvardin	1. Muharram
Urdibihisht	Safar
3. Khurdad	3. Rabi' I
Tir	Rabi' II
Murdad	Jumada I
6. Shahrivar	6. Jumada II
Mihr	Rajab
Aban	Sha'ban
9. Azar	9. Ramadan
Day (Daimah)	Shawwal
Bahman	Dhu 'l-Qa'da
12. Isfand	12. Dhu 'l-Hijja

It should be obvious from the context which calendar is being used, for instance in references to issues of newspapers or other Persian journals.

In England the change from Julian or Old Style to Gregorian or New Style dating took place in September 1752. We should note that use of the Old Style system in some of the European sources consulted, particularly Russian, persists up till as late as the first decade of the twentieth century.

1

Macroseismic data from historical sources

1.1 Description and evaluation of documentary source material

It is self-evident that any work based on documentary research can only be as comprehensive as its sources allow. Limitations on the factual data available determine the thoroughness of the investigation and the value of the conclusions that are possible. The objects of our concern, earthquakes, are specific events whose occurrence is significant: but so equally is their apparent lack of occurrence. The completeness of our information therefore assumes a great importance, and this in turn is the main burden on our sources and our responsibility in their interpretation.

Although seismologists are aware of the value of historical data and alert to their inherent limitations, the effect of these limitations is seldom examined systematically. Clearly a number of chance factors influence the survival of data, not least being the chance survival or destruction of documents containing information. Other factors are more constant and they must be investigated before we can assess how complete and representative a sample of seismic activity has been recorded, both in terms of its distribution (geographical and temporal) and its apparent intensity.

Fortunately, Persia has a relatively well documented history and a variety of source materials are to hand. The characteristics of these works, where relevant to their value as sources of macroseismic data, are noted in the course of this chapter, which aims to indicate the extent of the material made available and to discuss its suitability for our purposes. It is not necessary to describe works individually; their comparative merits and defects emerge from the use made of them. The same applies to secondary sources, such as specialist studies on Iran's history, geography and archaeology, or scientific publications. Many of the problems associated with the sources are dealt with in the endnotes to chapter 3.

1.2   A perspective on historical data

The transmission and survival of macroseismic data depend very largely on historical or geographical circumstances, which are not necessarily consistent for all regions over different periods of the past. We must therefore review our data in the light of the circumstances in which they were recorded. Despite its literary wealth, we are dealing with a long span of history of a society that has remained, until recently, essentially static in comparison with western Europe and sporting a low level of literacy: this is a factor for continuity. The vast extent of the country and its peculiar physical characteristics have served to make the different regions more or less isolated from each other, but linked by routes predetermined by natural features and thus of high antiquity. Similarly, the local urban centres have often played an important and independent role in the unfolding of events, in a region whose history has been turbulent, violent and subject to sudden change; earthquakes have only caused some of the scars on the battered features of the record of Persian history. In the analysis that follows some importance is attached to the role both of the cities and of the routes in the survival of data. This role may be formulated by analogy with modern seismographic stations: we need to be aware of their location and sensitivity and of how adequate is the publication of their records. Furthermore, we must look beyond the individual stations to the characteristics of the whole network, not only with regard to the distribution and sensitivity of the instruments, but to whether they report individually or transmit their data to a central organisation for processing. Finally, we have to know whether the stations have operated continuously, or only at certain periods.

As for the cities, it is an assumption, borne out by modern experience, that the larger towns are the main sources of information about earthquakes and that events occurring within the immediate vicinity of such a town are likely to be recorded, while those happening in remoter regions may well remain obscure. Thus it has been observed that the distribution of earthquakes reflected in historical sources is often closely related to the distribution and density of settled population, and not necessarily a function of the magnitude of the shocks themselves. The bias in volume of information available for the towns as against outlying rural areas may not only distort the picture of an individual event and the true location of its meizoseismal area, but also in more general terms, affect the apparent pattern of seismic activity throughout the whole region. For the period in which we rely exclusively on macroseismic data, this is clearly an important consideration.

The cities were linked by routes that were loosely defined, not restricted by a road surface but only by a series of fixed points. The links with the surrounding countryside, both at and in between stages, were close: the scope for exchanges along the route therefore correspondingly wide, unlike a modern motorway that is detached from the land it crosses. In addition, there were many sections of parallel or multiple routes, suitable for travel at different seasons, or for animals or activities of different types, for pasturage, trade or more rapid communications. This busy network facilitated the oral spread of news, albeit slowly, depending on its importance. In the case of an earthquake, which can have considerable impact on a local level, perhaps with wider repercussions, the spread of details reflects the number of people affected or interested. Whether or not the news was recorded in writing (and thereby given a better chance of survival) is a function of both the size of the earthquake and also its location, depending primarily on the geographical proximity of an urban centre to the epicentral region. The record may then become part of the local history of the district, provided a local historiographical tradition exists; this is generally associated with its political independence. In cases where details of an earthquake have not survived in local histories, but in more general works written elsewhere, this also reflects the relative importance of the town or region concerned and the ease with which news of it has travelled. This is why earthquakes near Ray (743) were recorded by authors in Constantinople, and why events near the Oxus (819) were noted in Baghdad or those in the Persian Gulf (1497) were of interest in Cairo.

We therefore expect macroseismic data to be more readily available for places situated along the major routes and particularly the cities at the termini or intersections of the route system. As the urban centres and the routes between them fluctuated in importance, their political and commercial fortunes responding to historical developments, we must be aware of these changing circumstances (and their influence on the distribution of recorded earthquakes) throughout the period under survey.

It is possible to divide this period into four sections of unequal length, on the basis of the predominant type of source material available. The first, from the seventh to the mid-thirteenth century, is defined by the fact that almost all our data derive from *Arabic* sources. The second division, up to the end of the sixteenth century, is in marked contrast to the first, *Persian* works becoming the main source of information. In the third division, covering the seventeenth and eighteenth centuries, *European* sources, particularly contemporary travellers' accounts, provide increasingly valuable evidence of earthquake occurrence in Persia; and finally, British *diplomatic archives* and European and Persian *newspapers* make available a very comprehensive sample of data from the nineteenth century onwards. The review ends around 1925, a convenient date marking the fall of the Qajars and the start of Iran's lurch into the modern world under the Pahlavis. Instrumental data are of course available before this date, but they are still unreliable and macroseismic data continue to be invaluable. There is obviously a measure of continuity between these periods, each standing for the addition of a new category of source material rather than the replacement of one type by another, so that ultimately all the different groups of sources contribute information. Nevertheless, it is convenient to maintain the divisions, as various characteristics of each can also be seen to influence the amount and quality of data that have survived. A brief description of these periods is intended to give a background to the occurrence of historical earthquakes, and also to illustrate some of the criteria conditioning an analysis of the raw material provided by the sources. There is in most cases a clear correspondence between the distribution of recorded earthquakes and prevailing historical circumstances; particular attention should be paid to the factors assisting or prejudicing the survival of data for those areas whose historical seismicity is inadequately known, as they throw light on the gaps in the record.

The state of the Muslim authors' understanding of seismic phenomena, which varied from a rational, though incorrect, scientific interpretation to one of superstitious ignorance, is mainly irrelevant to this study. Ample evidence exists of Muslim cosmologists' views on the causes and nature of earthquakes, which reflect the ideas of classical Greek writers, particularly Aristotle, but their discussion is mainly philosophical.[1] A rational viewpoint is seldom applied to the discussion of individual events, a rare example being some of al-Biruni's references to earthquakes (al-Biruni: 20–3). At the other end of the scale, earthquakes were regarded with primitive religious awe and were discussed, occasionally, in purely theological terms. There are hardly any works devoted exclusively to earthquakes, exceptions being the works of al-Suyuti and his continuators, al-Dawudi and al-Shadhili, who cover the period up to 1588, and of al-Jazzar, who was writing in 1576. A much later work by al-Qusi comprises events in all parts of the world up to 1907, but for the early period he gives no information that is not found in the better-known historical sources.[2]

A poor understanding of the nature of earthquakes does, however, inevitably lead to some irrelevancies or confusions in early accounts. This is particularly evident in the tendency to associate the occurrence of an earthquake with some other event, when such a relationship is in fact coincidental. The departure from Iran of Muhammad Riza Shah Pahlavi on 16 January 1979 and the occurrence the same day of a relatively large magnitude earthquake northeast of Qayin, killing a few hundred people, is such a coincidence. Similar associations occur in historical sources, particularly with the death of prominent people, and can often be used to confirm the accuracy of the dates given, though sometimes such correlations merely confuse the issue.[3]

In the same way, but more importantly, earthquakes are frequently reported along with other natural phenomena, such as an eclipse: a recent example of how this might arise is the coincidence of the Tabas earthquake of 16 September 1978 and a total eclipse of the moon later the same night. One more beneficial result of this type of association of events, particularly common in superstitious societies, is that earthquakes that might otherwise have gone unrecorded are mentioned in the sources. Heightened perception and recording of earthquake activity may thus extend to undamaging shocks or tremors that coincided with other natural phenomena or with important local political events. This factor has to be taken into account when assessing the gravity of the shocks themselves. The collective reporting of such diverse elements is particularly characteristic of Arabic chronicles, to which we may now turn.

It is emphasised that the discussion throughout is concerned only with sources that have actually been read, and not with works that may strictly speaking be available but have not in fact been used by the present writers.

1.3   The Caliphate period (622–1258)

This is more precisely defined as the early Islamic period, from year 1 of the Muslim era up to the sack of Baghdad by the Mongols in 1258 (the pre-Islamic period is treated separately, see below, § 3.2). The chief characteristic of this long period that allows it to be taken as a whole is the fact that Persia and Iraq were part of a unified empire, even if by the end the unity was only theoretical. Iraq being the heartland of this empire, almost all our information about earthquakes comes from Arabic sources, mainly historical chronicles. Very little has survived of native Persian works and their contribution to our data is small.

The systematic treatment of events in Arabic annals gives the data for these centuries a certain uniformity. Earthquakes are recorded factually and, because of the repetitive nature of the annalistic style, usually by a number of sources. The later chronicles generally provide an accumulated record of all previous events, certainly the most important ones. Of these works, the most notable is that of Ibn al-Jauzi (lived in Baghdad, d. 1200), who provides a comprehensive and invariably detailed record of events, forming the basis for most later compilations, such as that of al-Suyuti (of Cairo, d. 1505). The preservation of often summary data in a stereotyped format by generations of annalists promotes the survival of information, while removing much of its immediacy. Earthquakes are often reported baldly, along with eclipses, comets, shooting stars, floods, famines and plagues as 'events'. The only form of embroidery is provided by occasional suggestions of the supernatural at work, with stories of other freak phenomena, resembling much of the 'damned' data collected by Charles Fort (1973). The joint description of earthquakes along with other phenomena, such as meteorite falls,[4] strong winds, hail or thunderstorms, can give a confusing impression of the destructiveness of the shock itself;[5] nor is it always certain that the different effects were indeed simultaneous. Similarly some Arabic authors, such as Ibn al-Athir (of Mosul, d. 1233), often describe different earthquakes together in a collective account of all the events in a year, making it difficult to disentangle their separate effects, their sequence and the areas over which individual shocks were experienced. These defects are small, however, beside the overall thoroughness and regularity of reporting of earthquakes by Arabic historians; all positive statements, however inadequate, are of value and can be assessed critically (Melville 1978: 184–94).

There are three broad subdivisions in the period we are considering. Very little information has survived from the earliest period, partly doubtless because of its antiquity, but mainly for the lack of a pre-existing tradition of historical writing, which took time to emerge. The Byzantine model was adopted, as in many other fields, and Byzantine annals themselves have some data for this early period. The shift of capital from Damascus to Baghdad in 763 was of great importance for the re-emergence of the Iranian plateau from its comparative obscurity. At the same time, centralisation of the empire at

Baghdad made it the ultimate destination of all important news from the provinces, supplied by merchants or the official postal and intelligence system. From the end of the eighth century all major routes emanated from, or rather led to Baghdad, serving the commercial and political needs of the capital. Authors in Iraq were thus well placed for access to information; and Arabic was the dominant vehicle for all forms of expression and cultural evolution. This period of expansion, of comparative security and stability, encouraged the development of prosperous commercial centres and supported a large, predominantly settled population. All these circumstances were conducive to the survival of macroseismic data.

Political fragmentation of the empire began as early as the ninth century and was established fact by the eleventh, when the first wave of nomadic invaders swept from the east across Persia. From the mid-eleventh century onwards, various branches of the Turkish Saljuqs dominated Iran. There was

perhaps an increase in nomadism and a greater separation of the different regions of the country, but the underlying structure and coherence of the empire provided a thread of continuity until the Mongol invasions in the early thirteenth century.[6]

The distribution of earthquakes recorded in the Caliphate period closely reflects these conditions. Figure 1.1 shows the location of places mentioned as having been affected by earthquakes, the number of times this occurred and their relationship with the main routes of the period. It is most striking that almost without exception, the places named are directly situated on one of the arteries to the heart of the eastern Islamic world — or so near one as to be effectively within the route's catchment area of news and information. The network is drawn on the basis of details given by Muslim geographers of the ninth and tenth centuries, as summarised by Le Strange (1905). The figure also indicates the relative

Figure 1.1. The main routes under the Caliphate and the places affected by earthquakes during this period. The figure indicates how many times earthquakes were reported at each place and the close connection of these locations with the main lines of communication. Note the bias towards information for Iraq. The figure does not distinguish individual earthquakes nor their likely epicentral location, for which it should be viewed in conjunction with figure 5.2.

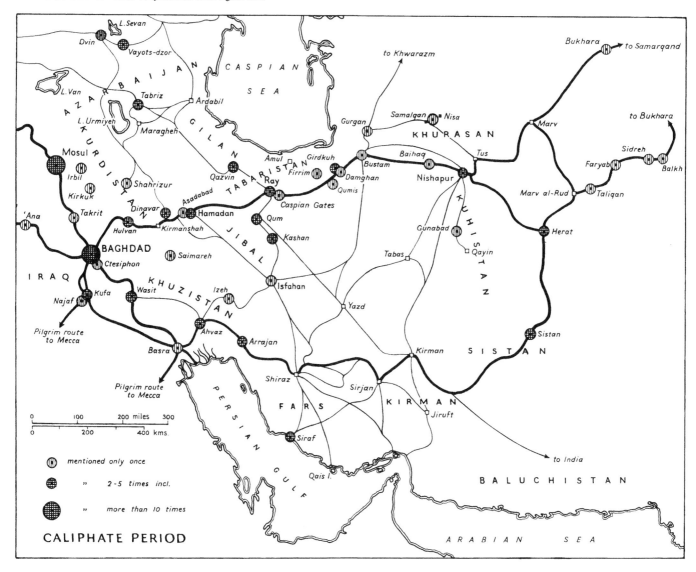

importance of the main regional centres along the way, as loosely defined in terms of their size and political or commercial influence. Such fluctuations in the state of the routes or towns as modify this general picture sufficiently to affect the survival of data, are noted below.

It will be observed that almost all the earthquakes recorded for Persia occurred in the region traversed by the main Khurasan highroad. This was the most important and most frequented of the five highways leading to Baghdad, the caravan route that brought products from China and India along the natural corridor between the foothills of the Alburz and the fringe of the central *kavir*, before turning southwest to cross the Zagros and descend into the Tigris–Euphrates valley. The importance of this route, both for trade and pilgrim travel, remained constant and news of the districts it crossed would be of current interest throughout the period. It seems reasonable to conclude that almost all the events of any significance in the places along the way would have been recorded. The details that have survived, a further stage of natural selection, must be of the most destructive earthquakes, particularly in the major cities of Ray and Nishapur: though not necessarily of the largest magnitude shocks in their respective provinces. Nishapur has a tradition of high seismic activity in the early period, but no details of these events have survived (Melville 1980).

A certain amount of information is available for the region between Ray and Azarbaijan. Qazvin was of some importance as a military centre for operations in the Caspian provinces and from its position on the ancient route across north Persia into Asia Minor. This route was eclipsed, at least till the late eleventh century, by the greater attraction of Baghdad and the Holy Cities beyond, to the southwest, and Qazvin although prosperous was not politically important. Seismic activity in the area is confined to events affecting Ray (864, 1177), the exception (1119) being recorded by a local source. It is unlikely that other events in the region would have been monitored, though the surviving record may give an accurate idea of the frequency of shocks seriously damaging in the town itself. The same may be said of Tabriz, which was of little size or importance until the tenth century. It became capital of Azarbaijan in the eleventh, but continued to share this position with Maragheh and Ardabil throughout the Caliphate period. The lack of macroseismic data for these towns may reflect an absence of genuinely destructive events there, for both were generally more important centres than Tabriz, though the sources for information on Azarbaijan are extremely poor for most of the period. The record of a destructive shock in Tabriz (in 1042) coincides with the passage of the traveller Nasir-i Khusrau along this route, which had again become more international with the spread of the Saljuqs west into Anatolia. Thereafter, though earthquakes in the city were undoubtedly more frequent than can be accurately determined,[7] the chances of other genuinely destructive events not being recorded are small. Another secondary route of some importance completed the triangle Baghdad–Ray–Tabriz, linking the latter with the Khurasan highroad between Hamadan and Kirmanshah, thus passing through Dinavar

(Minorsky 1964: 94). News from Azarbaijan and Tabriz would also reach Baghdad via Mosul (as in 1042), though the geographers do not mention such a route.

In contrast with the Khurasan road, those on the southern skirts of the desert are roundabout tracks, linking up regions of secondary importance (Minorsky 1964: 57). Nevertheless, the Gulf ports and the Tigris–Euphrates valley were busy sources of supply; the close connection of Wasit and Basra with the capital made detailed news readily available there. In such favourable circumstances, the few earthquakes recorded below Baghdad must reflect the low seismicity of the region at this time. In the Persian Gulf, details of events at Siraf (978, 1008) illustrate the influence of historical conditions on the survival of information. Described in the ninth century as the chief emporium for trade with China and India, the port reached a peak of prosperity in the tenth century, rivalling Shiraz and Basra. Al-Mas'udi (d. 956) refers to the high seismicity of this district, but such references cease in the eleventh century, when sources no longer throw light on affairs in the Gulf, which suffered a decline. This was occasioned in part by the collapse of Buyid authority in southwest Persia and also by the successful efforts of the Fatimids in Egypt to divert the Gulf trade into the Red Sea.[8] By around 1110, the island of Qais or Kish had become centre of the Gulf trade and Siraf was totally eclipsed. Although the local network of routes between Shiraz and the Gulf ports saw some changes in the eleventh and twelfth centuries (Aubin 1969: 36), these were of little consequence for the survival of news about the area. This would have reached Baghdad by the sea route, via Basra, as in the past; but the political upheavals and economic decline of southern Persia after the last half of the eleventh century would prejudice the transmission of all but the most extraordinary information.

In the southern Zagros, Shiraz had early on replaced Istakhr as capital of Fars and was developed in the ninth century by the Saffarids (Lockhart 1960: 43). Although the province, and by extension its capital, had no particular interest in the southeast, it did gravitate to the west.[9] It was the longest lived of the three Buyid capitals (the others being Ray and Baghdad), and particularly in the late tenth century, under 'Azud al-Dauleh, great importance was attached to developing land links between Shiraz and Mesopotamia, via Khuzistan. This involved building bridges and improving roads, and Arrajan (near modern Bihbahan) was developed as a major commercial city.[10] Abu Dulaf (*c.* 950) refers to the frequency of earthquakes at Izeh (Malamir), and though no details of these early events have survived, it is no accident that earthquakes there and at Ahvaz and Arrajan (1052, 1085) are recorded in Baghdad in the later Buyid period.[11] As for Shiraz, there is no reference to earthquakes in local sources and this, despite the decline of Fars in the late eleventh and the twelfth centuries, may be taken to indicate that none of any significance occurred in the immediate vicinity of the city. There is no record of any damage to the major buildings erected in the pre-Mongol period.[12]

On the western side of the desert, alternative routes connected Shiraz with Isfahan, whence roads led north to Ray and to Hamadan. From the head of the Gulf, routes through Khuzistan also reached Isfahan — one was followed by Abu Dulaf and another by Nasir-i Khusrau — while ancient tracks throughout Luristan linked the former Sasanian centres and continued much frequented in the Caliphate period (Siroux 1949: 2, 11). Despite this wide network of routes, no information has survived of a major earthquake in the central Zagros (which corresponds roughly to the Jibal province), with the exception of the Saimareh event of 872. It is possible, however, to qualify this apparent seismic quiescence. In the first place, Arab geographers of the tenth century refer to the high seismicity of the Jibal, especially round Hamadan.[13] To the west of the region, Baghdad, which we may regard as a very sensitive organ of perception, was frequently affected by shocks which caused little damage and are likely to have originated in the Zagros. Some of these may have occurred to the north of the Khurasan highroad, in Kurdistan, in which case they may also have been reported in Mosul.[14] For many of these Jibal events, there is no indication of a precise epicentral region or area of maximum damage. It is clear that the excellent record of earthquakes in Baghdad (seventeen in all) is due to historical factors rather than the high seismicity of its position (see figure 1.1). Shocks mentioned in Hamadan are similarly not always destructive there, and may be the result of more distant events. Further to the east, routes trending north—south were of less importance than those following the dominant axis of trade east—west, and their connection with Baghdad was clearly less direct. The towns of Qum, Kashan and Isfahan show little evidence of being affected by earthquakes; on the rare occasions they are mentioned, the connection is with events in the Alburz, not the Zagros (856, 958). Isfahan became particularly important under the Buyids (mid-tenth to mid-eleventh century), when it was on a par with Ray, and the Saljuqs later made it a capital city and created many fine buildings there. The absence of macroseismic data for Isfahan undoubtedly reflects a genuine lack of serious events there, while the chances of destructive shocks in the remoter regions to the southwest (such as Chahar Mahal) being reported either locally or in Baghdad during this period are negligible.

Other, less densely populated regions offer even less evidence of seismic activity. In southeast Iran there is a lack of data for the whole period, particularly noticeable in the Kirman region. This largely reflects the remoteness of Kirman (formerly Bardasir) and the earlier capital, Sirjan, from Baghdad, although under the Saffarids and to a lesser extent the Buyids, affairs in the province were fairly closely connected with those to the west.[15] The region however remained economically behind Fars until the collapse of the Buyids. Kirman then flourished for a long period under a branch of the Saljuqs (1041–1187), enjoying political stability and commercial affluence; it became the centre of a system of routes north—south from the Gulf (the Oman coast and Hurmuz were under Saljuq suzerainty) to the cities of Khurasan, and similarly eastwards to Sistan and Kuhistan (Aubin 1959). Reflecting this independence, local histories of Kirman are available from the twelfth century, the city being rather better represented in this respect than many other Persian towns. The lack of macroseismic data would seem to suggest that no significant event affected the city itself, as opposed to outlying regions: to the north, local oral tradition preserves the account of an earthquake in the twelfth century in the Kuhbanan district.[16] Local histories of Kirman concentrate on the deeds of the ruling families rather than purely local affairs and it may be that for some reason Kirmani authors were not interested in earthquakes.[17]

Eastwards to Sistan, such information as we have derives, significantly (and for this period almost alone), from a local source. Although apparently more prosperous and more populous in the middle ages than is now the case (Tate 1910), the area was nonetheless remote. Mediaeval geographers give few details about the province, which was connected to Herat and the towns of Kuhistan by local tracks, not comparable with the density of the network in the Zagros. The *Tarikh-i Sistan* records three early events (734, 805, 815) but is then silent on the subject of earthquakes, while continuing its coverage of affairs in varying depth up to the Mongol invasions.[18] The province came into wider prominence under the Saffarids (*c.* 870–911), who dominated much of the eastern Islamic world, and in these circumstances the absence of macroseismic data suggests that a period of prolonged quiescence followed the earlier burst of activity round the Hirmand (Helmund) basin. Mustaufi tells a fable of the destruction of a gold mine in Sistan by an earthquake, perhaps in the late eleventh century, and although worthless as a source of accurate information, legends emanating from such areas are clearly a valid indication of local seismic activity.[19] Local oral tradition is the source for the only earthquake recorded in Kuhistan in this period (at Gunabad in 1238), although others may be referred to under the general term 'Khurasan' — a suggestion made more likely by the assumption that had these events (763, 840, 1066) occurred near a main route, the locality would probably have been specified.[20] The routes in Kuhistan merely link up local centres, except where they connect with the main desert routes leading from Nishapur. Local sources are likely to be the only fund of macroseismic information in Kuhistan, and in the absence of such sources in the early period our data is clearly incomplete.

In Gurgan and the Kopet Dagh, traversed by routes north to Khwarazm and alternative itineraries from Bustam (or Shahrud) to Tus and Nishapur (e.g. those of Abu Dulaf and Nasir-i Khusrau), the survival of data remains fortuitous. The Gurgan shock (874) is recorded in connection with a specific historical incident by a unique source, while notice of the 943 earthquake, clearly of large magnitude, comes in the account of a contemporary traveller and other regional sources of information.[21] These conditions are not generally met, and the subsequent lack of data should certainly not be taken to reflect a total seismic quiescence. The direct route across the Sabzavar plain remained the dominant artery of travel after the eleventh century — the Saljuq caravanserai at Za'faraniyyeh was one of the largest in Persia (Siroux 1949: 16) — and news from further north would be unlikely to reach Baghdad.

These observations about areas of secondary importance

also apply to regions effectively off the route network altogether. Information is particularly deficient for the Caspian provinces of Gilan and Mazandaran, which were politically separate and commercially of minor interest in the Caliphate period. Historical circumstances did not facilitate the transmission and survival of macroseismic data and unfortunately there is inadequate contemporary local coverage of these areas beyond the Alburz. Al-Mas'udi (d. 956) states that Amul and many other towns in Tabaristan (Mazandaran) are subject to earthquakes,[22] but no details of these events have survived. Such information as we have (for Firrim, *c.* 1127) is fortuitous, although as in other remote areas, chance factors are more likely to operate in seismic regions than they are to illuminate relatively quiet zones.

Desert areas yield no information, for obvious reasons. Large shocks originating in the desert might be picked up by the main towns around its borders, but during this period the chances of this are slight. Regional termini such as Yazd and Kirman, but also places like Qayin, Na'in, Kashan and Isfahan were unlikely to record on a local level the feeble effects of a distant shock, which could pinpoint the epicentral region. The only chance of survival for earthquake data would be a traveller's account, or through direct transmission to Baghdad rather than a static local record. Well-worn tracks skirted and crossed the deserts of central and southeast Iran and news could travel with the caravans, especially if vital wells or water cisterns were destroyed. The volume of this traffic is hard to estimate; certainly the tracks from Yazd and Kirman through Tabas to Nishapur were important arteries in the late eleventh and in the twelfth centuries, flourishing under the Saljuqs of Kirman (see above), who greatly developed Tabas itself. Had an earthquake comparable to that of 16 September 1978 occurred in or around Tabas at this period, it is unlikely to have escaped widespread notice. The chances of a smaller earthquake, or one not affecting an important oasis, being recorded remain minimal.

1.4 The Mongol and Turkoman period (1258–1598)

This period is defined on one side by the Mongol sack of Baghdad and on the other by the transfer of the Safavid capital from Qazvin to Isfahan, which introduced a new era. The division exists by virtue of its complete contrast with the preceding Caliphate period. The most fundamental change is that a wide gulf developed between the Arabic world, now centred in Mamluk Egypt and Syria, and the former eastern provinces of the Islamic empire. Persia's affairs evolved separately as a function of internal conditions, with such outside influences as were important coming from the east. This is reflected in the fact that Persian works replace Arabic ones as the main sources of information. The difference is important, because the treatment of natural phenomena in Persian sources is far from systematic. Very few authors, even if covering the general history of long intervals, mention more than one or two earthquakes, and very few events are reported by more than one source. This does make for an individual account of each earthquake, with authentic distinguishing features, often embroidered with stories or other details of human interest.

Certain stories, such as events being predicted, with various results, are quite frequent (for example, the earthquakes of 858, 1042, 1549, 1593 and 1721). Allusion is also made to the behaviour of animals (as in 1485, 1608, 1695 or 1875; see below, § 3.4.3). One characteristic feature is the composition of poems about earthquakes which, apart from giving expression to the various emotions aroused by disaster, often contain useful information, such as the precise date of the event or of subsequent restoration work. But the fact that most of our accounts of earthquakes in the Mongol and Turkoman period derive from only one source means that it is generally not possible to confirm or supplement the details provided (Melville 1978: 194–8).

These characteristics of the Persian source material are in large part determined by a preoccupation with either straight political narrative or, more fruitfully, with purely local history, which may itself, however, have an entirely political emphasis. Inclusion of earthquake data in dynastic histories depended on a most favourable combination of circumstances, which rarely operated. Internal conditions in Iran after the Mongol invasions did not facilitate the spread and survival of macroseismic data on the general level. The country remained considerably depressed and depopulated after the invasions, many villages deserted and many towns greatly reduced. There was at the same time an increasing tendency towards nomadism.[23] While it may be argued that a greater mobility of population might encourage the spread of news, at least on a regional level, the decline of a settled, stable population would not assist its survival in written form (cf. chapter 2). Even the capital cities provided only temporary residences for the rulers, who in nomadic manner alternated between winter and summer quarters, or were away campaigning. Authors covering affairs at court were thus faced with a constantly changing geographical backdrop; the independent life of towns or regions at the centre of events was thus only sporadically brought into focus.

The Mongol Il-Khans dominated Persia from centres in the northwest (Maragheh, Tabriz and Sultaniyyeh) until 1335, during which time the entire length of the east–west trade route from China to eastern Anatolia was controlled by related Mongol states; most of this trade passed through Tabriz. Internal security remained poor and the Il-Khanid state quickly dissolved into factionalism before a new order was briefly introduced by Timur around 1380, from his capital at Samarqand. After his death in 1404, Timur's empire was effectively reduced to an eastern portion under his successors in Herat and a western portion under the Turkoman dynasties, centred in Azarbaijan and upper Mesopotamia. Both these succession states were eclipsed by the Safavids around 1502, whose capitals were again in the northwest; but by the end of this period, two of the four imperial cities (Herat and Tabriz) were in the hands of the Safavids' enemies (Uzbegs in the east and Ottomans in the west), while Qazvin itself was felt to be threatened.[24]

The frequent change of capital, lack of centralisation and relative insecurity prevailing in this period are important factors in the distribution of recorded earthquakes. In place of one long-term focal point, like Baghdad, where information could be accumulated and preserved, there were a number of more or less independent centres. In this respect, the role of the main routes as vehicles for the transmission of data is modified and other factors come into operation. Fragmentation of the country promoted the growth of local centres and local histories. These have often preserved information about areas that would otherwise undoubtedly have escaped notice in more general works. An increase in data for some regions off the main route network or rural districts not intimately connected to a major urban centre partly makes up for the unreliable reporting of earthquakes in dynastic chronicles. However, the amount of useful data found only in later compilations suggests that some sources of information have been lost or not yet identified. The Mosul annalist al-'Umari (d. 1811) is the sole source for about one-third of the events recorded in this period and our data would be seriously depleted without his work, which in many ways resembles Ibn al-Jauzi's and represents the continuing activity of Arabic historians in Iraq. He records events for several areas of Persia, his intermediary sources of information being unclear; he may be reporting oral news transmitted directly to Iraq as well as quoting documentary sources.

In addition to these indigenous histories, a small number of Muslim and European travellers have left accounts of their journeys in Persia (see figure 1.3). Their presence in the country was brief and intermittent, so that the likelihood of their coinciding with a major earthquake was small; furthermore, the accounts of their travels are generally meagre in geographical details about the areas they passed through, often confined to the vaguest indications of the author's movements or a bare list of places visited. Nevertheless, their passage through Persia introduces a further modification to the role of the routes they used, these becoming themselves potential sources of information rather than merely the channels along which news travelled.

Figure 1.2 shows the distribution of places mentioned as affected by earthquakes during the period up to 1600 and the number of times this occurred. The network of routes is based on details given by Mustaufi, who describes the situation at the end of the Il-Khanid period (*c.* 1340), with Sultaniyyeh as capital and the hub of five main highways; the picture is filled out for later periods on the basis of travellers' itineraries, and the traces are thus of the routes from which we would expect information to be available.[25] It is clear that although the distribution of recorded earthquakes is very different from that found in the previous period (figure 1.1), there is still a close coincidence of these places with the main routes.

Of these, two were of primary importance; one east–west across northern Persia, from the Oxus to Anatolia, the other diagonal from the northwest down to Hurmuz in the Persian Gulf. Both these routes went through Tabriz. Information on earthquakes in this city is available throughout the period, despite the fact that the events themselves (1273, 1304, 1345, 1459, 1503 & 1550) do not seem to have been too serious. These data must accurately reflect the seismicity of the time, for any large event should have been recorded had

one happened, given the international importance of the city.[26] The rest of Azarbaijan is similarly well covered: lack of information for Maragheh (capital till 1295) and Ardabil, which was much frequented and rose to a new prominence under the Safavids, suggests that no earthquake of any significance affected these places, while the Sarab and Miyaneh district in between does demonstrate some seismic activity. Data for the area west, round lake Van, also reflect the importance of this trade route as well as the high seismicity of the region.

A total lack of information from the regions of Sultaniyyeh (capital 1305–35) and Qazvin (1548–98) suggests a genuine quiescence for the periods of their importance, but not necessarily for the intervening two centuries. Although the routes through these cities were busy, our source material is inadequate to illuminate the apparent gap.[27] The same applies to Ray, which was superseded by Varamin, although the region remained populous; such details as we have are either dubious (1384) or reflect the effects of more distant events (1495). Despite the decline of the area, we would expect large destructive earthquakes there to be reported, though not with the same confidence as in the Caliphate period. If the occurrence of a shock around 1384 be admitted, a period of quiescence before and after it may account for the lack of further data.[28]

The Khurasan road east of Ray undoubtedly maintained its earlier importance, although we have few accounts of it.[29] The area was dominated between *c.* 1336 and *c.* 1380 by the Sarbadars, based on Sabzavar and Nishapur, whose intricate history receives some attention in the sources. The main routes passed to the north, through Gurgan, Jajarm and Juvain (Aubin 1971). Gurgan's importance as a winter pasture for the Turko–Mongol nomads is suggested by the record of three destructive earthquakes there in the fifteenth century (1436, 1470, 1498). Why similar information is not available for other intervals is not clear; perhaps a genuine seismic quiescence preceded and followed this concentrated burst of activity, though after the establishment of the Safavids at the beginning of the sixteenth century the region came under pressure from the Uzbegs and was only marginally under Persian influence.[30] A similar, though earlier paroxysm seems to have affected Nishapur, where three destructive earthquakes (1270, 1389, 1405) are reported in Persian sources. The subsequent seismic quiescence, during the period of activity in Gurgan, should *not* be seen as a function of the city's decline from the fifteenth century onwards or a corresponding dearth of information in contemporary histories (Melville 1980).

Kuhistan, peripheral to Nishapur and likewise dependent on Herat, yields perhaps the most consistent record of seismic activity during this period, with major events reported in 1336, 1493 and 1549. The two earlier earthquakes are mentioned by local historians, reflecting the vitality enjoyed by the area, in common with the whole Herat province, until the sixteenth century. The later event is recorded by a number of Safavid chroniclers.[31] Mustaufi relates a legend of a cypress tree at Kishmar, west of Turshiz, which protected the district from the earthquakes that frequently occurred all around it. The tree is said to have been felled in the ninth century and it may be that this action ended the seismic immunity of the Turshiz area. No details of specific events, however, are available before 1903.[32]

The route southeast from Sultaniyyeh down to the Persian Gulf was probably more frequented, certainly so by the few European travellers of the time. In view of the steady trickle of visitors, details should have survived of any destructive shock in one of the towns along this route. No such earthquakes are mentioned and the minor events recorded by al-'Umari for Isfahan and Shiraz are probably representative of the situation in those cities. The former was important throughout the period, more so, as a potential source of macroseismic information, than under the Caliphate. Shocks experienced in Isfahan (1344, 1459, 1495) all originated some distance away and can be used to form some idea of events in the Zagros. The infrequency of earthquakes in Isfahan itself is specifically referred to by Mustaufi (*Nuzhat*: 48). To the south, shocks in and around Shiraz in 1459, 1506 and 1591

leave a similar impression. Lack of information before the fifteenth century cannot be blamed entirely on unfavourable circumstances, for the city was visited and described by Ibn Battuta in 1327 and 1347 and a local history is extant, dating from the same time. Thereafter Shiraz was prominent under the Inju'ids and later the Muzaffarids, during which period the great poet Hafiz was active (d. 1390).

Beyond these two centres, routes to the Gulf reached Qais (chief emporium up to 1330) and Hurmuz (or Jarun, on Hurmuz island), the latter going via Lar by the end of the fourteenth century (Aubin 1969). Data for Qishm (1361), Lar (1400, 1593), Karzin (1440) and Hurmuz (1482–3, 1497) reflect the major commercial importance of the routes through this region. Descriptions of their itineraries are given by Ibn Battuta and various Europeans, such as Nikitin in 1471, Newberie in 1581 and Teufel in 1589. Information recorded for these areas, by a variety of sources, must be a fairly complete sample of seismicity of the southern Zagros.[33]

Figure 1.2. The main routes under the Mongol and Turkoman dynasties and those for which details are available from travellers' accounts. The figure shows the places affected by earthquakes during this period. Note the absence of data for Iraq compared with figure 1.1. For the epicentral location of events, see figure 5.2.

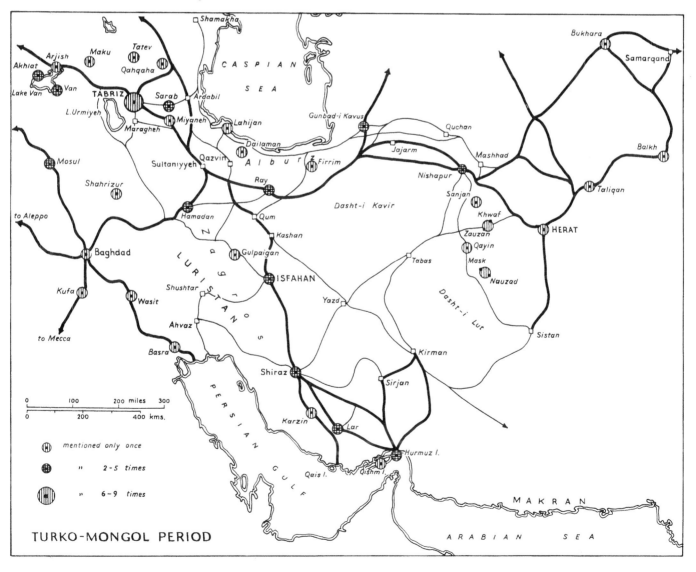

In contrast with this region, data are almost totally lacking for southwest Persia, the Tigris–Euphrates valley and the western Zagros, a gap which, compared with the preceding Caliphate period (figure 1.1), can only be seen in terms of historical and geographical circumstances. In Luristan, a local atabeg dynasty maintained the security and upkeep of roads to Isfahan up to the early fifteenth century, but increased nomadism in the area would prejudice the survival of macroseismic data.[34] The latest available account of this overland route is Ibn Battuta's; by the end of the sixteenth century, at the height of the Aleppo trade, accounts are to hand of merchant's voyages from Baghdad down the Tigris or Euphrates to Basra and so by sea to Hurmuz (Steensgaard 1974: 37), but none of these mention earthquakes. Two shocks in this area (in 1430, 1457), at a time when we have no travellers' accounts, are given by al-'Umari, which may point to epicentres in the Zagros. Isolated shocks to the north, in Kurdistan, are mentioned by Arabic sources,[35] but it is clear that the perception of events in the whole of this western zone is greatly reduced from its previous level. Only the Hamadan–Gulpaigan earthquake of 1316 is mentioned by Persian sources; the area was crossed by a secondary route of some importance at this time (Mustaufi, *Nuzhat*: 171–2). Later indications of possibly comparable events in the region are provided by al-'Umari's reports of shocks in Hamadan (1430?, 1495) and Isfahan (see above).

The southeast of Iran again presents a blank. In some respects, this gap is harder to account for than in the Caliphate period. Although by virtue of its location and terrain much of the southeast may be considered remote, its removal from the main stream of events in the Mongol and Turkoman period was by no means as great and its distance in relation to Baghdad is no longer relevant. Kirman witnessed a succession of rulers, notably the Qutlugh Khans and Muzaffarids up to the end of the fourteenth century, who attract attention in the main sources of the period. The former capital, Sirjan, again achieved considerable importance at this time, diverting the main flow of traffic to Hurmuz from the more easterly route through Jiruft (Sabzvaran) and at the same time benefitting from its position on the route from Shiraz to Kirman.[36] Thereafter, details of events in the province are more intermittent, but local dynastic histories continued to be produced. The existence of such works does not of course guarantee their reporting of earthquakes, but on the other hand it is likely that destructive events in Kirman itself would have been noted had they occurred. Information about the trans-desert routes is insufficient to form a precise idea of the frequency of traffic they maintained. Yazd and Kirman, with other desert towns, seem to have remained comparatively prosperous, as noted by Marco Polo (in 1272), Friar Oderic (*c.* 1325), 'Abd al-Razzaq (in 1442) and Nikitin,[37] but the trade that filtered down to the great emporium at Hurmuz was probably of a lesser order than that going via Shiraz and Isfahan, certainly during the sixteenth century.

Sistan is similarly served by local histories throughout the period under review; and furthermore, until the advent of the Safavids in the sixteenth century, was sufficiently within

the orbit of Herat for news to be available to the late Timurid historians of the region, as was the case for Kuhistan (see Tate 1910). It seems probable that the implied absence of large earthquakes, particularly before 1500, is genuine, although it may be that hints of such events may be found in local oral legends.

The Caspian provinces, finally, maintained their isolation from the wider circle of affairs in the rest of Persia. The few details of earthquakes issue from purely local sources, of which a number have survived, covering the whole of the period. The southwest corner of the Caspian was traversed in the 1560s and 1570s by British merchants of the Muscovy Company, plying between Shamakha, Ardabil and Qazvin, with excursions to Rasht and Lahijan, the main town of the area.[38] About a century earlier, Barbaro and Nikitin also penetrated the Alburz. These brief voyages have left no information about the seismicity of the Caspian provinces, however, most of the traffic keeping to the south of the Alburz. It remains probable that had any other event comparable to that of 1485 been experienced in Gilan or Mazandaran, it would have received attention in the sources available.

1.5   The seventeenth and eighteenth centuries

The period inaugurated by Shah 'Abbas's transfer of the capital to Isfahan saw increased stability and prosperity in the Safavid dominions as a result of his rule, with a greater degree of centralisation than had been present for centuries. In 1722, this relative tranquility was abruptly disturbed by the Afghan invasions of Persia and the quarter century that followed, embracing the career of Nadir Shah, saw the collapse of political stability, depopulation of the countryside and deterioration of the economic life of the region. Nadir's capital was Mashhad, and after his death in 1747 Persia was divided into separate spheres of influence. His Afsharid successors and the rise of an independent Afghanistan dominated affairs in the northeast and east, while after more than a decade of violence and anarchy order in the south and west was largely restored by Karim Khan-i Zand, whose metropolis was Shiraz. After his death in 1779, there was a protracted struggle for power between his successors and the ultimately triumphant Qajars, who were based on the Caspian provinces and assumed control in 1794; the new capital was Tehran. Superimposed on this political background was the heyday and gradual decline of Persia's position in the world of international and intercontinental commerce, the overland routes through the country slowly yielding to the ocean routes opened up by the Dutch and the English in the early seventeenth century (Steensgaard 1974).

Travel books have been treated as a separate source of information for the whole of the Middle East. Accounts of Persia by about 650 travellers in the period prior to 1900 have been read, of which 160 are from trips that skirted Persia along the peripheral routes of Mesopotamia, the Persian Gulf or Transcaspia, or from European residents who remained static in one place. The itinerary of each of the remaining 490 travellers has been drawn on a 1:8 500 000 scale map (unpublished), with details of the period of each traveller's stay in the country as well as observations relating to seismicity and geography. This corpus of itineraries, which is still in progress, has been used to identify routes, to sample the state of the country and to check for descriptions of damage, a method that has been found invaluable for other regions such as Turkey, Syria and Palestine; the volume of a similar collection for these latter countries is more than ten times greater.

The contribution of contemporary travellers to our knowledge of Persia's seismic history before 1600 has been seen to be small — isolated indications of earthquake-prone regions and the occasional valuable account of a specific event stand out from a generally sparse coverage of regional affairs and conditions, despite the comparatively long period spent on the journey or in individual towns. More particularly, only a few records of travellers' journeys are available. After the turning of the seventeenth century, this situation alters quite considerably (see figure 1.3). Before the fall of the Safavids, numerous Europeans visited Persia, either as missionaries, diplomats, merchants or simply for curiosity. They have left a great volume of material that is not only valuable as a record of earthquakes (about one-quarter of the events listed for these two centuries are mentioned by contemporary European authors), but also as a fund of negative information. The steady passage of merchants and others down well-worn routes permits the assumption that areas through which they passed are adequately documented: brief statements on the condition of places they visited, and what might be called any positive lack of information about earthquakes, have some value. It remains true, however, in common with earlier periods, that their coverage of the region is neither consistent nor continuous. Some routes may have been constantly busy, but individual records are still intermittent and in addition generally cover only a brief span of time (seldom more than two years). Not long was usually spent in any one place and few authors

Figure 1.3. Number of travellers through Persia (*N*) per two decades, whose itineraries have been read. Solid lines show number of non-Russian travellers; height of horizontal bars shows number of Russian itineraries.

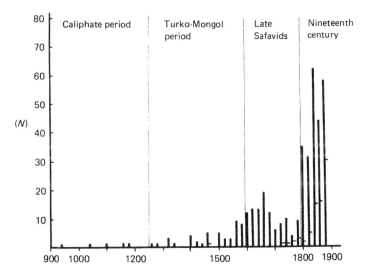

were concerned to enquire very deeply into the recent or past history of the places they passed through (this is less true of visitors in the nineteenth century, see below). On the other hand, some Europeans were resident for long periods, particularly in Isfahan: Père Raphael du Mans, for example, from 1644 to 1694 and Chardin from 1664 to 1670 and 1673 to 1677. They were thus well placed to hear such news as was available, i.e. of the most important or destructive earthquakes; smaller shocks would only be reported if a traveller happened to pass through the area affected. After the Afghan invasions and particularly the death of Nadir Shah in 1747, European presence in Persia was dramatically reduced (figure 1.3), few penetrating beyond the shores of the Caspian in the north and the Gulf littoral in the south. Those who went further inland did not stay for any length of time and the precise intervals for which their reports are relevant are thus correspondingly more limited.[39]

Erratic recording of earthquakes in Persian sources, already noted, continues to be typical. Safavid chronicles are almost devoid of references, even when it is known from other sources that events did occur in regions that one could reasonably expect would be covered, such as Bandar 'Abbas (1622) or Tabriz (1641, 1717). This trait is even more marked in the period before the establishment of the Qajars, despite numerous contemporary histories of Nadir Shah and the complex sequence of events in the unsettled years that followed; the emphasis remains on purely political matters.

While Persian contemporary sources yield little data of interest, almost as much again is provided by modern Persian works, particularly local histories, which suggests that original source material, oral or written, existed or still exists to be explored. In addition, poems and inscriptions contribute a certain amount of information, not only about the occurrence of earthquakes, but also giving an insight into their effects on buildings, which is relevant to an assessment of the Intensity of the shock.

Other oriental sources also provide information, making the available material less homogeneous than in previous periods. A number of useful Armenian data have survived, often in the form of manuscript colophons or marginal notes.[40] Historical sources of the Ottoman Turks, who were in constant competition with the Safavids and their successors, also throw light on affairs in Persia. Among Ottoman sources can be included the Arabic chronicle of al-'Umari, referred to above (§ 1.4), effectively a contemporary record for most of the eighteenth century and certainly a repository of earlier information.

These sources inevitably benefit our knowledge of events in northwest Persia rather than any other region, and this bias is reflected in figure 1.4, which shows the number of times various places are mentioned as having been affected by earthquakes during the period. The concentration of Turkish and Armenian authors on Ottoman—Persian frontier regions of Azarbaijan guarantees that the record of high seismic activity in the area between Van, Erivan and Tabriz is probably complete in all essential details. Tabriz itself was a great mart throughout the Safavid period, despite an exceptionally destructive attack by the Ottomans in 1635, as is attested by most visitors to the city. Even though in common with many other regions it suffered a decline in the eighteenth century, details of large magnitude events, such as that of 1780, continued to be recorded as they occurred, even by the generally reticent Persian sources.[41]

Figure 1.4 also illustrates the routes most frequented by European merchants and other visitors during these centuries, on the basis of information given by Tavernier, who made nine journeys into Persia in the course of six voyages between 1632 and 1668. This scheme is supplemented by details of other itineraries and thus represents the towns and the routes between them that are relatively well documented. Little comment is necessary on the coincidence of data being available for districts crossed by these routes wherever this occurs. The journey most commonly made by traders to the Safavid capital led from Tabriz through Sultaniyyeh and Kashan, an alternative route from Shamakha through Ardabil or Rasht to Qazvin and so to Isfahan also being much used. This is

reflected in the distribution of recorded earthquakes. Qazvin, entrepot of the Gilan silk trade throughout the period, is mentioned only in connection with shocks elsewhere (1608, 1721) and seems to have been free from destructive earthquakes itself. Kashan, continuously visited throughout the Safavid period, was apparently unaffected by earthquakes till the latter part of the eighteenth century (1755, 1778). The previous low level of seismic activity in this area is probably authentic. It is likely that only small or non-destructive shocks would escape attention in the region between Qazvin and Kashan, both of which were still flourishing in 1796.[42] Equally representative of the situation is the absence of any significant data for Isfahan, whose role during the Safavid period is somewhat comparable to that of Baghdad under the Caliphate. Both the state of preservation of monuments in the city and the silence of various authorities residing there are eloquent indications that no major earthquakes were felt (Ambraseys 1979). On the other hand, its perception of events in the central Zagros was no better than in previous periods. It

Figure 1.4. Main itineraries followed by travellers in the seventeenth and eighteenth centuries and the places affected by earthquakes during this period. Note the bias towards information for the northwest of the region. For the epicentral location of events, see figure 5.2.

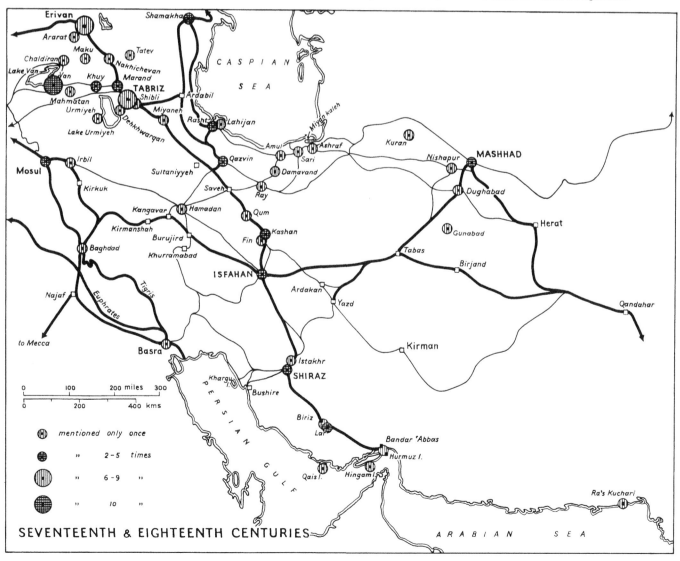

is unfortunate (except for the inhabitants of Isfahan) that the city is so well removed from a genuinely seismic area that it is unlikely to record distant events even when all other factors would tend to facilitate the reporting of shocks.

The extension of the route from Isfahan down to Bandar 'Abbas, via Shiraz and Lar, was a major thoroughfare. Absence of reported earthquakes in Shiraz throughout the seventeenth century must be significant in view of the regularity with which the city was visited by observers who should have mentioned any such event. Those recorded for the eighteenth century (1752, 1765 and 1784) were all minor shocks and caused very little damage. It is significant that such small events are mentioned at a time when Shiraz was at its peak of commercial and political importance as centre of the Zand state: the two later shocks are noted by visitors who happened to be in the city at the time. It is unlikely that a large earthquake in Shiraz would have gone unrecorded at this time; on the other hand, this observation only applies to the city itself and to the places along the main tracks to and from it. On the route between Shiraz and the Gulf, Lar (1677) and Bandar 'Abbas (1622) both suffered earthquakes in the seventeenth century, as recorded by European travellers. Pietro della Valle (p. 480) remarked that the numerous shocks in Bandar 'Abbas in 1622 were considered unusual, but when La Boullaye-le-Gouz was there in 1648 he noted (p. 121), that shocks were so frequent that the English factor preferred to sleep in a tent: no details of these earthquakes survive, but it is likely that they were small tremors not normally damaging. Absence of information for the eighteenth century must largely be a function of the fact that this route declined drastically after the Afghan invasions and more particularly after the death of Nadir Shah. Merchants from Mashhad, Nadir's capital and temporary hub of communications throughout Persia, were reluctant to trade to Bandar 'Abbas after his death and the European merchants soon moved to the top end of the Gulf in response to Karim Khan's attempts to restore trade from his capital at Shiraz (Perry 1979: 258). Throughout most of the eighteenth century the main flow of traffic from east to west and back by-passed Iran, the so-called overland route from India involving a passage in the Persian Gulf, perhaps touching at Khargu, Bandar Rig or Bushire (Bushahr), but with Basra the main port and terminal of various routes across Iraq to the Levant and Asia Minor.

Despite the frequency with which routes through Iraq were used, we have very little data on the seismicity of the Tigris–Euphrates region throughout this period. In the seventeenth century, the Mosul area was affected around 1620 and again, more seriously, in 1666 (Longrigg 1925: 37; Wilson 1930: 115), but that is all. In the eighteenth century, similarly if not more fully covered by the works of European travellers, the few details available must be taken to represent a genuinely low seismic activity. Events recorded by Europeans in Basra (1705) and Baghdad (1769) were not severe. To the north, the route between Baghdad and Mosul via Kirkuk and Irbil was much used in the latter half of the eighteenth century by employees of the East India Company and others, avoiding politically unstable Iran, and a severe earthquake in this area

is most unlikely to have gone unnoticed. Al-'Umari records shocks in Mosul (1764 and 1782) and Irbil (in 1714),[43] but again, none of them was severe and the control provided by his chronicle allows a fairly positive interpretation of the lack of data in European sources for this particular region. The slight seismic activity reported in Iraq may not only represent the true situation in the towns mentioned, but serve to indicate more serious activity to the east in the Zagros proper.

Figure 1.4 reveals a large area, stretching from south of lake Urmiyeh down the whole length of the western Zagros, for which no earthquakes have been identified, with the one exception of the 1666 shock, mentioned by al-'Umari as affecting Hamadan, Isfahan and Shiraz; but the details of this event are highly uncertain. We have seen that valuable indications of the frequency or intensity of earthquakes can be given by European authors, but their coverage is fortuitous in that it depends on their physical presence in the country (or any specific place), which was ephemeral. There is no substitute for local sources of information and it is both fortunate that al-'Umari's chronicle has survived and unfortunate that Persian sources do not provide a comparably reliable reporting of events to complement European sources in other areas. In the Zagros, the European record is not full. Despite the importance of the direct route between Isfahan and Baghdad, which generally by-passed Hamadan and went through Khurramabad or Burujird before joining the Khurasan highway at Kangavar, few accounts of this journey are available.[44] European presence north or south of this comparatively narrow band of routes was non-existent and the chances of a severe earthquake coinciding with the passage of a traveller who left a record are small. While al-'Umari illuminates the situation in Kurdistan, no similar control is given by Persian sources. Although Hamadan and Kirmanshah were of some political and commercial importance, they have not produced local histories. Furthermore, the fact that the pilgrim route from Mashhad to Najaf and Mecca went right through the region, and it was a matter of urgent domestic policy that it should be kept open and secure, was clearly insufficient to ensure the transmission and survival of information in general chronicles from places along the way. This highlights the contrast between Persian sources and Arabic annals written under the Caliphate. In short, the apparent seismic quiescence of the western and central Zagros in the seventeenth and eighteenth centuries must be seen in terms of a deficiency of source material; at the same time, lack of data probably reflects the remoteness of areas that may have been affected by relatively small events, and the limited radius within which such shocks may have caused serious concern.

By contrast, a certain amount of information is available for the Caspian provinces, details of events in 1608, 1665, 1678 and 1687 being provided by Persian sources.[45] Our record of this area is certainly incomplete, for Chardin (1811: iii, 285) states that earthquakes in Hircania, i.e. Gilan and Mazandaran, were frequent and furious, though seldom fatal. Nevertheless, the largest shocks have probably been reported, particularly during the Safavid period; the Caspian was the Shah's favourite resort after Isfahan and the area was also important as the centre of silk production. The lack of details of earthquakes after the fall of the Safavids may reflect historical circumstances, for penetration of the central Alburz by Europeans remained marginal and after the death of Nadir Shah in 1747 the British position round the Caspian was totally absorbed by the Russians. Russian works, however, do not contribute much information about the region, which saw the early build-up of Qajar power.

The distribution in time of those earthquakes recorded for Khurasan is influenced by similar historical factors, but also perhaps by the relatively superficial level at which the available literature has been examined, particularly for the eighteenth century. Shocks are reported for 1619, 1673, 1678, 1687 and 1695, mainly by Persian sources, and demonstrate the general importance of the area and especially its capital, Mashhad. The city was developed as a religious centre by the Safavids and became capital of Nadir Shah's empire. On the other hand, the importance of Khurasan as a corridor for international trade declined throughout the period; Tavernier (p. 625) says that the route to Qandahar via Khurasan was not much used. Turkoman encroachment threatened the security of the routes even under the Safavids and conditions deteriorated considerably in the last three-quarters of the eighteenth century.[46] After Nadir Shah's death, as already mentioned, events in northeast Iran progressed independently of developments in the southwest. Coverage of affairs in the Afsharid zone is not comparable, in the sources that have been studied, with that of affairs under the Zands. This bias is reflected in the survival of macroseismic data for the west and southwest of the country and its almost total absence for Khurasan: one event in 1780 is recorded in isolation, but there is no indication of its location, itself suggestive of the poor level of available news. It was reported by an author writing in India, where many other historians sought refuge from events in Persia (see Perry 1979: 303). Europeans too were even rarer in Khurasan than in the western half of the country. Thus although the routes from Mashhad to the Caspian, to Herat and especially across the desert, through Tabas to Isfahan or Yazd and Shiraz, continued to carry trade and pilgrims, information about earthquakes in the northeast would have been of minor concern and limited currency in the areas and sources from which most of our data derive. Unfavourable circumstances are therefore responsible for the lack of details about events in Khurasan throughout the eighteenth century.

How far similar explanations can be applied to the lack of macroseismic information for the whole of the rest of the country is hard to judge. We find an apparently quiescent area stretching from south of Gunabad to the Indian Ocean, embracing the two deserts, Sistan and the districts of Kirman and Yazd. The east—west route from Qandahar to Isfahan via Birjand, Tabas and Ardakan was certainly much used in the seventeenth century and probably continued to be so thereafter, on a lesser scale. Equally, both Kirman and Yazd were major marketing and production centres, retaining their prosperity, or comparative importance, throughout the fluctuating conditions of the period.[47] But remarks already offered about these regions continue to be pertinent and we

cannot be sure that even larger magnitude events would have been recorded had they occurred. Few Europeans have left detailed accounts of southeastern Persia, where earthquakes would probably not have been recorded on a local level other than in oral tradition (as in 1838). The more general Persian sources would also be unlikely to refer in detail to events in these remote regions, with which communications were probably even less regular than in previous periods. We may note, however, that until the middle of the eighteenth century, news from Yazd or Kirman reached and was occasionally reported by agents of the East India Company at Isfahan and Bandar 'Abbas, so that there was some chance of a large magnitude event being recorded had it affected one of the more important centres of population.

1.6   The Qajar period (1794—1925)

With the advent of the nineteenth century, there is an immediate and obvious improvement in the volume and quality of macroseismic data, which becomes more complete as the gap before our own time narrows. This is largely due to the availability of additional sources of information, namely Persian press reports and British diplomatic correspondence, which will be treated separately below. Previously identified categories of documents nevertheless continue to provide much useful material.

Iran itself remained a strongly traditional society under the Qajars, only slowly responding to developments taking place in Europe, though unable to remain isolated from their influence. The capital was established at Tehran, symbolising among other things the greater economic and strategic importance of the northern provinces as opposed to the south, which took far longer to recover from the disorders attendant on the demise of the Zands.[48] Both Azarbaijan in the northwest and Khurasan in the northeast were sensitive areas throughout the period, not only because of continuing hostilities with the Ottomans on one side and the Turkomans and Uzbegs on the other, but also because of the new menace posed by Russian expansion southwards. The comparatively superior availability of news from these areas is reflected in the distribution of recorded earthquakes, in as much as the northern half of the country seems to have been more frequently affected (see figure 5.3). The significance of historical factors alone is hard to assess, however, as this is also broadly speaking the most seismically active zone in the country (see figure 5.13), nor is the south inadequately documented.

Qajar Persia was thoroughly penetrated by foreigners, particularly Russians and British, who brought with them their inquisitiveness, thoroughness, mobility and technology. One important product of their presence was the direction of these talents, for strategic or diplomatic ends, towards the accumulation and transmission of information about the country. Numerous travellers have left accounts of their journeys in Persia during the nineteenth century, containing many valuable details about earthquake occurrences. Two things distinguish the quality of their data from those provided by earlier generations of visitors. First, a large number of people went off the beaten tracks to regions previously considered remote and little known, so that their potential perception of shocks extends over a greater territorial area. Although their passage through the country was generally more rapid than in earlier centuries, they were present in far greater numbers than before and the record they provide is more continuous (see figure 1.3). Secondly, even though intervals of varying length still occur between the passage of visitors through certain areas, this is in large part compensated for by a tendency to describe in some detail places visited, not only as they were at the time but also with reference to their past history and legends: this applies to the smaller as well as to the more important towns and villages. Geographical observations and details of inscriptions or the state of preservation of monuments, provide useful indications of earthquake effects in their local context. This aspect of the record is particularly noticeable in the accounts of travellers on diplomatic service, both French and Russian; the books of British or Indian government officials overlap, as a source category, with official archives (see below), where many of their observations were originally reported.

It is probably fortunate that we do not need to rely heavily on Qajar historical literature, for although this is extensive, it has not proved to be of great value for our purposes. Relatively little of this material has in fact been exploited, one notable exception being the works of Saniʿ al-Dauleh. His compilations of historical and geographical information are of some value for his own period and contain many brief notices of earthquakes not readily found elsewhere; on the other hand his work is rather too inaccurate to be reliable for the more distant past (Melville 1978: 199). Other local geographical histories or biographies provide some data, for example about repairs to architectural monuments, while poems, inscriptions and oral legends preserved in modern local histories, as for earlier periods, complement our information with occasional useful pieces of evidence. The value of field studies as a source of macroseismic data of this type is discussed in chapter 2.

An important change in the character of the monarchy occurred with the accession of Nasir al-Din Shah in the mid-nineteenth century (1848–96), when a policy of modernisation and greater centralisation of government was introduced. This further enhanced the primacy of Tehran over the provincial cities and was facilitated by improved internal communications, brought about by the introduction of the telegraph between 1860 and 1870, as a link in the line between Britain and India (Wright 1977: 128). This service was clearly crucially important to the transmission and availability of information. Figure 1.5 shows the way in which the network developed and the dates when various centres were connected to the system, based on data and a map found in the India Office.[49] Although the running of the line was by no means straightforward or uninterrupted, it remained until the end of the Qajar period the only well-developed means of internal and international communications in Iran; the introduction of other facilities, such as railways or surfaced roads, was greatly hampered by the clash between British and Russian interests (Kazemzadeh 1968).

In view of the partial opening up of the country and the greater ease with which news could travel, it is no longer relevant to assess the information at our disposal mainly as a function of the accessibility or importance of the areas involved. Nor is it necessary to view available macroseismic data as products of varying local or regional conditions. It is indeed not feasible, in contrast with previous periods, to show on one map all the individual places mentioned as affected by earthquakes under the Qajars. The extent of information available is reflected by the fact that over 400 shocks are recorded for the nineteenth century, as opposed to only 23 in the eighteenth (see figure 5.1).[50] Events are reported for all parts of the country and their distribution can be seen from the zoning of these earthquakes on figure 5.3. This wealth of data is largely contributed by two categories of source material which must now be described.

The Persian press. Regular production of newspapers in Iran began in the middle of the nineteenth century with the official newspaper *Ruznameh-yi vaqaʾiʿ-yi ittifaqiyyeh*, which first appeared on 7 February 1851 (Hashimi 1948: iv, 330). The complete collection of this paper in the Majlis library, Tehran, preserved with very few gaps, provides a large volume of information, in the form of quite lengthy reports of events in all parts of the country. The paper appeared more or less regularly about once a week, news of the provinces coming from various written sources and often therefore being published some considerable time after the events concerned took place. News from Tabriz appeared quickest, on average two weeks after the event, but intervals of up to six weeks were common, as from Bushire. In cases where earthquakes themselves are not dated, therefore, only an approximate indication can be taken from the date of the issue in which they are reported. Some thirty shocks are reported in the decade up to 1860, a third of them occurring in Azarbaijan. It is worth remarking in this context that Tabriz was the second city of Persia and official residence of the heir to the Qajar throne.

After issue 470, dated 9 August 1860, the name of the paper was changed to *Ruznameh-yi daulat-i ʿaliyyeh-yi Iran* or later, *Ruznameh-yi daulati.*[51] This was also produced for a decade and is preserved in the Majlis library. The amount of data contained in this set is considerably smaller than for the previous decade: only a handful of events from different parts of the country (with none from Azarbaijan, despite the fact that Tabriz was the first place to be connected to the telegraph network, see figure 1.5). This paucity of information may in part be explained by the altered character of the paper, which laid greater emphasis on affairs at court, hunting expeditions, military reviews and so on, illustrating these occasions with engagingly naive lithographs. It appeared at irregular intervals and is of less value as a reliable source of information.[52]

The paper was again renamed when it came under the control of a newly created press department, later a ministry, supervised by Saniʿ al-Dauleh. The first issue under the name *Iran* appeared on 2 April 1871 and it continued to be produced for about forty years.[53] Fragmentary collections of this publication in the Majlis, Melli and Tehran University libraries combine to preserve an almost complete set. In the early years of the paper, the emphasis remains on court domestic affairs or foreign news; coverage of news from the provinces is poor. This tendency is reversed by the 1880s, when internal news dominates the contents of the paper. A reflection of this is the fact that more than twice as many shocks are recorded for the 1880s than for the 1870s.[54] Issues continued to be irregular in their appearance, generally coming out at intervals of ten days to a fortnight; there often remains, furthermore, a large gap between the occurrence of an event and the publication of the details. The paper's sources of information are not specified, but news received by telegraph probably formed the main contribution from an early stage. Despite this improved facility, intervals between the occurrence of an event and its publication remained anything up to three weeks, as in the case of the Quchan earthquakes of the 1890s, when *Iran* specifically cites telegraphic sources for the first time. Apart from making searching for details of known earthquakes more arduous, the importance of this point is simply that some news may not have been reported because it arrived too late for one issue and too early, unless it was considered very important, for the next.

Figure 1.5. Telegraphs in Iran in the Qajar period. The figure shows when various centres were connected with Tehran; a = telegraph stations; b = wireless telegraph stations, opened 1915–17; c = Persian Gulf submarine cable of 1864. *Notes*: 1. This section was dismantled in 1905; 2. The Persian Government line Tehran–Firuzkuh–Sari–Astarabad lasted only 1864–6; 3. This section is shown on the map in Preece 1879, but not on the India Office map of 1906; 4. Preece shows a Persian Government wire connected to Kirman in 1879; this was extended to Hurmak after the convention of 1901, Kirman being connected in 1903. At the same time the Persians built a line from Mashhad to Sistan; 5. The link Sistan–Rubat opened for international traffic in 1915, but the end of the new Central Persia line had previously been connected with the Indian Telegraph Department line from Duzdap (Zahidan) to Quetta. In the First World War, the Indian system was expanded to connect Panjgur and Gwadur; 6. The section Kirman–Bandar 'Abbas was completed in 1917 and the Baft-Sirjan extension shortly afterwards; this was later taken over by the Persian Government; 7. The Hamadan–Qazvin line was built in the war to supply Dunsterforce and later given up to the Persian Government. Based on data in Houtum Schindler 1877b, Harris 1969, *FO* 60 650 and *IO* 10 210, 729. Redrawn from *IO* W/L/P & S/B 37 (1906) and the map in *IO* 10 326, which sketches the situation in 1922.

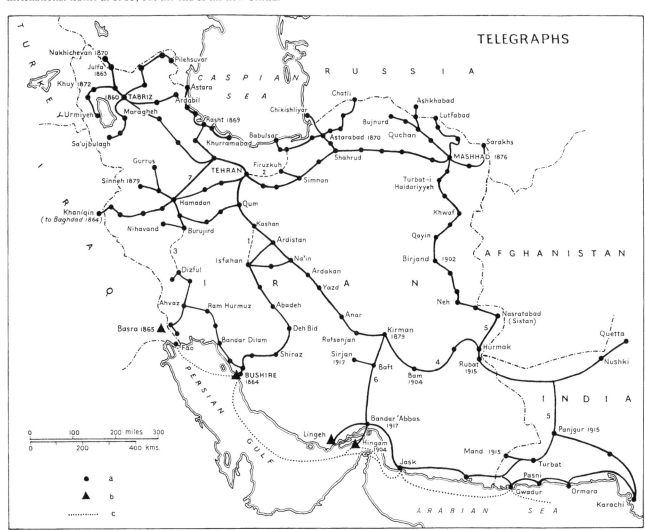

The set of *Iran* has been read up to 1901. Around this time numerous other papers become available, of which *Khulasat al-hawadith*[55] and *Habl al-matin*[56] have been used for the period bridging the end of the nineteenth and beginning of the twentieth centuries; both yield some macroseismic data. A large number of publications appeared in the constitutional period (1906–11); they were generally of brief duration and topics of greater concern than earthquakes were in the air. Nevertheless, a study of the newspapers *Muzaffari*, *Iran-i nau* and *Istiqlal-i Iran*[57] proved profitable for the retrieval of information about earthquakes. For the period 1912–16, the paper *Aftab*, kept in the Melli library, is the source of only a small amount of data, despite the large space given to home news, events at the capital and internal telegraphic news.[58] With the start of a new series of the paper *Iran* on 1 November 1916, details of earthquake occurrence are once more fairly steadily reported in the press.[59] This paper has been read up till late 1923, during which time the amount of space devoted to telegraphic news from the provinces, and the amount of information about earthquakes, fluctuates considerably, with particularly noticeable gaps in 1918 and 1921; the significance of this is discussed below.

Of other foreign papers, the most informative have been the Tiflis newspaper *Kavkaz* and for the later period the *Turkmenestanskie Vedomosti* of Ashkhabad. The use made of these sources has not been comprehensive, papers being read only for specific events. The same applies to the *Neologos Constantinoupoleos* of Constantinople (Istanbul), which occasionally gives detailed accounts of earthquakes in eastern Anatolia and Azarbaijan. Of European papers, the *Times* has been checked through its detailed Index from 1801 to 1948. Besides this, many other papers, such as the *Augsburger Allgemeine Zeitung* (1843–71), the *Leipziger Illustrirte Zeitung* (1859–77) and the *Kölner Zeitung* (1861–77), have been used but yield little information not found in the Persian and Russian press.

The Persian newspapers that have been read provide a more or less continuous coverage for the period 1851–1925; only the years 1904–9 have been inadequately investigated and we have no details from the press about earthquakes between 1905 and 1908 inclusive. During this time several destructive shocks occurred and should have been reported in the papers. The years 1924–5 are similarly under-represented. Otherwise, the scope of press coverage, as noted above, is limited to some extent by the long intervals between issues, a feature that is not eliminated until the turn of the century. In territorial terms, however, the papers give a comparatively full coverage. Although there are gaps in the news about certain areas for certain periods, a large number of different places are named as affected, from most parts of the country, including eastern Khurasan, the Kirman region, southern Fars and the Persian Gulf, the central Zagros and Azarbaijan. The only areas for which press coverage is consistently deficient are in the southwest corner of the Caspian: several events in the Anzali (Pahlavi)–Linkuran–Mughan region in the late nineteenth and early twentieth century are not reported, but these districts were under the direct domination of Russia and for-

tunately the Russian press gives a full account of seismic activity in the region.

Between 1851 and 1899, about 135 shocks are recorded in the Persian press, of which just over half were merely tremors, felt locally and causing no damage. Information about such small events marks a new departure in the quality and range of our macroseismic data. The large number of shocks mentioned facilitates identification of aftershock sequences and permits a better control of classifying earthquakes than was previously possible. At the same time, the greater number of places mentioned gives a better idea of the radius of perceptibility of the events. Even at their most inadequate, with perhaps only the main town of a region reporting details of effects, newspapers at least draw our atten-to an event having occurred and this can usually be tied in with independent evidence from other sources, particularly in the nineteenth century from travellers' accounts or local information from the field. All the major earthquakes identified for this period were mentioned in some form in the press; and on the other hand, only very few events that are known to have occurred escaped coverage in the papers, mainly in the 1870s when the amount of provincial news available is small.

Between 1900 and 1925, about fifty shocks are recorded in the Persian press, a much lower average per annum than in the previous half century, with a wider disparity between known events and those actually reported. This is partly explained by the political upheavals of the Persian revolutionary period, followed by the Great War, which profoundly affected the contents of the papers. In addition, the number of known events increases, with more earthquakes starting to be reported in other sources or even recorded instrumentally.[60] On the other hand, many small shocks *are* picked up in the papers, while some of the events apparently omitted were damaging earthquakes. It is often not clear whether gaps in the press coverage reflect a genuine absence of seismic activity or merely a lack of interest or available information on the part of the papers. In 1918–19 and again 1921–2, when there is little information in *Iran*, the gap coincides with a period of very restricted operation of the seismograph network (see figure 4.7), but also apparently of genuinely low seismic activity, for other sources do not make good the omission. Destructive earthquakes in 1923 once again command interest in the press.

It is further to be observed that although the Persian press does not seem to have any systematic blind spots (apart from the strongly Russian sphere of influence around the Caspian), some pronounced patterns can be discerned in the distribution of earthquakes recorded at different periods. Between 1916 and 1923, eighteen of the twenty-four shocks reported in *Iran* occurred east of Damghan, in either Khurasan or eastern Persia. In 1900–15, however, only five of the twenty-four events mentioned in the press occurred in this area and the bias was rather to the northwest and north-central part of the country. Low accuracy instrumental recordings for 1900–25 do not reveal any such obvious groupings, so the implied migration of seismicity through different regions at different times does not seem to find confirmation. Changing emphases must thus be influenced by local circumstances and the internal workings of news communication, the presence of reporters and methods of press output — influences which only minutely detailed investigation might elaborate. More generally, it is possible that the bias of data for the northern half of the country in the Tehran press reflects the proximity and importance of these regions to the capital. Fortunately, any such bias is corrected by information found in British diplomatic correspondence, which limits the detrimental effects of gaps or omissions in the newspapers; the two combine to give an excellent overall coverage of earthquake occurrence.

British official archives. British correspondence concerning Persia preserved in archive collections goes back to 1602 with the general correspondence of the East India Company and the Persian factory records from 1620, now in the India Office library. In 1778, the headquarters of the Company in the Persian Gulf was established at Bushire, where the post of Factor ultimately became that of Political Resident, an office that continued throughout the Qajar period (Wright 1977: 62; Tuson 1979). The Residency Diaries are a valuable source of information not only about the Gulf but also of events inland.

Permanent diplomatic relations between the British government and Persia were officially established in 1809 with the mission of Sir Harford Jones. Between 1824 and 1835 the East India Company and its London Board of Control were responsible for the mission in Tehran and after the supersession of the Company by the India Office in 1858, the latter was again briefly responsible for Persian affairs. Although thereafter the Foreign Office resumed control, the Government of India contributed a large portion of the costs and a majority of the personnel. A practical result of this is some duplication of records preserved at the India Office and the Public Record Office, particularly important in cases where documents have been weeded out or are otherwise missing from either collection. The most informative classes of documents are the General Correspondence or Despatches from Tehran and the Embassy and Consular archives, which often contain drafts of despatches not otherwise preserved.

In the present work, use made of the available archive sources has not been comprehensive. The method generally adopted has been to search files for specific events, mentioned in the first place elsewhere (i.e. mostly in contemporary European works), thus restricting reading to certain years of the correspondence from particular areas. Very little correspondence before 1830 has been read,[61] and only intermittent files thereafter. This is mainly because such files as have been used have not been very forthcoming with macroseismic data. Only the largest or most repercussive events are noted (such as earthquakes in Tehran in 1830, Shiraz in 1853 and Quchan in the 1890s). The limited nature of the information reported is explained by the fact that British representatives were not widely distributed about the country until towards the end of the nineteenth century. Besides the Tehran legation, it was not until 1837 that a consulate was established in Tabriz, followed by another in Rasht in 1858. Only after the establishment of

the next, in Mashhad in 1889, did the appointment of British officials proliferate, to the extent that by 1925 there were over twenty different consular posts (figure 1.6). Thus for more than three-quarters of the century, consular news was restricted to Tehran, Tabriz and Rasht in the north and Bushire in the south, with occasional bulletins being provided by native agents. Furthermore, the practice of compiling consular diaries and news summaries was not developed until the last quarter of the century.

For the nineteenth century, therefore, the use made of archive sources is very limited; while some additional information might be available for this period, it seems unlikely that anything of great importance remains to be found, even assuming that the relevant documents are accessible: they are often difficult to identify. On the other hand, a number of works by British or Indian government officials provide much useful information for the whole period and cannot be regarded purely as travel books, since they make readily available data that could only be extracted laboriously from their political correspondence. One may cite as examples the works of Morier (1812, 1818), Ouseley (1819) and later Napier (1876) and Yate (1900), or the confidential publications by Vaughan (1890), Sawyer (1891) and Maunsell (1890). In the north of Iran in particular, Russian missions were equally active in topographical and geographical research.

Very little correspondence of the Russian consular services has been studied, and this only in connection with a few late nineteenth and early twentieth-century events in the *Sbornik Konsul'skikh Doneseniy*. However, information provided by Russian travellers for northern parts of Persia and for Azerbaijan is considerable, particularly after the first quarter of the nineteenth century. Detailed reports in the voluminous *Sbornik Materialov po Azii*, published by the Imperial General Staff, in the *Zapiski Pusskogo Geograficheskogo Obshchestva* for geography and ethnography, and in individual publications, were of particular value.[62] For the Qajar period, Russian sources and press reports constitute a significant part of the material searched for data.

It is regrettable that time did not allow the retrieval of information from French, Portuguese and Belgian archive sources. Occasional reference to these sources is mainly due to information supplied to us by colleagues.

In the twentieth century, British diplomatic archives contain considerably more news of regional events. The files have been searched systematically, yielding references to some sixty shocks, of which ten were damaging earthquakes and the rest minor tremors or aftershocks. The distribution of these events, as one would expect, closely reflects Britain's political preoccupations and presence in the country, which centred round the three nodal points of Tehran, Mashhad and Bushire. The Anglo—Russian Agreement of 1907, which divided Iran into three spheres, Russian, British and Neutral, emphasised a long-standing position of Russian encroachment in the north and British interests in the south (figure 1.6). Of the sixty-odd earthquakes reported by British consuls, over forty occurred in the southern half of the country, in the so-called Neutral and British zones.[63] The rest are all located in northeastern

Khurasan, reports coming from Mashhad, Turshiz and Turbat-i Haidariyyeh, where a strong British presence was maintained by the Government of India, to balance Russian activities in this sensitive area bordering Afghanistan and the northern approaches to India. No shocks have been reported from the northwest or north central part of the country, the region more particularly covered by the Persian press at this period. The Tabriz consulate does not seem to have produced news diaries and little of its correspondence has been read. Despatches from Tehran itself are naturally more exclusively concerned with political events, and although news summaries were prepared, incorporating the more important items from provincial consular correspondence, there is not the same flavour of local news available for northern districts. One could reasonably say that only large or destructive earthquakes in the Alburz would be reported from Tehran (as after 1925),

and not the smaller tremors that are put on record in the other zones.

Within these territorial restrictions, particularly in the region of the Gulf and Kirman (the only town of note directly within the British sphere of influence), the coverage of events provided by political correspondence is surprisingly full, considering the difficult and inaccessible nature of much of the country in comparison with northern provinces. This is largely due to an initiative taken by the British Association for the Advancement of Science (BAAS). In 1909 Milne wrote to the Foreign Office with a circular for consuls, requesting their co-operation in the collection of a register of destructive earthquakes, which 'will possess not only scientific but also practical utility' (*FO* 248 955). On 21 September 1909, this circular was passed on by Sir George Barclay (British Envoy to Tehran 1908–12) to his consuls, asking them to report on any

Figure 1.6. British consular posts in Iran in the Qajar period. 1 = key British centres, with date established; 2 = consulates maintained in 1925, with date established; 3 = consulates closed before 1925, with the period they were open; 4 = places with a Consular Agent before the establishment of the full post. Towns with Native Agents are not shown. *Sources*: Rabino 1946, Wright 1977 and *Foreign Office Lists*. See also Browne 1910: 172, who sympathetically condemns the Agreement of 1907 as tantamount to a partition of the country.

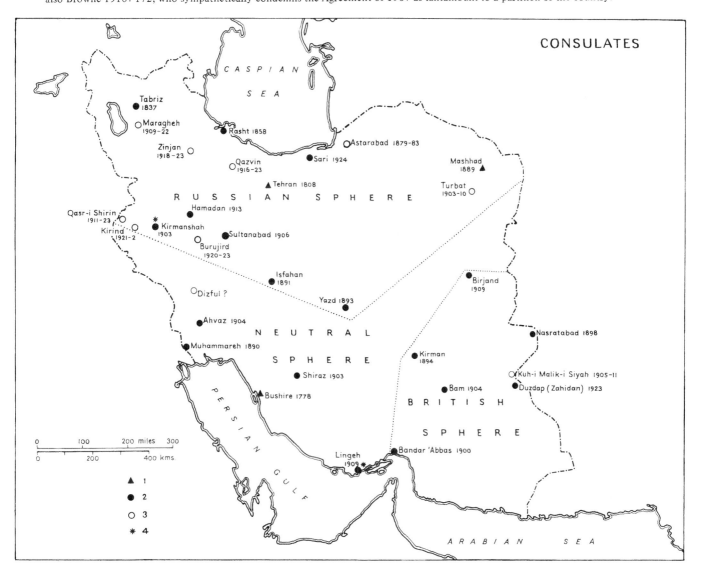

important earthquakes occurring in their consular districts as follows:

> You should be careful to divide earthquakes into these classes: 1. shocks which have damaged a few buildings; 2. shocks which have damaged buildings within a limited area; 3. shocks which have caused destruction over a large area. No unimportant earthquakes need be reported.[64]

The British Association's request came in the wake of the Silakhur earthquake of January 1909, which had attracted world-wide scientific interest (see below, § 4.3). It is not clear how long Barclay's instructions remained in force,[65] but earthquakes do seem to be mentioned in the consular diaries as a matter of course, under such headings as 'Meteorological' or 'Miscellaneous', even if they were only slight tremors.

As with the press coverage, a number of earthquakes identified from early instrumental records as occurring in Iran during this period have not been traced in diplomatic correspondence, but there are cases when the reverse is true (as with the series of destructive shocks in Lar in 1911). Either the relevant documents are missing or have not been found, or else the shocks were not felt or reported in the nearest consulate. This naturally applies almost exclusively to smaller events, or to those in remote areas. But it is also true, in view of the generally sensitive perception of shocks and their systematic recording by provincial officials, that apparent failure to record a local earthquake may indicate that it did not in fact occur there. Given the inferior accuracy of instrumental determinations of earthquake locations at this period, such apparently negative evidence can be of value for relocating certain events (see chapter 4). Certainly, from early in the twentieth century, consular correspondence provides much valuable information, even about important events that are imprecisely located by the early seismograph network and inadequately covered by the Persian press, such as the Silakhur earthquake of 1909 or the 1923 shock near Turbat-i Haidariyyeh.[66]

1.7 Recent period (since 1925)

After 1925, the availability of documentary information continues to improve as the century progresses and retains its value for modifying the location of instrumentally determined epicentres, particularly up to around 1950. A large number of known shocks still pass unrecorded both in the press and in diplomatic correspondence, generally only so far as minor events are concerned, but many obscure earthquakes have been successfully identified. We have made a fairly systematic resort to files up to around 1940: access is governed by a 'thirty-year rule' and documents are presently available up to 1949.[67] A search for information about specific events has also been referred to the newspapers, particularly *Iran* and *Ittila'at*, which first appeared on 11 July 1926 (Elwell-Sutton 1968: 77, 79). These papers are found in the Tehran University and Melli libraries; similar resort to the local press has been made right up to the present. Documentary data are now available for almost anywhere in the country and for shocks

of very minor importance and there is no need to pursue our examination of these literary sources any further.

One new source of information emerges during this period in the form of unpublished technical reports associated with the construction of large engineering structures such as railways, roads, bridges and dams, as well as with the geological mapping of the country and oil exploration. Almost all these reports contain invaluable first hand information collected *in situ* by engineers or geologists after an earthquake or in special surveys made to assess local seismic hazard.

Earthquake relief reports prepared by the Red Lion and Sun Organisation and by other international agencies, damage statistics and reconstruction projects prepared by various ministries and institutes, as well as numerous technical papers in the second half of this century, add to the macroseismic material at our disposal.

This period has also seen the publication of numerous regional and global macroseismic catalogues; more than twenty such lists of earthquakes have appeared since 1930, quite often containing useful information about earthquakes in Persia. A few of these catalogues are fresh and pertinent, quoting their sources;[68] some are out of date or compiled at second hand, containing many inaccuracies;[69] and some are oversimplified and misleading, containing gross errors and duplications in entries, which are occasionally difficult to disentangle.[70] It is only during the last few years that the cataloguing of historical and more recent earthquakes has become more systematic and reliable.[71]

The present study supersedes all these catalogues for references to events in Persia and no attempt is made to draw attention systematically to previous misconceptions, although individual inaccuracies are discussed as seems relevant or important.

1.8 Conclusions

In the foregoing review, we have seen the extent of the survival of information and noted some of its main characteristics. Arising out of this discussion, a general point to be made is that seismicity, in terms of large earthquakes, can sometimes appear exaggerated and sometimes under-estimated. Whether, and how, a historical event is recorded remain subject to many factors that are irregular and operate fortuitously; if it is mentioned at all, it may be dismissed in a brief phrase, or described in great detail with poems and stories: neither method in itself being necessarily indicative of the size of the earthquake. At the same time, the role of the cities in distorting the location of early events makes macroseismic information available for densely populated areas rather than remoter regions, thus also potentially distorting the true size of the earthquake. The same is true today; the Buyin Zahra earthquake of 1962 was consistently referred to as the Qazvin earthquake by the Persian press, similarly, the Dasht-i Biyaz event in 1968 as the Qayin earthquake. This is a natural and convenient generalisation, but important because it operates against the availability of adequate information about more seriously affected villages. Conversely, evidence that a major town *was* affected destructively, with details of extensive damage and heavy casualties, may still not be a sufficient indication that a large magnitude earthquake was involved: the Moroccan earthquake of 29 February 1960, which killed 12 000 in Agadir and attracted world-wide concern, was barely felt forty kilometres from the city and had a magnitude of only 5.7. The earthquake of 11 July 1927 in the Holy Land, affecting Jerusalem, caused relatively little damage but was even so, widely reported purely for the importance of the place concerned.

Such obvious limitations in historical accounts of earthquakes have to be treated as they arise on the merits of each individual case. In more recent periods, particularly since 1800, the position is improved by increased coverage of news from remoter areas and details about small as well as large settlements. One major contribution to this improved knowledge comes from studies made in the field, which are an important source of macroseismic data. The additional, particular value of field studies is that they permit correlations between the effects of historical earthquakes and those of this century. This is possible despite shortcomings in the documentary evidence such as those discussed above, for one can, in many cases, determine from internal or contextual evidence such details as the duration of the shock, the size of the area affected and the extent and type of damage caused, the duration of aftershock sequences and the association of the event with ground deformations, such as surface faulting, landslides and drastic changes in the underground water regime. All such details, ideally to be found in or inferred from historical sources, when calibrated against similar information derived from field studies of modern events, permit assessment of the magnitude of the earthquake and of the different Intensities experienced throughout the affected area. Such a correlation is helped by the fact that local conditions in Iran, such as building materials and techniques, have not until very recently changed greatly over the whole period under review, and some factors thus remain constants. In the chapter that follows, we look more closely at this area of investigation.

2

Macroseismic data from field studies

2.1 Scope of field studies

The site of a damaging or destructive earthquake constitutes a full-scale laboratory model from which significant discoveries can be made by keen observers, be they seismologists, geologists, engineers, archaeologists, historians or sociologists. As our knowledge of the complexity of earthquakes has increased, we have become more aware of the limitations nature has imposed on our capacity to understand, in purely theoretical terms, the effects of these events, the behaviour of a community and of the ground itself, or the performance of man-made structures.

Our experience shows that any advancement of knowledge about earthquakes must be based on reliable observational data with which we may approach the ideal situation of using our theoretical knowledge to the fullest extent, whilst lessening the risk of being occasionally misled by it. Collection of these data is best achieved by field study of earthquakes, which not only offers a unique opportunity to develop an intimate knowledge of the actual situation created by an earthquake disaster, but also promotes an understanding of the real problems that it creates. For instance, it is only in this way that ground deformations or faulting associated with earthquakes can be discovered and studied. Also, the resistance of local methods of construction and century-old building techniques, as well as the efficacy of new methods, can only be gauged after an earthquake. The social and economic repercussions of the event can also be studied at first hand, so that the picture obtained is not distorted by non-scientific considerations, such as those of local politics, which may otherwise obscure the evidence.

Field studies of earthquakes which occurred before the advent of modern seismology are even more important. The historical sources discussed in the previous chapter describe the occurrence of a multitude of destructive earthquakes in the past whose effects were partly or totally unknown. These accounts proved useful in guiding field studies to sites of early events, some of them associated with faulting representative of very recent tectonic activity. These indications of past activity are essential for an accurate assessment of local earthquake risk, for global estimates of seismicity may be only approximately compatible with local tectonics and of little practical value. Surprise earthquakes, such as Buyin Zahra in 1962, Dasht-i Biyaz in 1968 and Chaldiran in 1976, show how little we know about the risk involved in areas which are seismically quiescent at the present time.[1]

Our field investigations of earthquakes in the Middle East began in the late 1950s and continued until early 1978. After several experiences of the difficulties created by accepting second-hand information, our scheme has been to make a fresh start, based on original sources, in which both documentary and field evidence are presented in their historical and regional context. The size (magnitude) of historical earthquakes can be assessed from the degree of damage caused in the epicentral area of the event (epicentral Intensity), the extent of the area over which the shock was felt (radius of perceptibility), and from other factors which can then be calibrated against macroseismic information of similar twentieth-century earthquakes. Much of our effort, therefore, was concentrated on identification in the field of the meizoseismal area of all significant events, both historical and modern, for which we had literary or instrumental data, and on uniform assessment of their epicentral Intensities. Local tectonics were also noted, but at this stage only in a cursory manner. Our scheme envisaged a final stage, once the local seismicity studies were completed, in which the tectonics of the whole region were to be studied in detail. However, recent political developments in this part of the world are likely to delay the accomplishment of this work for some years.

2.2 Topographical material

The topographic maps available during most of the period of our field studies left much to be desired. Complete coverage of the region was initially available only in the India Survey 'quarter inch' series of the Geographical Section of the General Staff (GSGS series, pre-1942), in the United States Air Force Aeronautical Approach Charts, scale 1:2 250 000 (USAF, base information 1951–6), and in the Farsi versions of these series. For eastern Anatolia we had in addition the 1:200 000 scale maps of the Harta Genel Müdürlügü. Only later did maps of larger scales become available, but only for a limited area (AMS Farsi series, 1:100 000 and 1:50 000).

The triangulation control of the GSGS, USAF and AMS series differ considerably. Comparison of the former two with the AMS series and photomosaics shows location errors in GSGS and USAF of five to ten kilometres, particularly in Mazandaran, central Iran and Khurasan, where many areas are furthermore left blank or marked 'unsurveyed'. The *Village Gazetteer* of the Statistical Centre (27 volumes, 1968–71), which is based on the national census of 1966 and the USAF series, does fill in these gaps of country, but quite often only by hearsay. This becomes apparent to anyone who follows a track across the Kopet Dagh, the Chahar-dangeh in Mazan-

daran, or the region between Kuhbanan and Khur or Khabis, along which he will come across almost all the settlements shown in the *Gazetteer*, but in a haphazard sequence.

For most parts of interest we were able to consult vertical aerial photographs of the late 1950s vintage, usually on a scale of about 1:55 000. Photographs on a larger scale were also made available and proved extremely useful in mapping certain regions of recent tectonic activity. In a few cases photographs on a scale 1:5000 to 1:7500, taken immediately after an earthquake, were provided.

The geological maps available for most of the period of our studies only covered a very small part of the country and we had to rely on small-scale maps too general for our use. At a later stage, the issues of the Geological Survey of Iran and the Tectonic Map of Iran by Stöcklin and Nabavi proved very useful.

One irritating feature of all maps is that they do not follow any common system for the spelling or transliteration of place names. For example, Aivaneki = Eyvanaki = Eyvankey = Eivanekey = Eiwan-e Keif = Ivan-i Kay. This has led to some difficulties in the production of our own maps to illustrate individual earthquakes (see chapter 3).

The problem is not acute for the period before 1800, because few places are mentioned in early events, and these are mostly well known. We spell these places as they are written in the Arabic or Persian texts, in accordance with the transliteration system described above (p. xiv). Places mentioned by early European travellers, where identified, are also spelt uniformly with this system and where names have changed the modern name is indicated. After 1800 a much greater number of places are mentioned, particularly villages, not only in a variety of written sources, but especially on the basis of information obtained from field studies or oral reports. While the majority of these small localities can be found or identified on the Farsi maps described above, and so transliterated consistently, the Farsi spellings themselves are often not consistent. Furthermore, names sometimes appear on one map but not on another, so that it is impossible to keep to one set to use for a standard guide. In general, however, reference is made whenever possible to the maps in the *Gazetteer*,[2] which contains both Farsi and (rather erratic) English spelling, supplemented where necessary by the other series.

2.3 Local source material

In the field study of twentieth-century earthquakes, our method has been to seek out local information about earlier events in the area at the same time. Quite often, local knowledge was brought to the front of local memory by the recent earthquake, even in areas without a strong literary output, where oral tradition provides the only reliable source of information (for example, the work of Tabandeh 1969). Information on past events derived from local sources varied. It was found to be much poorer in detail and accuracy in eastern Anatolia, Luristan, Fars and Baluchistan than that acquired from other parts of the region. Very few people could accurately remember and relate what had happened one generation ago, and even fewer would agree among themselves on where or when an earthquake had occurred. One of the reasons for this is that in the border areas of eastern Anatolia there are few sedentary people left, the wars of the late nineteenth and early twentieth centuries having caused wholesale population movements and mass emigration, with the almost total disappearance of the Christian monasteries that in the past had helped the survival of information. Similarly, in the tribal areas of Persia, there were few permanent settlements whose damage or destruction by earthquakes would be preserved in legend. Only the larger events have left a trace, some recorded in unpublished family documents, and occasionally in western archive material. Since these areas were (and some still are) almost totally uninhabited during the winter (*sardsirs*) or summer (*garmsirs*), only those events that occurred during the rest of the year would have been witnessed and perhaps recorded by the nomads (Bakhtiyaris, Luris, Qashqa'is, Mamasanis etc.). In contrast, information was found to be richer in detail and more accurate in Gilan, Mazandaran and Khurasan, even for areas that are today considered to be almost totally deserted. Quite often we came across instances of handed-down stories of recent or earlier earthquakes, the veracity of which their narrators were able to prove to us in the field.[3] Some of the most isolated settlements, along old and abandoned caravan routes and in small mountain valleys, provided the most valuable information on early and recent earthquakes. The tradition about the 1838 earthquake on the borders of the Dasht-i Lut in Sistan is perhaps the most interesting case on record. It mentions the rare event of large-scale liquefaction of desert sands and slumping of clay flats, which is also associated in oral tradition with the earthquakes of 1903 west of Turshiz and of 1953 in the *kavir* near Turud.[4]

The use in local toponymics of terms such as *zalzaleh kharabeh*, *zalzaleh kuh*, *gaud-i zalzaleh*, *zalzaleh sang* or *kharab-deh*, which are to be found particularly in Mazandaran, Khurasan and Sistan, also implies local earthquake effects.[5] We have not always been able to substantiate the true significance of these appellations. The fact that they seek to explain the curious appearance of geological formations, the frequent occurrence of landslides or the drying up of water springs, as being the result of earthquakes, is interesting even if difficult to verify.

Information was usually collected from local educated elders or people who because of their trade were familiar with local history. Quite often they led us to others in the region who possessed additional information and occasionally family records. In many instances two or three independently-interviewed individuals gave basically the same details, thus confirming the reliability of the information, and we seldom had to deal exclusively with a village fool. Without family records, oral information seldom went back for more than three generations, a much shorter period than was noticed in central and western Anatolia, with the information becoming vague and rather fabulous. It was also difficult to assess local information about early events in areas more recently affected by a large earthquake, when what we were told was often heavily contaminated by statements derived from popularised articles in the local press.

Thus, up to the end of the eighteenth century, both oral and documented local information contributed very little beyond allusions to legendary events. Field studies were, however, valuable in confirming that sites of some of the earlier earthquakes were closely connected with tectonically active regions. Under the Qajars (1794–1925) local information improves rapidly as the nineteenth century progresses, to the extent that by the end of the period, not only could almost all the events determined instrumentally be identified *in situ*, but also new events for which previously only unassociated instrumental readings were available.[6] In fact, for the first half of this century, lack of local oral or documentary evidence for the occurrence of an earthquake located instrumentally was generally found to mean that its instrumental location was in error, this seemingly negative evidence being of great value for relocating certain events.

2.4 The effect of earthquakes on local houses

One of the most striking conclusions that can be drawn from field observations of recent earthquakes is the comparative ease with which local types of dwellings in Iran are damaged or destroyed. The average house is so badly built, indeed, that some early European residents in Persia supposed that it was the prevalence of earthquakes that deterred people from building good houses or dwellings with more than one storey.[7]

Persian houses were, and to a great extent still are, of mud-wall or adobe-brick construction, covered with domed or vaulted brick roofs. The clay available is often too silty and produces walls and bricks of surprisingly low strength. Where timber for rafters and beams is available, roofs are flat and heavy, consisting of a rough boarding covered with twenty to eighty centimetres of tamped earth. Timber beams are rarely straight and are often re-used and weakened by overloading. Flat roofs are generally preferred as they provide additional space for sleeping in the hot seasons and for drying crops. In mountain villages, where houses are usually built with rubble-stone masonry laid in clay mortar in terraces, the flat roof of one house is often the yard of the house above. Thus, apart from the Caspian provinces and some regions in the Kopet Dagh and the Zagros where houses are built chiefly of wood, in the rest of the country they are all made of adobe or low-quality brick, one storey high. In villages they are built close together in clusters, separated by narrow, winding alleys. Occasionally, they are partly below street level, the ground inside having been used for making bricks with which the house was built, enlarged or repaired. Better houses on the outskirts of villages and in towns are often detached and surrounded by a garden and a high wall.[8]

Typically, these houses are very vulnerable constructions. Even houses built in the last few decades with non-traditional materials such as reinforced concrete, have little extra resistance. As a matter of fact, the introduction of new materials in the absence of proper building codes and enforceable regulations has produced a new class of highly vulnerable structures.[9] But what really makes these structures, old and new, so defenceless before the forces of nature is the *fons et origo* of all evils, neglect.[10] Repair to damage is rarely carried out with the intention of strengthening the structure. Thick plaster or glazed tiles, sometimes with inscriptions recording repairs, often conceal damaged structural elements which have not been strengthened and sometimes not even repaired.[11]

The large number of adobe houses and public buildings that collapse every year without help from earthquakes is evidence of the inherent weakness of this type of construction. The rains of May 1769 in Baghdad destroyed 4000 houses in a few hours. The earthquake of 26 February 1894 in the *dihistan* of Kirbal caused no damage in Shiraz beyond the tumbling down of some old walls. The heavy downpour of rain that followed the earthquake, however, destroyed 2000 houses in the city.[12] Such associations have often given rise to an exaggerated impression of the destructiveness of the shock itself. In 1677, Fryer noticed that 'after the snow melts it proves fatal to the houses which are built of mud; for whilst they seek to secure the roof, many times by sweeping it thence, the snow melts at the bottom, and undermines their foundations, that oft-times they become mixt with the dirt in the streets'. During the nineteenth century, rains in Tehran, Tabriz and other parts of the country washed down a multitude of walls, houses and public buildings, with casualties that could be numerous when the rains happened at night. Entire neighbourhoods were ruined every year and washed away, leaving soaked up mud houses on the verge of collapse.[13] The detrimental effects of rain or snowfalls before or after an earthquake were well recognised by the people. The snowfall that followed the earthquake of January 1924 in Kirman totally destroyed the damaged roofs and walls in the villages around Gughar. After the Bandar 'Abbas earthquake of 1902, the main concern of the people was that if it rained, all the buildings in the town would have collapsed. Similarly, the shocks that persisted for some time in Bushire in 1911 terrified the inhabitants, not so much because of their strength, but because of the state of their houses, which were already weakened by continuous rains.[14] Flash floods are often responsible for the destruction of whole villages and the breaching or bursting of small dams (see for instance Bellew 1874: 431).

In view of this, it is surprising that in contrast with Anatolia, even in the most seismic regions of Iran there is little or no conclusive evidence that local building materials and methods of construction were adapted to produce earthquake-resistant structures, or that traditional techniques for building against earthquakes ever existed. In a few instances, in the aftermath of large earthquakes short-lived attempts were made to use more adobe with wood and to reduce the height of houses by going partly underground. For example, for a few years after the 1780 earthquake in Tabriz, all new houses were intentionally built as low as possible with thicker mud walls and more wood, and the bazaars were covered only with light wooden roofs. However, no attempt was made by better-off citizens, including the governor, to build more resistant houses for themselves. Instead, they opted for the easy solution of erecting temporary timber-framed pavilions, the so-called *takht-i push* shelters, in their gardens in case of emergency and

slowly rebuilt their palaces in very much the same manner as before.[15] Similarly, temporary improvements and innovations in building houses more resistant to shocks were made after the Quchan earthquakes of 1871 and 1893,[16] and the larger earthquakes of this century, but with little impact on traditional methods of construction. Plate 1 shows a twentieth-century *takht-i push* type of earthquake-resistant dwelling, built in northern Tehran in the mid-1960s, a spherical house, totally unsuitable for poising on sloping ground.

A general observation about a typical house in Iran is therefore that its inherent strength is very low and extremely variable, and its vulnerability to earthquakes extremely high. The degree of damage or destruction observed in the field after earthquakes is proportional to the size of the housing unit or village affected. It was found that the larger the unit or settlement, the heavier the apparent damage. This is understandable when one considers how housing units and villages are built and expanded, i.e. simply by joining new adobe elements to the existing ones. As a single detached adobe house develops into a cluster of houses, and as the size and age of the group increases, it becomes increasingly probable that weaknesses and flaws will be present. This produces weak units, and the resistance of a cluster, be it a single large house or a settlement, tends to become that of its weakest unit. In many instances it was found that the collapse of a unit affected others attached to it, the damage spreading out for a considerable distance from the original failure. This is particularly clear on aerial photographs taken immediately after an earthquake, which give the impression of damage caused by bombing (plates *2a*, *2b*). Weak units were found to be due to improper repairs, ageing, the addition of a second storey which was supported by the old walls, the removal of bearing walls when rooms were added on the same floor, the collapse of *qanats* and the saturation of the ground with water escaping from the *hauz* or village pond.[17] These observations apply more to the *qal'eh* (fortress) type and less to the open, loosely knit mountain village, where in addition the use of timber reduces the vulnerability considerably. They do not apply at all to nomad camps, where local dwellings are literally indestructible by earthquakes (plate 3).[18]

2.5 Assessment of Intensity

As a consequence of these defects, the maximum Intensity in any destructive earthquake in Iran appears to be effectively the same; that is, at Intensity VIII on the Modified Mercalli (MM) scale, all adobe houses are destroyed and any Persian town or village would thus appear equally, but no more, devastated at so-called higher Intensities (plates *4a*, *4b*). Higher Intensity earthquakes can only be assessed from the behaviour of timber-framed or other types of construction with greater inherent resistance. In our field studies, wherever adobe was found in the epicentral area, it was totally destroyed, while timber-framed houses suffered comparatively less. The fact that in practice only one of these types of construction was normally available for observation in any one area made it practically impossible to assess epicentral Intensities greater than VIII (MM). The degree of damage to a

Plate 1. A modern attempt to produce earthquake-resistant constructions: one of the few spherical houses built in the early 1970s. This one is in northern Tehran.

Plate 2(*a*). Damage to adobe houses built in clusters: Aerial view of the village of Kariskh (9 kilometres west of Dasht-i Biyaz) taken after the earthquake of 31 August 1968. Notice that with the exception of a few houses southwest of the mosque (M), the village was totally destroyed (80% of the houses collapsed killing 20% of the people, see Ambraseys & Tchalenko 1973).

Plate 2(*b*). Another victim of the Dasht-i Biyaz earthquake: no domes remain intact.

Plate 3. Turkoman *yurt* in northeastern Iran after the destructive earthquake of 30 July 1970.

Plate 4. Typical behaviour of adobe constructions in an earthquake of Intensity about VIII ± (MM): (*a*) Panoramic view of Buyin after the earthquake of 1 September 1962; (*b*) Cumulative damage at Dilman (Shahpur) caused by the earthquake sequence of May 1930 in the Salmas plain: the Armenian church and a general view.

(*a*)

(*b*)

Plate 4(*a*) (continued)

Plate 4(*b*) (continued)

single adobe house was generally an indication of the weakness of the structure rather than of the strength of ground shaking, making it very difficult to assign Intensity as the *mode* rather than the mean value observed at a particular site with relatively few dwellings.

We soon realised that for the study of earthquakes in the Middle East and in particular Iran, conventional Intensity scales were too subjective and quite often misleading, especially when they were designed to describe conditions in other parts of the world. A comparison of epicentral Intensities assessed independently by different investigators of the same event showed unacceptable differences, from one to three grades. For the Dasht-i Biyaz earthquake of 1968, for which we have seven independent assessments, Intensity values ranged from VIII to XI, with eight to twenty-six kilometres as radii of the epicentral region.[19] For the Ashkhabad earthquake of 1948, our data did not permit a reliable assessment,[20] while for other events differences remained significant, particularly when Intensities were assessed by seismologists from questionnaires and press reports, without visiting the region.[21] This shows how sensitive the method is to the experience of the investigator and the means and time at his disposal.

In contrast with other more developed parts of the world, the ease with which earthquake damage in rural areas seems to have been made good suggests both a tendency to exaggerate Intensities of old and recent events, and also that local materials permit not only easy destruction but quick reconstruction. Additional evidence, such as whether or not a site was abandoned because of loss of its water supply,[22] decimation of its labour force, or decline of its commerce, is necessary for an assessment of Intensity of historical earthquakes. Details of destruction of public buildings, mosques and minarets, as well as the number of casualties, all contribute to a better understanding of the effects of the event. Survival of information about earthquake damage over a very long period, a millennium or so, though not in itself conclusive, can also be supposed to suggest comparatively high Intensity effects.

Natural exaggeration in the sources, historical and modern, is not difficult to detect. The authenticity of the source, the style of its narrative, internal evidence in the account and experience gained from processing local information, all combine to permit a realistic assessment of the gravity of the situation. For instance, statements such as that after a destructive earthquake everyone was staying on the roofs of their houses, or that they took refuge in mosques, suggest that the shock was not in fact all that destructive. There are also cases when European travellers exaggerate damage unintentionally, mistaking the many dilapidated old ruins typical of a Persian town, for signs of recent destruction (e.g. in *FO* 246 611). Comparison between Persian and European versions of an event, when possible, is also revealing. One example of how differently the same event can be reported is provided by the earthquake in Shushtar in 1928. According to Shushtari (1952: 30), the most fearful of all earthquakes in Khuzistan this century occurred during the night of 28 August 1928, when the earth shook with great force eighteen times,

causing the people of Shushtar to flee into the open; the shock recurred even more strongly the next morning. All the houses were damaged and the inhabitants left town, only the police remaining. On 2 September there were two more strong earthquakes, becoming more and more terrible and horrifying. Light shocks recurred at the end of the month, causing people to camp out; there was no great loss of life because of the strength of the buildings. This event is recorded in the *Ahvaz Consular Diaries for Khuzistan* as follows: 'On 27 and 28 [Aug.] several light earthquake shocks were felt at Shushtar. A few houses have been slightly damaged but no lives have been lost. The police have taken precautions to safeguard the place against possible disorders; on the 3rd [Sept.] several light earthquakes were felt at Shushtar' (*FO* 371 13069). The substance of these accounts is the same, but the impressions given of the gravity of the event (of magnitude $M = 5.0$) are very different.

In contrast, statements referring to fatalities among leading citizens mentioned by name, remission of taxes, famine or emigration and poems written after an earthquake all suggest a destructive event of high Intensity. Destruction of post stages, water cisterns (*ab-anbars*) and the disruption of communications also imply the occurrence of serious events.

Casualty figures are rather difficult to check and are anyway not necessarily indicative of the magnitude or the Intensity of an earthquake. In the past, because of the emphasis on effects in major centres, reported figures depended to a large extent on population distribution and density; even today, with fuller coverage of events, the same bias is maintained, although this may reflect the genuine situation: quite often the largest number of people killed were in towns or villages outside the meizoseismal region, within which semi-abandoned settlements were totally destroyed but with little or no loss of life. Precise figures, such as those given for the 856 and 1344 earthquakes in Damghan and Isfahan (45 096 and 20 casualties respectively), or for the 1485 earthquakes in Gilan, seem to be unexaggerated and most probably authentic. Others appear to be of comparable accuracy with those estimated for large nineteenth- or twentieth-century events. Considering that before 1956 there was no national census, these casualty figures may not be all that unrealistic.[23] For the whole period studied, the same figure is rarely repeated more than once and they do not seem to be connected with the favourite multiples of thirty, forty and seventy.[24] They are much lower for events in regions where timber houses are used, such as round the Caspian, in the Kopet Dagh and the high Zagros, and they are on average higher for events that occurred at night,[25] particularly in the winter or after heavy rains. Figure 2.1 shows the density of population in Iran in 1869 and 1956.

The present survival of historical buildings is not necessarily an indication that early earthquakes to which they were exposed were of low Intensity. Most of the buildings that are still standing have in fact, during their lifetime, been subjected to a number of destructive shocks and have survived through a process of natural selection. They are a very small fraction of the total number of structures that existed in early times and they represent a sample of buildings of the best final design and construction, achieved through the ages by experience, trial and error techniques, or by chance. Furthermore, frequent references occur in the sources to extensive restorations after earthquakes (or for other reasons), and even if the opportunity was rarely taken to introduce changes in design and construction to ensure a more resistant structure, these restorations, sometimes very numerous, prolonged the life of the buildings. Thus the mere existence of early monuments should not necessarily be taken on its own to mean that their sites have been free from high Intensity shaking in the past. Rebuilt or repaired *masjid-i jami's* on sites devastated by recent earth-

Figure 2.1*a*. Density of population in 1869. (After Fredy Bémont, *Les villes de l'Iran*, Paris 1969. See also *Camb. Hist. Iran*: i, 469.)

Figure 2.1*b*. Density of population in 1956.

quakes, such as Kakhk, Gunabad, Bidukht, Quchan or Damavand hardly give the right impression today of the situation created by earthquakes a few decades ago.

Similarly, the existence of tall, solitary free-standing structures such as a minaret, tower or belfry, made of brickwork, does not necessarily indicate an absence of damaging earthquakes. These special structures are far more flexible and better built than ordinary dwellings and their response to an earthquake is quite different.[26] A near-by, strong earthquake, which may cause houses to collapse, may affect a minaret less, by merely whipping off its top. On the other hand, a large distant shock, not damaging to buildings, will overturn some minarets and damage the tops of others, but this is not an indication of high Intensity shaking. These effects are determined not simply by the distance of the minaret from the epicentre, but by a variety of factors, such as the resilience, flexibility, height and aspect ratio of the individual structures. Without details of these characteristics of minarets that no longer exist, early statements of damage (see § 3.3, under 1273, 1364, 1483 and 1641) are not a reliable measure of the Intensity of shaking experienced by the structures affected, nor are these sufficiently numerous to allow any valid conclusions. Although the survival of towers, minarets and belfries over the centuries is not surprising, their disappearance nevertheless suggests the occurrence of large shocks somewhere in the immediate or distant vicinity (cf. figure 5.10). The same applies to the evidence of free-standing columns. Columns consisting of rigid cylindrical sections of stone can fail, not in bending as brick minarets do, but by rocking on their base and joints, a process that for homologous columns makes the highest the most stable. But because this type of structure has little reserve strength and totally lacks the benefit of being structurally redundant, the Intensity of early events can be assessed only from their observed effects on a large number of such structures.

Secondary effects, such as landslides, rockfalls and soil failures, as well as faulting, are also of limited value in assessing Intensity. The destruction of a village perched on a slope, caused by sliding of the ground on which it stands, and damage to houses from rockfalls or other ground deformations can, and often does, occur without the assistance of an earthquake. It is practically impossible to determine how strong or light a shock would be necessary to produce these secondary effects: earthquakes, like heavy rainfalls, act as the last straw to trigger landslides prematurely, particularly in the Zagros where mountain faces crumble rapidly without assistance from earthquake shocks. It is very probable that in the Zagros, the pre-historic slide in the Saimareh valley, which has a debris volume of 20 000 million cubic metres, perhaps the largest recorded in the eastern hemisphere, as well as the much smaller slides at Shimbar and Irene, were triggered by earthquakes. Historic events are known to have triggered slides in these regions and legends of similar occurrences in the Zagros persist into the present century.[27] Elsewhere in Persia, earthquake-induced slides and rockfalls happen more often in regions of unstable topography. Landslides from the flanks of mount Sahand, for example, often occur with or without earthquakes, in the process destroying old troglodite dwellings that have been excavated in the tuff.[28] In Mazandaran and the Atrak valley, landslides often dam tributaries of the Tijan, Nika and Atrak rivers, while in the south, parts of the Gulf coast slump into the sea.[29]

As we have seen, assessment of Intensities in Iran as well as in eastern Anatolia and Iraq presents serious problems, particularly if a scale is sought which will also be applicable to the historical period. To resolve this problem, therefore, a very simple Intensity scale was devised, describing in a most uncomplicated manner the overall effects of an earthquake in this part of the world. The scale has only five grades, which fit most descriptions met with in the sources or observations made in the field. The five grades are:

$i = 1$ Total destruction of all man-made structures, including stage posts and water cisterns (*ab-anbars*), with a large number of people killed, including leading citizens; total loss of livestock.

$i = 2$ All dwellings destroyed and many public buildings ruined, with numerous casualties and some loss of livestock.

$i = 3$ Many houses ruined and a few people killed.

$i = 4$ Few dwellings ruined, public buildings cracked, without fatalities.

$i = 5$ Shock widely felt, causing concern and in places panic. Damage to isolated public buildings, survival or collapse of minarets and free-standing walls, faulting, landslides and rockfalls, as well as liquefaction and slumping of the ground were not used as criteria for grading Intensity. As we are not in a position to be completely rigorous in our definitions of various ratings, our intention has been that the investigator should have enough leeway to use his own judgement without being hemmed in by a scale that is too specific. Provided that the same investigator assesses Intensities for events over the whole period, his ratings will reflect differences in Intensity more faithfully than absolute values.

A large and representative sample of the material from which these Intensities are assessed is presented in the case histories in the following chapter; the same detail is not available for all events, in some the balance being towards historical information, in others towards technical. For all events up to the present for which documentary or field evidence of damage is available, the authors assigned Intensities (i) independently; occasionally a compromise was reached by adding or subtracting a fraction of an Intensity grade, either to satisfy differences of initial assessment or to account for special 'contaminating effects' such as the presence of timber structures, which moderated damage, or the occurrence of aftershocks, landslides or rockfalls that added to the damage. At the same time, Intensities (I) on the Modified Mercalli scale were assigned for events, particularly of this century, for which there was sufficient data.

Using the simplified Intensities (i) thus assessed, a map of the meizoseismal region of each earthquake was prepared, usually on a half-million scale, on which were plotted the Intensities assigned to each place identified or visited. The meizoseismal or epicentral region of an event was then defined

as the area within the highest isoseismal, which was drawn an elliptical shape to contain the bulk of the places to which the highest Intensity (i_0) was assigned. To simplify matters, the relief of terrain within this region is depicted approximately by the river courses and low-lying areas. Some of the maps thus prepared accompany the case studies in chapter 3, on which other features associated with the earthquake, such as landslides or faulting, are also shown.

The macroseismic 'epicentre' of an earthquake was therefore defined by the geographical co-ordinates of the centre of the ellipse, and the size of the epicentral region by the radius (r_0) of the equivalent circle. For small, shallow depth earthquakes this radius was only a few kilometres, but for large events could reach up to thirty kilometres. However, because the epicentral or higher Intensity (i_0) or (I_0) was taken to be the value assigned to the largest number of observations in the meizoseismal region, the value of (r_0) becomes a function of the higher Intensity. Thus for a small earthquake, the meizoseismal region may be an area where Intensities perhaps did not exceed $i_0 = 4$, while for a large event, meizoseismal Intensities may exceed $i_0 = 2$.

2.6 Macroseismic epicentres

Data provided by historical sources and field studies are generally adequate to permit accurate locations of the meizoseismal region of earthquakes of the last two centuries. For events before about 1800, however, it is often difficult to ascertain the true epicentral region of a given event.

When several places or areas are mentioned, without any apparent indication of a meizoseismal region and no evidence of serious damage at any one location, the epicentral region can be estimated fairly closely by reckoning its position from the various places named. This technique was used to identify potential epicentral regions, not only of historical events but also of early twentieth-century earthquakes for study in the field, particularly in sparsely populated areas. The meizoseismal region of shocks felt in towns and villages round the borders of deserts or mountainous regions was often located successfully in this way, with an error of a few tens of kilometres. For instance, using the places at which the earthquake of 22 July 1927 was felt (see chapter 3), the location error is about forty kilometres.

In cases where only the name of a province or district is given as being affected, no precision is possible, although in the early period at least, the mention for instance of the Jibal province as the epicentral region and the simultaneous reporting of the shock in say, Baghdad and Mosul, suggest a large area with an uncertainty of location of 100 to 300 kilometres.

It is often the case, particularly after the Caliphate period, that only one place is named as affected, and then a great deal depends on the extent of detail available about the effects in that place and in its dependencies. Internal evidence often suggests that destruction was not particularly severe and that outlying villages also suffered damage, indicating that the meizoseismal region should be sought some small distance from the town itself. Alternatively, mention of a town but with no details of serious damage may suggest either the occurrence of a local earthquake of small size, or that of a larger event some distance away.

The problem of location is intimately connected with that of assessing the size of the earthquake. One measure of this magnitude is the radius of perceptibility (r') of the event, or the average epicentral distance at which the shock was felt. Our sources are generally adequate to permit an assessment of this parameter. Long-period effects, such as the collapse of free-standing high walls, or tops of minarets, the sloshing of water out of ponds and the swinging of curtains and screens in houses, all help assess the relative size of an event, and thus the location of the epicentre in relation to places where these effects are reported.

Each macroseismic epicentre has been graded according to the quality and quantity of the macroseismic information from which it has been derived.

Grade (*a*), or good determination, implies the existence of good isoseismal maps or sufficient information describing the meizoseismal region to help define its limits. This may include evidence of faulting, or significant changes in the water supply and other factors deduced from written sources or field studies, or both. Such precision is rarely encountered in early historical documents.

Grade (*b*), or moderate accuracy, has been assigned to epicentral locations deduced from evidence of the locality or district (*dihistan*) where the Intensity was maximum, with indications of places or areas affected at lower Intensities, that help assess the limits of the meizoseismal region.

Lower grades, all denoting poor determination, are assigned according to the types of macroseismic data available:

Grade (*c*) has been used for locations based only on the evidence of high Intensity shaking reported at one locality or town, and also for locations calculated only from lower Intensity isoseismals, from radii or perceptibility or from a combination of the two. Indirect evidence, such as abandonment of sites, may also help assess the location of the epicentral region.

Grade (*d*) implies that the shock was felt locally; it may indicate the approximate location of a small shock, or the occurrence of a large magnitude event some distance away, for which data are still incomplete.

Grade (*e*) denotes a very approximate location of events that were probably large but for which we have insufficient details; such epicentral locations are not shown on figures 5.2–5.4 and are of value only on a global scale. Similar locations but not based on reliable data are deemed to be associated with dubious events, which are not listed. They are referred to in the following chapter.

These assessments are shown in tables 5.1 and 5.2 in the column 'q'.

3

Case histories

3.1 Selection of case histories

This chapter contains descriptions of some of the largest
and most interesting earthquakes to have occurred in Persia.
Over two hundred in number, these events have been selected
partly because of their individual destructiveness, but chiefly
for their general illustration of the effects of earthquakes in
this part of the world. Our accounts of them are based on
information retrieved from the various sources described in
the previous two chapters. These accounts contain the essential data available and an assessment of this material in the
light of the relevant factors already discussed, thus illustrating
some of the problems associated with evaluating the seismicity
of the region under study. A quantitative evaluation of these
earthquakes is incorporated in tables 5.1 and 5.2, which list all
the historical events that have been identified up to the early
nineteenth century and the larger earthquakes thereafter (see
p. 156).

From the tables it will be observed that a number of
earthquakes have not been described in the case histories that
follow, some of them large events. In the earlier period, such
omissions are rare and occur generally when the information
available is small; most of the references that have not been
discussed are apparently to minor events. Similarly, case
studies of some historical earthquakes have been given very
little space, either for lack of details or in cases where the
sources quoted provide all the information without compli-
cation or need for further comment. In most cases, discussion
of the earthquakes is as full as possible, particularly when
there is a wealth of data or when it is necessary to demonstrate
characteristic problems of handling the source material. Our
accounts of these events are written with historians, seismologists, archaeologists and researchers in related fields in mind.
The main burden of textual analysis, critical documentation
and support for interpretations offered falls on the notes to
the chapter, which may be ignored by those prepared to

accept statements as they stand without more ado. The notes may also be consulted for information pertinent to particular aspects of the events, such as their historical or geographical context, or their effect on monuments situated in and around the epicentral area.

In the later period, from the mid-nineteenth century onwards, pruning has perforce been more ruthless and a large number of shocks listed in tables 5.1 and 5.2, themselves a selection from the total data available, are not discussed in this chapter. Omissions are either large, well documented events for which a complete list of references is given, or less well-known earthquakes, the accounts of which perhaps do not give rise to any points of interest, or which occurred in areas already identified and discussed in connection with other more important events. These omissions, dictated by lack of space or occasionally lack of conclusive information, can be followed up in the sources, to which the most appropriate references are quoted.

All the earthquakes discussed in this chapter, as a matter of principle and for the sake of completeness, are documented as fully as possible; inevitably, however, there are cases where such thoroughness is misplaced and references are omitted when they add no relevant details. Sources which give inaccurate information are generally mentioned for comparison. As many as possible of the works that have been found to contain information are referred to in the course of these various accounts, particularly when they contribute valuable or original material; any less useful statements they contain, which may not have been referred to, on examination may easily be seen in perspective by the reader himself.

Maps accompanying about a quarter of the events described were originally mostly drawn on a half-million scale, of an accuracy consistent with that of our data. They help the reader identify the location of the earthquakes, their extent of destruction and their associated ground effects. The maps show places for which details of effects are available and the location and number of other villages in the area, suggesting the differing densities of population in affected regions. Place names are spelt as uniformly as possible, ideally on the basis of a transliteration of the name as written in Farsi. Maps illustrating earthquakes in closely adjacent regions are drawn with some overlap, to produce a composite map if so desired. Reference to these maps should indicate the method used by the authors to assess macroseismic epicentres, the equivalent radii of destruction and the basic pattern of the spread of damage (cf. § 2.5). Symbols used on the maps are given in the caption of figure 3.1, but are only employed systematically as the information becomes fuller, towards the nineteenth century.

Items of information on separate topics connected with the occurrence of earthquakes in the region are brought together at the end of this chapter.

3.2 Pre-Islamic period

There are not many statements concerning earthquakes in the pre-Muslim period, and the little we know about the seismicity of the region at that time is based on archaeological evidence and on a few early historical accounts.

The earliest event may be placed in the district of Buyin Zahra. Archaeological excavations in the graveyard and mound of Sagzabad show that a devastating earthquake in the third millennium B.C. must have put an end to the settlements in the district (unpublished information and Negahban 1971). Archaeological evidence also suggests that an earthquake in the second millennium B.C. totally destroyed Ak Tepe in the region of Ashkhabad (Atlas). Similar evidence suggests that Godin Tepe near Kangavar was affected by an earthquake in the seventeenth century B.C. (Young 1968: 160).

Much of the information for earthquakes for the next millennium is based on the royal correspondence of the Assyrian Empire, and it concerns either the region of Nineveh, which is just outside the area of our interest, or localities of the empire which remain unknown. Thus, letters record the restoration of the temple of Emašmaš and of the tower gate in Nineveh after a damaging earthquake in the thirteenth century B.C. The same event is referred to in later correspondence, which adds that in the twelfth century the tower gate at the front of the temple of Ishtar in Nineveh was damaged again by an earthquake and that it was repaired. This damage to the temple may possibly have been confirmed by archaeological excavations (Thompson 1937; Mallowan 1966). Royal correspondence of the same period also mentions another earthquake that 'lasted for a whole day' and which was the only shock recorded in three generations (Thompson 1937; Waterman 1929: 136).

An Assyrian letter written in one of the outlying towns of the empire some time in the eleventh century (plate 5) says 'On 21 Elul [Syrian calendar; see Grumel 1958: 174] an earthquake took place. All the back part of the town is down; all the wall at the back of the town is preserved (except) 30½ cubits therefrom being strewn and fallen on the near-side of the town. All the temple is down . . . Let the Chief [architect] come and inspect' (Thompson 1937). Four other Assyrian letters from the Sargonid Dynasty, i.e. probably to be dated 750–612 B.C., refer to earthquakes. One mentions an earthquake (rebu) at Dur-Šarruken (i.e. Khursabad, northeast Nineveh) on 9 Addar. Another quotes an omen 'if the earth quakes in the month Simanu settlements in abandoned outlying regions will be settled again' (Waterman 1929: 22, 27, 79, 80, 127, 136, 247, 251). However, it is not possible to identify the localities that were affected. It is quite clear that in some parts of the empire earthquakes were rare and in others they occurred more often. One omen text appears to believe in periodicity of earthquakes 'for in twenty-one years an earthquake will correspond with an earthquake' (Waterman 1929). The earthquakes of the sixth century B.C. at Susa given by Sieberg (1932: 803), we have not been able to substantiate.

The first earthquake in Persia recorded in history is that of the fourth century B.C. which devastated the region of Ray. According to Duris of Samos, the author of a history of Greece and Macedonia who flourished about 350 B.C., 'Rhagae[1] in Media has received its name because the earth about the Caspian Gates had been rent[1] by earthquakes to such an extent that numerous cities and villages were destroyed, and the rivers underwent changes of various kinds'

(Strabo: i.3.19). Poseidonius of Apameia, in his history written in the middle of the second century B.C., adds that in this earthquake 'numerous cities and 2000 villages were destroyed' (Strabo: xi.9.1). Later writers, for instance Apollodorus of Artemita who flourished about the middle of the first century B.C., add that Rhagae was rebuilt by Seleucus Nicator (312–280 B.C.), and that it was renamed Europos (Strabo: xi.13.6). This major earthquake must have occurred after Alexander's passage through Rhagae in 330 B.C., probably during the reign of Seleucus, and it is likely that these accounts refer to more than one destructive event, similar to those occurring in A.D. 743 and 855.

For the next earthquake we have only archaeological evidence. It occurred during the first decade B.C. and totally destroyed old Nisa and the settlements east of Bagir, west of Ashkhabad (*Atlas*). The degree of destruction suggests the occurrence of a large earthquake in this part of the Kopet Dagh.

Some historical evidence has survived of an earthquake in the region of Ararat in A.D. 139 (Stepanian 1942, *Atlas*). It is not clear, however, whether this event is the same as that of the fourth century A.D. in Sipandag (?) mentioned by Dzhanashvili (1902) and Alishan (1882) for 338–40.

From the fifth century A.D. earthquake notices become more numerous but they refer to events not further east than Van, Mosul and Kirkuk. We hear that late in the fourth century Mosul was seriously damaged (Nau 1911) and that in A.D. 461 the district of Malazgirt, north of Van, was destroyed (Hovhannes Avagerets, in Hakobyan 1956: 27). It is also said that at the birth of the prophet Muhammad (traditionally in November 570) an earthquake destroyed thirteen towers of the palace of the Khusraus at Ctesiphon south of Baghdad (Ya'qubi, *Tarikh*: ii, 5). While the occurrence of the earthquake may be admitted, its symbolic coincidence with the birth of Muhammad need not be taken seriously.

Another destructive earthquake in Taraun, between Kigi and Mush in Armenia, is mentioned by contemporary writers (Yohannes Mamigonian: ii, 369; cf. Michel: x, 23) for April 601. Seismic activity in this part of Anatolia seems to have continued intermittently for about forty years (Yohannes Mamigonian: iii, 370, v, 381; Michel: xi, 5) with a damaging shock occurring in 628 in the region of Kirkuk (?) (*Chronicon miscell. ad annum 724*: 139; Hoffman 1880: 77).

Places mentioned as the sites of earthquakes in the pre-Islamic period are shown in figure 5.2.

Plate 5. Letter from Nineveh describing an Assyrian earthquake (BM. 123358: TH. 1932-12-10, 301). For a translation, see Thompson (1937).

3.3 Earthquakes in the Islamic period

The first event recorded in our area of interest was in
658 **Basra**. A shock in 37–8 H. that caused no damage.[2]
Another shock in Basra is recorded for 714, but it seems
to have been a spurious event.[3]

c. **Sistan**. An earthquake 'such as no one had experienced
734 before' devastated Sistan in the period 111–120 H. It is
probable that Zarang, capital of the province, was particularly affected.[4]

735 **Vayots-dzor**. A locally destructive earthquake in
Armenia killed 10 000 people in the valley of the upper
courses of the Arpa-chai. Aftershocks continued for
forty days.[5]

743 **Caspian Gates**. Late in the spring of 743 there was a
destructive earthquake east of Ray in the region of the
Caspian Gates, that is, the valley of Tang-i Sar-i Darreh,
which runs through the Kuh-i Namak.[6]

763 **Khurasan**. In 763 there was a catastrophic earthquake in
Khurasan which threw mountains from their place and
left neither trees nor rocks in position. The area involved
is impossible to identify but the shock, which triggered
landslides or was associated with ground deformations,
does not seem to have affected any centre of population.[7] Possible locations, traversed by important
routes, would be the Kuhistan district (Khwaf, Qayin,
Tabas) or the Jajarm, Juvain, Nishapur area.

815 **Sistan**. In 199 H. the third destructive earthquake in
eighty years occurred in Sistan.[8]

819 **Balkh–Taliqan**. In Dhu' l-Hijja 203[9] a catastrophic
earthquake in eastern Khurasan destroyed a quarter of
the city of Balkh and ruined the *masjid-i jami'* there.[10]
Other places severely affected were the towns of Faryab
(Daulatabad) and Taliqan (Qal'eh Vali) and the districts
of Juzjan in the west and Tukharistan in the east. Many
houses were destroyed, with heavy casualties in these
areas. The shock was also felt in Marv and Transoxania.[11]
As a result of the earthquake the desert at Sidreh
between Shaburqan and Balkh was flooded by an
excessive rise of the water table, which turned the
country into a fertile area.[12] Aftershocks lasted for a
long time.

840 **Ahvaz**. In 225 H. there was a destructive earthquake in
the Zagros. In Ahvaz many houses, including the *masjid-i
jami'*, were destroyed and the people left the city. The
hill overlooking Ahvaz was fissured. Aftershocks continued for some time.[13]

849 **Herat**. A damaging shock in Herat in 234 H. caused some
houses to collapse.[14]

855 **Ray**. A major earthquake in Ray destroyed many houses
and caused a large number of casualties in the district in
241 H. The shock was strongly felt, perhaps with some
damage, in Qum and Kashan. Aftershocks continued for
more than a month.[15]

856 December 22 **Qumis**. On Tuesday, 18 Sha'ban 242, there
was a catastrophic earthquake in the eastern Alburz
which devastated the district of Qumis and the region of
western Khurasan dependent on Nishapur (figure 3.1).[16]

Along a fertile tract of land running for 350 kilometres
between the Alburz and the Dasht-i Kavir, from Khuvar
to beyond Bustam and in parts of Tabaristan and Gurgan,
200 000 people were killed and practically all villages
were ruined. In the district of Qumis, the earthquake
was worst at Damghan, which was half destroyed with
45 096 casualties. In the mountain regions there were
extensive ground deformations, including probably surface faulting. The old city of Shahr-i Qumis, former
capital of the province, was also destroyed and was probably finally abandoned after the earthquake.[17] One third
of Bustam collapsed and the region between this town
and Damghan still showed the effects of the earthquake
two generations later.[18] Tabaristan and Gurgan were also
affected by the shock,[19] which had a disastrous effect
on the water supplies of the district of Qumis, either
causing the drying up of springs and *qanats*, or triggering
landslides which dammed the streams flowing down to
the plain.[20]

Outside the meizoseismal region, Nishapur[21] to
the east and the Jibal province to the west and southwest of Qumis were also strongly shaken, the earthquake
being felt in Ray, Qum, and as far as Isfahan, where it
caused great concern.[22] Aftershocks continued for some
years, probably causing damage in western Khurasan.

858 **Tabriz**. An earthquake in 244 H. almost totally
destroyed the growing town of Tabriz, which was rebuilt
on the orders of the Caliph.[23]

859 **Khurasan**. This was presumably a damaging aftershock
of the 856 earthquake in the region of Qumis, towards
Khurasan.[24]

859 **Baghdad**. Mada'in and Baghdad were shaken by an earthquake in 245 H., probably causing some damage.[25]

863 February 13 **Dvin**. A locally destructive shock occurred
in the region of Dvin, with casualties. The town of Dvin
was badly damaged. The walls and public buildings,
together with many houses, were shattered and for three
months continuing shocks kept the survivors camping in
the open.[26]

864 January **Ray**. Ray was affected by an earthquake in Dhu
'l-Hijja 249 which destroyed many houses and killed a
large number of people. The survivors left the city and
stayed in the surrounding plains. The shock seems to
have been experienced in Qazvin as well.[27]

872 June 22 **Saimareh**. On 11 Sha'ban 258 an earthquake
devastated the region of Saimareh, following a damaging
foreshock the previous day. Most of the town was
destroyed. The walls fell down and about 20 000 people
were killed. It is probable that the town of Sirvan was
also affected. The shock was possibly felt in Iraq at
Wasit and Basra, and was also responsible for large-scale
landslides in the Saimareh valley.[28]

874 **Gurgan**. Late in 874[29] a locally destructive earthquake
killed 2000 of the troops that had taken refuge in
Gurgan. The damage was so serious that many of the
Gurganis emigrated to Baghdad.[30] Violent shocks continued for three days.[31] It is unlikely that the destruc-

tive effects of the earthquake extended beyond the limits of Gurgan (see figure 3.1).

893 December 28 **Dvin.** In the night of 15 Shawwal 280 an earthquake in the district of Ararat devastated Dvin (figure 3.2). Houses, palaces and churches were destroyed and many of the inhabitants perished in the ruins.[32] The area worst affected was that of the city of Dvin and its immediate surroundings, which had already suffered considerable damage in an earthquake thirty years before. All but about 100 houses were destroyed, together with the great church and palace of the Catholicos, and 30 000 people were killed, Damage extended over the plateau of Artashat where landslides added to the destruction, and Grigor, Bishop of R'tshunik and some of his followers who happened to be in retreat in the mountains, perished.[33] This was a locally destructive earthquake affecting a rather small but densely populated region. Shocks recurred for five more days, adding to the damage.[34]

902 June **Baghdad.** A series of earthquake shocks and thunder storms in Rajab 289 caused some concern in Baghdad.[35]

906 April **Vayots-dzor.** An earthquake in the region of Vayots-dzor in Armenia destroyed a number of settlements and monasteries, among them those of Karakop' and Khotakerits. A long sequence of aftershocks contributed to the abandonment of the region for some time.[36]

912 May **Kufa.** In Ramadan 299 a violent hailstorm accompanied by an earthquake destroyed many houses in Kufa, causing many deaths.[37]

914 April **Bukhara.** It is possible that an earthquake in Ramadan 301 was responsible for the destruction of the *masjid-i jami'* in Bukhara in which many people were

Figure 3.1. A.D. 856 (22 December), Qumis.
Symbols used on maps illustrating case histories:
■ Sites destroyed or heavily damaged with many casualties.
● Sites heavily damaged or partly ruined with some casualties.
□ Sites that suffered considerable damage, without casualties.
• Other sites for which there is insufficient information of small damage.
✧ Instrumental epicentre for twentieth-century events. If epicentre is not shown its location is outside the area covered by the map.
Dashed elliptic contours or shading shows average extent of meizoseismal area.
Heavy solid lines show location of earthquake fault.
Heavy dashed lines show inferred location of fault break.
○ New springs of water appeared.
Dates are given in local time; when different, GMT is shown in brackets.

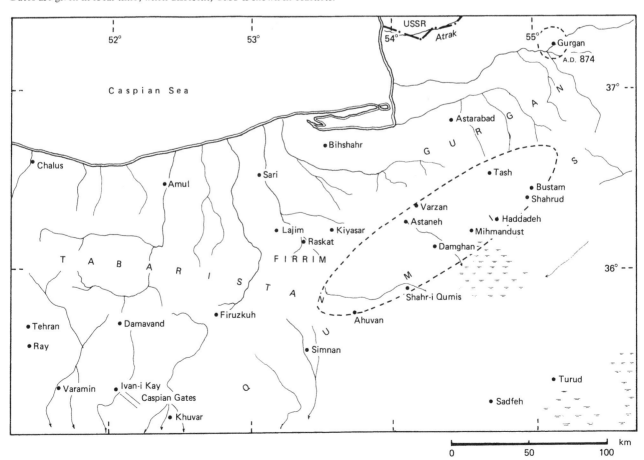

killed. In the whole city many people died, leaving it deserted.[38]

943 August **Atrak–Nisa**. In Dhu' l-Hijja 331 a catastrophic earthquake in the district of Nisa destroyed many villages, killing more than 5000 people.[39] In the Samalqan valley more than thirty villages were overwhelmed by landslides and ground deformations appear to have had significant effects on the valley streams, damming and disrupting the flow of water.[40]

956 **Hamadan**. Asadabad and Hamadan were seriously damaged by an earthquake in this district in 345 H. Many houses were destroyed, including the government office in Hamadan, killing a large number of people.[41]

958 February 23 **Ray–Taliqan**. On 1 Dhu 'l-Hijja 346 there was a catastrophic earthquake in north-central Persia. It destroyed all villages in the districts of Ray and Taliqan, both in the plain and in the mountains, and much of the city of Ray was totally ruined, heavy casualties being reported from both districts.[42] In Taliqan there were only 30 survivors and in the district of Ray 150 villages were destroyed, one village in the mountains being overwhelmed by landslides. A mountain near Ray was fissured and water poured out of the ground.[43] In the mountains of Ruyan to the north of Ray, large-scale landslides blocked the course of a river whose waters receded to form a lake.[44] Damage extended northwest into Dailam and south to Qum and Kashan. The shock was possibly felt in Isfahan and as far as Baghdad.[45] Damaging aftershocks continued for forty days, and were felt throughout north-central Persia. It is possible that the earthquake was connected with an abnormal drop in the level of the Caspian Sea which, however, seems to have occurred before the event (figure 3.3).[46]

958 April **Hulvan**. In Nisan 347 H. an earthquake destroyed Hulvan, modern Sar-i Pul-i Zuhab, killing many people in the Jibal.[47] The shock, which was felt in Baghdad, and its aftershocks, which continued intermittently throughout the early part of the year, affected the underground water supplies in the Zagros.[48]

977 November **Baghdad**. Several shocks were felt without damage in Rabi' I, 367.[49]

978 June 17 **Siraf**. A locally destructive earthquake occurred at the port of Siraf in Fars on Sunday 7 Dhu' l-Qa'da 367.[50] Most of the houses in the town were damaged or destroyed and more than 2000 people were killed. Aftershocks continued for a week and the inhabitants took to the sea.[51]

1008 April 27 **Dinavar**.[52] On the night of Sunday 16 Sha' ban 398 there was a destructive earthquake in the central Zagros. Damage was concentrated in the important city of Dinavar which was totally destroyed with the loss of more than 16 000 people, apart from those who were overwhelmed by landslides.[53] Survivors built shelters outside the town, where the loss in personal property was more than could be estimated.[54]

1008 **Siraf**. In the Spring of 398 H. there was an earthquake at Siraf and along the coast of the Persian Gulf. Many people were killed and a number of ships were sunk, probably by a seismic sea-wave (tsunami).[55]

1042 November 4 **Tabriz**. A catastrophic earthquake occurred in Tabriz late in the evening of Thursday 17 Rabi' II, 434. Part of the city was totally destroyed and part was undamaged. 40 000 people are said to have perished.[56] There is no evidence that destruction extended far beyond the region of Tabriz, but the city was finally ruined by the violent aftershocks that continued for

Figure 3.2. A.D. 893 (28 December), Dvin.

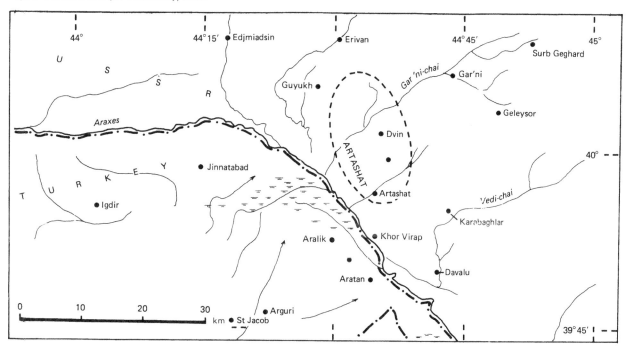

some time.[57] The citadel, walls, houses, baths and markets of the city, as well as most of the ruler's palace, were destroyed.[58]

There seems to have been no long-term detrimental effect on Tabriz and no decision to move the site of the city, which was rebuilt a few years after the earthquake.[59]

1052 June 2 **Baihaq.** A destructive earthquake took place in the region of Baihaq and particularly around the capital, modern Sabzavar. The main shock on 1 Safar 444 and continuous aftershocks for more than a month reduced the city and its walls to ruins. The earthquake was widely felt, and the year in which it happened was still recalled as the year of the earthquake a century later. It killed many people and the city walls remained in ruins for twenty years.[60]

1052 **Khuzistan.** A destructive earthquake shook Arrajan and the regions around Ahvaz in 444 H., causing walls to collapse and battlements to be thrown down from castles. The main shock was felt throughout Khuzistan, affecting, among other places, Izeh (Malamir) and particularly the region around Bihbahan, where many villages were ruined. A large mountain there was fissured, possibly by landslides. Aftershocks were repeated several times during the year.[61]

1058 December 8 **Western Zagros—Hamrin.** At dusk on 18

Shawwal 450 a widely felt earthquake in the western Zagros caused considerable damage, killing many people. In Baghdad slow ground movements caused panic and the collapse of many houses. The shock caused great concern in Takrit and it was felt as far as 'Ana, Mosul, Hamadan and Wasit.[62]

1063 December **Wasit.** A prolonged earthquake shock was felt without damage in Wasit in Dhu' l-Hijja 455.[63]

1064 **Ani.** An earthquake at Ani in Armenia is said to have caused considerable damage to the fortifications of the city.[64]

1066 May **Kuhistan.** There was a series of earthquake shocks in Khurasan which continued for some days in Jumada II, 458. In Kuhistan, mountains were split and a number of villages were totally destroyed. Many people perished and the survivors remained out in the open.[65]

1069 **Jahrum.** An earthquake at Khurshah near Jahrum allegedly caused a loss of water in the castle, thus forcing its defenders to sue for peace.[66]

1085 May **Arrajan.** In Muharram 478 Arrajan and the districts bordering it were shaken by an earthquake, causing many deaths and the collapse of the *masjid-i jami'*. Crowds of men and beasts were buried beneath the wreckage. The shock was felt in both Khuzistan and Fars.[67]

1087 November **Hamadan.** Hamadan and the nearby region of

Figure 3.3. A.D. 958 (23 February), Ray–Taliqan.

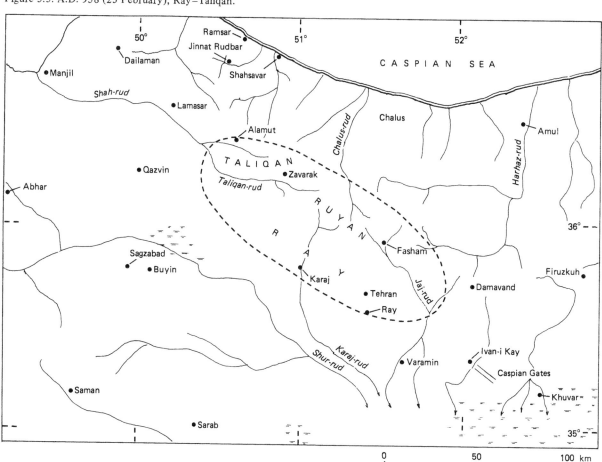

the Jibal were shaken by an earthquake in Sha'ban 480. Two towers of the castle of Hamadan fell down and two districts on the outskirts of the town were destroyed. Many houses collapsed and a number of people were killed. Strong aftershocks lasted a week and the inhabitants went out into the open country until the shocks subsided.[68]

1094 February **Baghdad.** An earthquake shock was felt in Baghdad in Muharram 487.[69]

1102 **Damghan.** An earthquake was experienced at Girdkuh in the Damghan region.[70]

1102 February 28 **Herat.** A number of houses and other buildings in Herat were destroyed by an earthquake which heavily damaged the *masjid-i jami'*, with some casualties.[71]

1118 April 3 **Western Zagros.** An earthquake in the western Zagros on 9 Dhu' l-Hijja 511 was widely felt in Kurdistan and in Iraq. In Baghdad slow ground movements ruined a number of houses, particularly on the western side of the city, a few of which collapsed without casualties.[72]

1119 December 10 **Qazvin.** A severe earthquake on the night of 5 Ramadan 513 in the region of Qazvin killed many people and caused extensive damage. The walls of Qazvin and one-third of the city was destroyed, with serious damage occurring to the mosque of Abu Hanifa,

which needed subsequent restoration work. Aftershocks continued for a year.[73]

1127 **Firrim–Chahar-dangeh.** A major earthquake in the Hazar Jarib district of southern Mazandaran caused wholesale destruction of the villages of the Firrim district, which is a wide valley in the mountains east of Pul-i Safid (figure 3.4). The villages of Kunim and Zarim were more or less completely ruined, and Daulat was carried by a landslide to the other side of the stream on which it was situated. The whole of the Hazar Jarib must have been affected by this earthquake, for Firrim is in the Dau-dangeh division of the region and Kunim and Zarim are in the Chahar-dangeh.[74]

1130 February 27 **Jibal.** Early in the evening of 16 Rabi' I, 524 there was a destructive earthquake in the western Zagros. The shock caused widespread damage in al-Jazira, in Iraq as far as Mosul and in the Jibal. In Baghdad slow ground motions that persisted for a long time caused the collapse of houses in the eastern and western parts of the city, without casualties. Aftershocks continued for some time.[75]

1135 July 25 **Kurdistan.** On 11 Shawwal 529 a destructive earthquake in Kurdistan caused a large number of casualties. The shock was strongly felt in Iraq, Mosul, the Jibal region between Hamadan and Maragheh, as well as

Figure 3.4. A.D. 1127, Firrim–Chahar-dangeh.

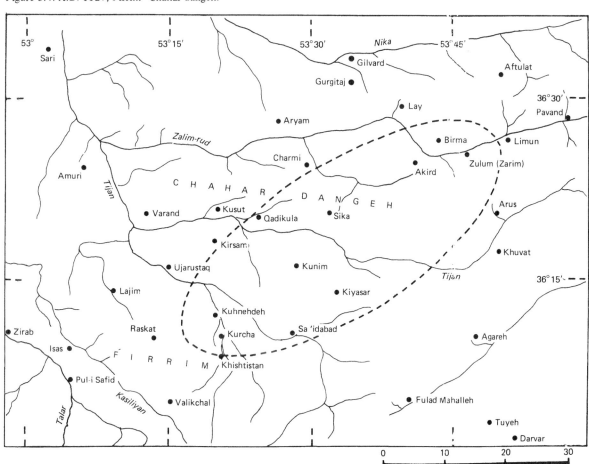

in Baghdad. Violent aftershocks continued in succession until after the end of the month, causing additional damage.[76]

1135 August 13 **Kurdistan.** A violent shock on 1 Dhu' l-Qa'da 529, possibly originating from Kurdistan, was widely felt. In Baghdad ceilings cracked and walls collapsed. Tremors continued for some time, causing panic. This was probably a strong aftershock of the previous earthquake.[77]

1144 May 29 **Baghdad.** A strong shock on 24 Dhu' l-Qa'da 538 was felt without damage in the city.[78]

1145 **Nishapur.** In 540 H. a locally destructive earthquake tempted some of the inhabitants of Nishapur to emigrate.[79]

1150 April 1 **Hulvan.** On the morning of 1 Dhu' l-Hijja 544 there was a destructive earthquake in the region of Hulvan (Sar-i Pul-i Zuhab), causing ground deformations in the mountains. Ribat al-Bahruzi was destroyed and a large number of Turkoman nomads were killed. The shock was strongly felt in Baghdad, where the ground moved in waves a number of times, cracking some walls.[80]

1156 February **Northern Iraq.** An earthquake shock in Dhu 'l-Hijja 550 was felt in northern Iraq and possibly in the mountain areas bordering with Kurdistan. Baghdad is among the places said to have felt the earthquake.[81]

1177 May **East Buyin Zahra.** In Dhu' l-Qa'da 572 an earthquake destroyed many towns of Persian Iraq, along the southern slopes of the Alburz up to the region beyond Ray. The cities particularly devastated were Qazvin and Ray, where many people were killed. Internal evidence suggests that the Ray area, eastern Buyin Zahra and the Karaj settlements were the worst affected.[82]

1179 April 29 **Irbil.** A destructive earthquake in the Great Zab on 12 Dhu' l-Qa'da 574 ruined castles and villages in the region of Irbil, killing a great number of people. To the north of Irbil, large-scale landslides dammed the river for two years. The shock was felt in Armenia and it was perceptible in Baghdad.[83]

1191 **Hamadan.** A strong shock was felt in Hamadan without damage.[84]

1194 March **Najaf.** A shock, which was widely felt in Iraq in Rabi' I, 590, caused some damage at Najaf.[85]

1209 **Nishapur.** A catastrophic earthquake, felt throughout much of western Khurasan, almost totally destroyed the district of Nishapur in 605 H.[86] Few of the buildings in Nishapur withstood the shock, exceptions being the mosque of Mani'i and the main square. The rest of the city collapsed killing a great number of people, despite the fact that foreshocks warned many of them to take flight into the open. Damage was equally heavy in the countryside, where in several villages not a soul escaped alive.[87] In all, about 10 000 people were killed (figure 3.5). Aftershocks continued for two months and the city of Nishapur was rebuilt on the same site.[88]

1226 November 18 **Shahrizur.** A destructive earthquake in Kurdistan on 25 Dhu' l-Qa'da 623 ruined Shahrizur and destroyed six other castles in the area and many villages.

Figure 3.5. A.D. 1209, Nishapur.

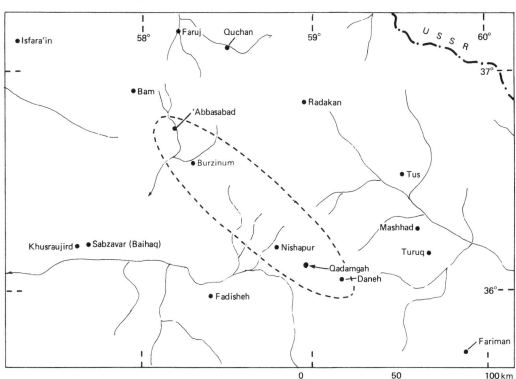

The shock was widely felt in Persia and Iraq as far as Mosul, where it caused some concern. Aftershocks continued for over a month.[89]

1238 **Gunabad.** A destructive earthquake occurred in Gunabad around 635 H.[90]

1251 **Nishapur.** An earthquake in Nishapur in 649 H. totally ruined part of Shadyakh.[91]

1270 October 7 **Nishapur.** On the morning of 19 Safar 669,[92] a catastrophic earthquake occurred in Nishapur, ruining the former suburb of Shadyakh and a number of villages, with the loss of 10 000 people. All the major buildings were affected, including the minaret of the *masjid-i jami'*. Aftershocks continued for two weeks almost without a pause. Finally, Nishapur was rebuilt some distance away from its previous site.[93]

1273 January 18 **Tabriz.** On the night of Wednesday 18 Kanun II, 1584 Sel. there was a large earthquake in Azarbaijan.[94] In Tabriz, many houses collapsed and the tops of the minarets fell off, but damage was not excessive and only 250 people were killed. The shock caused no damage to the Christian church but some of the mosques were overthrown.

This earthquake probably occurred some distance from Tabriz, but no information is available for the destruction caused outside the city, which was the capital of Persia at this period. Although the damage to Tabriz was comparatively small, the event stands out as an important earthquake in the context of a long period of subdued seismicity since the 434/1042 disaster.[95]

Eighteen shocks in the first twenty-four hours were followed by intermittent shocks for about four months.

1291 **Shiraz.** It is alleged that an earthquake in 690 H. caused some damage to the Masjid-i Nau in Shiraz.[96]

1301 **Firrim.** In 700 H. an earthquake totally destroyed many villages in southern Mazandaran, causing the decline of the district of Firrim (figure 3.6).[97]

1304 November 7 **Tabriz.** On the night of 7 Rabi' II, 704, a strong earthquake in Azarbaijan caused much damage in Tabriz.[98] In Sarab buildings swayed to and fro, causing panic.[99] Aftershocks continued for a few months.

1305 April 16 **Azarbaijan.** On 20 Ramadan 704 there was a strong earthquake in Azarbaijan.[100]

1310 **Shahrizur.** In Kurdistan, an earthquake destroyed many houses, killing a great number of people in Shahrizur in 710 H.[101]

1316 January 5 **Gulpaigan.** On 8 Shawwal 715 more than twenty villages in Siya . . . (?), one of the districts of Hamadan, and in Gulpaigan, were totally destroyed. Everything in the affected region was flattened to the

Figure 3.6. A.D. 1301, Firrim.

ground and many farmers were killed and their possessions buried.[102]

1319 **Maku—Taddeus.** In 1319 an earthquake in the region of Maku destroyed the twelfth-century monastery of St Taddeus (the Qara Kilisa or church of Tata'us) in the vicinity of Siyah Chashmeh, killing seventy-five people.[103]

1320 **Ani.** In 1320 an earthquake in the district of Ararat totally destroyed the city of Ani.[104] Damage extended to the countryside and the shock was felt in the districts of Siunikh, Gegharquni and Shirak. The earthquake ruined tens of thousands of houses and hundreds of churches in the city, but there is no evidence that destruction extended beyond the immediate surroundings of Ani.[105] The earthquake did, however, hasten the decline of the city.[106]

1336 October 21 **Khwaf.** Early in the morning on Monday 14 Rabi' I, 737 there was a catastrophic earthquake in Khurasan in the district of Khwaf. The town of Jird was totally destroyed and 20—30 000 people perished in the villages between Jird and Zauzan. In the latter, the local ruler was killed in the collapse of his palace as a result of the shock. An epidemic, probably cholera, broke out after the earthquake and a further 11 000 people in the region between Sanjan-i Zaveh (Sangun) and Dughabad died as a result (figure 3.7).[107]

All the indications suggest that this was one of the largest earthquakes ever to occur in Kuhistan. Jird, Zauzan, and all the villages between them[108] were destroyed, and Khwaf itself was almost certainly seriously affected. The meizoseismal area of the earthquake, therefore, should run along the valleys of the Rud-i Dunakh and Rud-i Fanuk up towards Sanjan (Sangun). The long axis of the area is about 110 kilometres and it aligns in places with Quaternary faults, mostly in alluvium, which are visible on aerial photographs and can be followed on the ground discontinuously for some kilometres, striking N-140° to N-150°-E. The large size of the meizoseismal area, its alignment with what seem to be features of recent tectonic origin,[109] the effects it probably had on the water supplies of the district and the long duration of shaking,[110] all suggest a large magnitude event.

1344 **Isfahan.** An earthquake in Isfahan in 745 H. destroyed the walls and a number of houses, killing about twenty people.[111]

1345 **Tabriz.** An earthquake was felt in Tabriz in 746 H. without causing damage.[112]

1361 **Qishm Island.** An earthquake caused much damage to the Island of Qishm.[113]

1364 February 10 **Herat.** On 6 Jumada I, 765 there was a destructive earthquake in the region of Herat.[114] Most of the buildings in the city were ruined, particularly the tall structures. The shock caused the battlements to fall from the ramparts and several metres fell off from the top of the Falak al-Din minaret.[115] The *masjid-i jami'* in the city was also damaged.[116] The main arch collapsed although its two supporting pillars remained intact.

The details of this earthquake suggest that the shock originated some distance from Herat, where the main damage caused was due to long-period ground movements, possibly from the Gulran district.

1384 **Ray.** A destructive earthquake allegedly occurred at about this time.[117]

1389 February **Nishapur.** Preceded by four days of strong foreshocks, a catastrophic earthquake again affected Nishapur in Safar 791. The main shock totally destroyed the city and killed all but a handful of its inhabitants, striking with great violence and causing almost instantaneous destruction. Some months after the earthquake, the survivors erected buildings roofed with poles and tents near the ruins of the city. Ground deformations, probably landslides, caused major damage to some of the villages and led to subsequent dispute over land ownership in the areas affected.[118]

1400 **Lar.** An earthquake in Lar destroyed 500 houses.[119]

1405 November 23 **Nishapur.** There was a catastrophic earthquake in Nishapur and its dependencies on 30 Jumada I, 808. The city was completely destroyed and only those out in the fields survived. Destructive aftershocks continued for several days and in all more than 30 000 lives were lost; no building remained standing. The city was rebuilt, perhaps on the site of its present location.[120]

1406 November 29 **Tatev.** A strong earthquake in Tatev caused some concern.[121]

1410 **Balkh.** In 813 H. earthquakes were felt in the towns of Afghanistan (?) affecting Balkh and Bukhara. Landslides in the mountains dammed streams to form a deep lake.[122]

1428 **Taliqan.** There was a destructive earthquake in Taliqan in 831 H. Shocks continued for ten days and a great many people perished.[123]

1430 **Hamadan.** In 833 H. there was an earthquake in Hamadan. Some places in the region slumped into the ground; houses and walls were destroyed, causing a great loss of life.[124]

1430 **Wasit.** A damaging earthquake in the region of Wasit occurred in 833 H.[125]

1436 **Gurgan.** In 839 H. a damaging earthquake killed a number of people in Gurgan.[126]

1436 **Azarbaijan.** In the same year there was a locally destructive earthquake in Azarbaijan. A village was completely overwhelmed and not one of its inhabitants survived, not even the animals.[127]

1440 **Karzin—Qir.** A destructive earthquake in southern Fars in 844 H. caused considerable damage and loss of life at Karzin, and also in other parts of the district of Shiraz, killing nearly 10 000 people.[128]

1457 **Tigris.** An earthquake, perhaps originating from the vicinity of Amara on the Tigris, was felt in Baghdad, Kufa and Basra.[129]

1459 **Zagros.** In 863 H. a shock was felt in Shiraz and also in

Isfahan, where it did not cause any damage. The meizo-seismal region of this event must be sought near the southeast foothills of Kuh-i 'Alijuq.[130]

1459 **Azarbaijan.** In the same year an earthquake in Azar-baijan, associated with large-scale landslides, was felt in Tabriz.[131]

1470 **Gurgan.** In 875 H. there was an earthquake in Gurgan and one of its villages, possibly near Abasku, sunk into the ground. Gunbad-i Kavus was only lightly affected.[132]

1483 February 18 **Western Makran.** On 21 Ramadan 887/3 November 1482, a series of foreshocks began, culminating three months later on 10 Muharram 888/18 February 1483 in a destructive earthquake in the Strait of Hurmuz. In Jarun the earthquake damaged or threw down certain tall buildings, the minarets of the mosques and the ventilation chimneys (*badgirs*) of the houses.[133] About this period northeast Oman was also affected by an earthquake.[134] The details of the effects of the shock

Figure 3.7. A.D. 1336 (21 October), Khwaf: A.D. 1979 (14 November), Karizan: A.D. 1979 (27 November), Khuli.

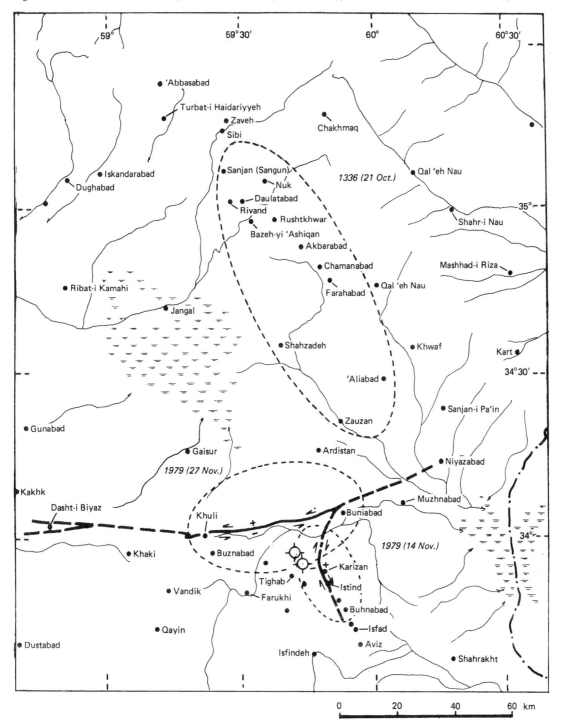

in Hurmuz suggest that the island was some distance from the epicentral region of a large magnitude earthquake. The absence of data from the Persian mainland, and the reference to the earthquake in Oman, suggest therefore the further possibility that the shock originated offshore from the western coast of Makran.

1485 August 15 **Mazandaran–Gilan.** Just before sunset on Sunday 3 Sha'ban 890,[135] there was a catastrophic earthquake in Gilan, particularly affecting Dailamistan,

a large area between Gilan and Mazandaran to the east (figure 3.8). In Tanikabun the shock demolished substantial buildings such as castles, mosques, shrines and *hamams* and what was left was damaged beyond repair.[136] In Gurjiyan[137] and Gulijan[138] damage was equally severe, with casualties, and a strong castle in the region was levelled with the ground. Also in Shakur[139] many villages were ruined and old buildings were destroyed, with casualties. Further to the south, in

Figure 3.8. A.D. 1485 (15 August), Mazandaran–Gilan.

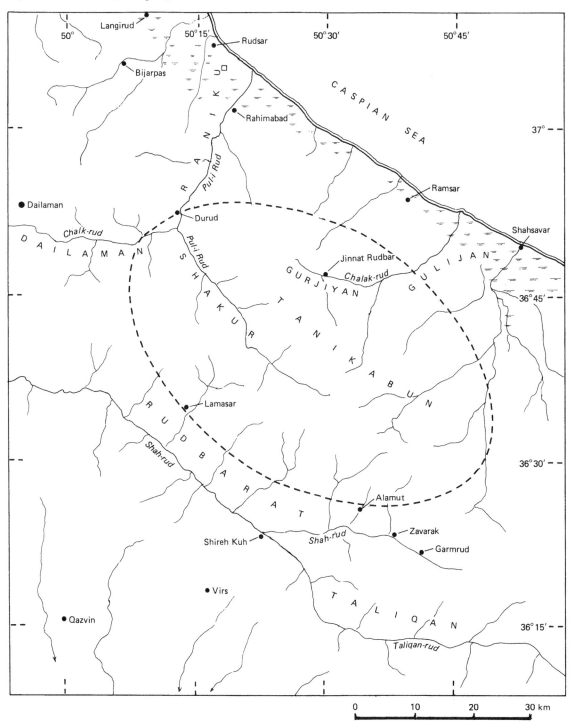

Rudbarat, many people perished, but the number is not known exactly.[140] In Taliqan other castles were ruined and at Palisan the fort collapsed completely, killing 78 of its inhabitants.[141] Throughout Dailamistan the shock triggered rockfalls from the mountains and many animals perished.[142] To the northwest, in Dailaman, many old buildings collapsed but the inhabitants and the local ruler, who was at prayers, escaped. Part of the palace at Raniku[143] fell down, but further to the north and northwest, at Lahijan, Gukeh, Kaisum, Pashija and Lashtinshah,[144] there was little damage and no one was killed, although the shock was strongly felt in these areas.

Aftershocks continued for six weeks until the end of Ramadan, or early October, keeping the survivors camping out in the open. However, another strong aftershock occurred on Monday 1 Rajab 891/3 July 1486, but it was not as destructive as the main shock.

1493 January 10 **Mu'minabad.** This was a major earthquake, which affected the mountainous district of Mu'minabad to the east of Birjand. It occurred around midday on Friday 21 Rabi' I, 898. Most of the inhabitants of Nauzad perished and many also died in Mask. Much damage was done and houses were levelled to the ground, but there were no casualties in Dar Miyan (figure 3.9).

The earthquake was associated with a fault break that extended between Nauzad and Mask, the trace of which is still visible on the ground (plate 6).[145] The trace follows the contact between flysch and the valley alluvium and near Mask this is overthrust, the fault plane dipping sharply to the south–southwest. This Quaternary fault passes south of the villages of Sargaz, Mask, Mughdam and Kalateh Mazar, bearing approximately N-95°-E, and near Nim-i Rah it turns, bearing N-125°-E. It continues by passing southwest of the villages of Khunik, Nauzad, Chak, Takhrij and Tashman. Up to this point the valley alluvium is downthrown to the north and northeast. Thereafter the trace is discontinuous. It can be picked up again near Barkandan and Chah Tuk, about five kilometres north of Dar Miyan, where it turns south, bearing N-170°-E. At Nim-i Rah a series of faint surface traces of terraces facing northeast were found traversing the valley alluvium, heading northwesterly. They can be followed to the southwest of Gask and Malikabad and they are thought to be a clue to the continuation of the Nauzad fault in a northwesterly direction under the valley alluvium.[146] The earthquake happened not far from Durukhsh which was destroyed early in the twentieth century.[147]

1495 **Jibal.** There was an earthquake in the Jibal in 900 H. It was felt in Hamadan, Isfahan and in the district of Ray. In the region of Hamadan the shock caused a large landslide.[148]

1497 **Hurmuz.** A whole town in the vicinity of Hurmuz, most probably Gambrun, was totally destroyed and its inhabitants perished in the ruins.[149]

1498 **Gurgan.** In 903 H. a destructive earthquake caused the collapse of most of the houses in Gurgan (Gunbad-i Kavus), killing 1000 of its inhabitants.[150]

1503 **Hakkari.** In 908 H. there was a major earthquake in the Hakkari region. In the district of Mosul many houses were destroyed and some buildings collapsed in the town. The shock was felt in Azarbaijan as far as Tabriz and in Akhlat on lake Van.[151]

1506 **Shiraz.** An earthquake in Shiraz in 912 H. caused the collapse of the roof of the library in the Shah Chiragh mausoleum. It is probable that other buildings were also affected.[152]

1549 February 15 **Eastern Qayin.** On the night of Wednesday 17 Muharram 956 there was a major earthquake in the region of Qayin.[153] The shock completely destroyed five villages, possibly in the Zirkuh district,[154] with the loss of 3000 lives. Qayin itself, presumably some distance from the epicentral region, does not seem to have been seriously affected by the earthquake. The event was predicted by a local astrologer, who was himself killed.[155]

1550 **Tabriz.** A damaging earthquake in Tabriz in 957 H. caused many casualties and extensive landslides in the mountains. Aftershocks continued for six days, possibly affecting the region of northwest Sahand.[156]

Figure 3.9. A.D. 1493 (10 January), Mu'minabad.

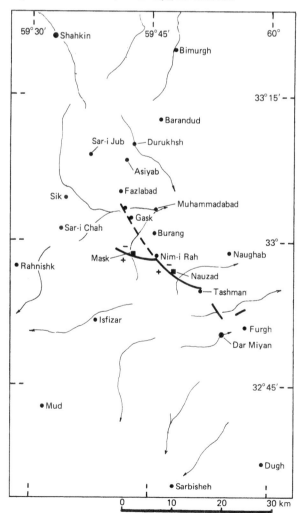

Plate 6. Mu'minabad fault and extant villages along it.

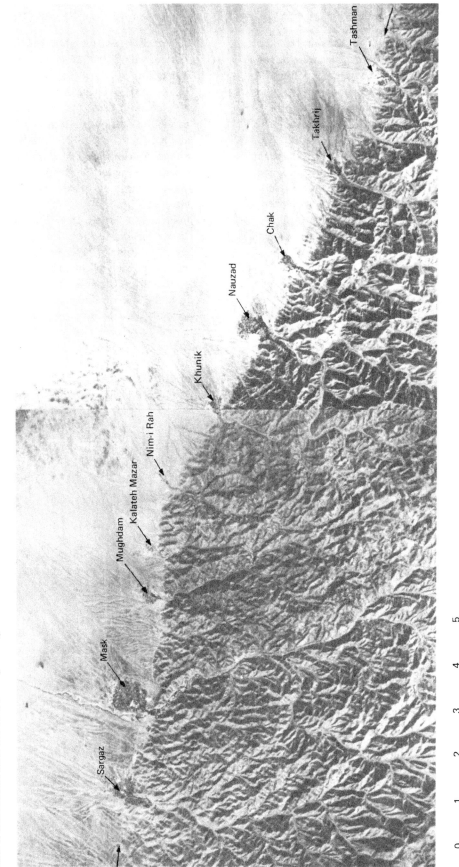

1567 **Arasbaran.** In 974 H. a damaging earthquake in the region of the Araxes caused the collapse of part of the castle of Qahqaha in the Baft district of Qaracheh Dagh. Among the casualties were Sam Mirza and two of his nephews who were imprisoned there.[157] The continuing use of the castle suggests that the structure was not totally destroyed.[158]

1591 **Shiraz.** In 999 H. there was a damaging earthquake in the region of Shiraz.[159] The mountains outside the city were fissured and many houses were destroyed in outlying settlements to the northwest of Shiraz. The city itself does not seem to have been seriously affected.[160] Nevertheless, the shock damaged the Shah Chiragh and the Masjid-i Nau so that extensive restoration was necessary.

1593 **Sarab.** In the summer of 1001 H. Sarab was totally destroyed and its district was levelled with the ground. Damage extended to the district of Miyaneh, where two villages were overwhelmed by landslides.[161]

1593 September **Lar.** A destructive earthquake affected Lar in the late summer of 1001 H. Three shocks occurred in forty-eight hours. The second of these, in the early evening, destroyed all the large houses and major buildings. The ground opened up, in some places by as much as two to three metres, causing the collapse of many houses and the destruction of the cisterns for rain water. The third shock dislodged great boulders from the mountains at the foot of which Lar is situated, and landslides overwhelmed the rest of the town. The earthquake destroyed more than 1200 houses and caused the death of 3000 people. The greater part of the walls were destroyed and the castle on the east side of the town collapsed, although solidly built on rock.[162]

    The earthquake, which was preceded by foreshocks, does not seem to have affected a large area.[163]

c. **Mashhad.** Allegedly, an earthquake near Mashhad dam-
1598 aged the cupola of the Masjid-i Gauhar Shad.[164]

1604 June 18 **Basra.** On 20 Muharram 1013 a great earthquake at Basra destroyed most of the houses both inside and outside the city. Many thousands perished in the ruins.[165] The destruction was certainly the result of an explosion of the arsenal in Basra and not because of an earthquake.[166]

1608 April 20 **Taliqan.** On 4 Muharram 1017 there was a major earthquake in southern Gilan, causing great damage over a large area.[167] Many houses were destroyed in Taliqan, Rudbarat-i Alamut and in the region of Qazvin. The castle of Darband was also ruined; a recently built tower fell down, shattering the inside of the castle, and three towers over the gateway also collapsed. Further east in Amul, Sari and Ashraf, 280 kilometres away from the epicentral area, houses were cracked and chimneys collapsed. The shock was associated with violent ground movements more than 300 kilometres away, at Miyan Kaleh.[168] It caused large waves in the Caspian Sea which crashed up the coast and resulted in great alarm among men and animals (figure 3.10).

1619 May **Dughabad.** A destructive shock in the district of Zaveh-o-Mahvalat in Khurasan utterly destroyed Dughabad in 1028 H. Despite the fact that most of the people were out in the fields, the earthquake killed about 800 in and around the town. This event was considered greater than anything that had previously occurred in the district.[169]

1622 October 4 **Bandar 'Abbas.** A damaging earthquake in Bandar 'Abbas and on the island of Hurmuz ruined many houses and caused the collapse of a tower of the fortress. Consecutive shocks for two days added to the damage.[170]

1623 **Marv-dasht.** An earthquake in the Marv-dasht destroyed Qal'eh-yi Shikasteh and Qal'eh Istakhr, among other places. It is probable that this shock was responsible for the collapse of a number of columns in Persepolis and for damage caused in the area of Naqsh-i Rustam.[171]

1624 **Tabriz.** In 1033 H. there was allegedly a strong earthquake in Tabriz, but this is almost certainly a spurious event.[172]

1639 **Qazvin.** An earthquake in Qazvin in 1049 H. is said to have killed 12 000 people. This is possibly an incorrect reference to the earthquake of 1608.[173]

1641 February 5 **Dehkhwarqan–Tabriz.** On Friday night, 5 February 1641, there was a destructive earthquake in the region between Tabriz and Lake Urmiyeh in Azarbaijan.[174] The districts of Usku and Khusraushah on the northwest slopes of mount Sahand, as well as Dehkhwarqan (modern Azarshahr), were totally destroyed with great loss of life.[175] In Tabriz many houses and public buildings collapsed, among them most of the famous historical monuments. The half-ruined complex of buildings of Sham-i Ghazan, situated about five kilometres west of the city, fell in leaving only the four corners of the main structure standing. The Masjid-i Ustad–Shagird was partly destroyed, as was most probably the 'Alishah mosque which also stood in a ruined state in the centre of Tabriz. The shock particularly affected the domes and minarets of mosques, which fell shattered to the ground, and also the main buildings in the city, although some of them were already largely in ruins by this time.[176] Damage extended to neighbouring districts and in all the earthquake caused the deaths of 1200 people (see figure 3.11).

    In the mountains the shocks triggered large-scale rockfalls and landslides that added to the destruction, and in places the ground opened up.[177] Elsewhere, presumably in the Talkheh-rud plain, water began to flow from fissures in the ground, only to be cut off on the third day by a violent aftershock, when it emerged elsewhere before drying up. There is no evidence that these ground deformations were of tectonic origin. The region is prone to landslides.[178]

    The main shock, which was felt in the region of Van, was followed by many aftershocks which gradually died out over a period of five months, being especially violent in the first three days, when they probably added to the damage. In Tatev, about 160 kilometres away, aftershocks were felt for three months and in the meizoseismal region it was a month before all the possessions and corpses could be unearthed.

1648 March 31 **Van–Hayots-dzor.** A destructive earthquake occurred in the region of Van in the Hayots-dzor. To the north of the Hoshap' river the densely populated region of Van was almost totally destroyed. The Armenian monasteries and cloisters of Srkhuvanq, Noragivt, Berdak, Kr'nkuvanq, Kendenanits, Kur'ubash, Varagavanq, Susans, Salnapativanq, Ardjakuvanq and Aleruvanqer were either destroyed or damaged to the extent that some of them were abandoned after the earthquake. The town of Van was also damaged. The walls of the lower citadel from Tavrizu-dargali to Khani-burts collapsed together with many houses and a number of churches in the town. The shock caused twelve springs of water in Avants, near Van, to dry up and at Noragivt it triggered a landslide that carried away and destroyed the village and nearby cloister (figure 3.12).

To the south of the Hoshap' river in the less densely populated valley of Hayots-dzor, damage was equally heavy and more extensive. All the cloisters were ruined and the monastery of Khegavanq was destroyed. At Ab-i Ghner–Liarn, near Hogeatsvanq, landslides dammed the stream and caused the abandonment of a number of water mills. In this region rockfalls killed a number of people. At Hermerugivt the ground deformed and slumped creating ponds in places, while at Eghnaberd and Lower P'aghakh springs dried up causing six water mills to be abandoned and the inhabitants to remove to Kasrik. The shock was felt strongly in Tabriz and in Armenia. Aftershocks continued for about three months.[179]

1650 **Tabriz.** It is alleged that an earthquake did much damage in Tabriz.[180]

1657 **Tabriz.** It is said that Tabriz was destroyed by an earthquake.[181]

1659 **Tatev.** An earthquake occurred at Tatev in the province of Goris which was responsible for a large landslide.[182]

1664 **Tabriz.** Many places, including Tabriz, were ruined by an earthquake in 1074 H.[183]

1665 **Damavand.** In 1075 H. there was a destructive earthquake in Damavand and its dependencies. The earthquake destroyed many houses and buildings in Damavand. An inscription in the *masjid-i jami'* refers to the earthquake damage and records the restoration work done in 1081/1670.[184]

1666 **Zagros.** A destructive earthquake in the upper reaches of the Karun river to the northwest of Haft Tanan decimated the local tribes in 1076 H. The shock, which brought down massive rockfalls, blocked mountain passes and dammed streams. It was felt in Hamadan and Shiraz and caused some concern in Isfahan.[185]

Figure 3.10. A.D. 1608 (20 April), Taliqan.

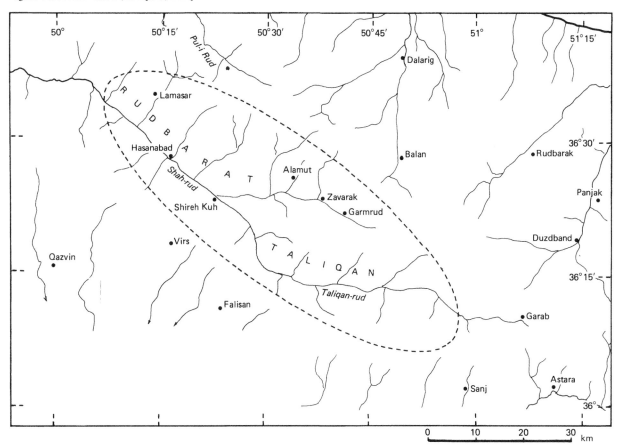

1673 July 30 **Mashhad.** On 15 Rabi' II, 1084 a destructive earthquake occurred in Khurasan. Two-thirds of Mashhad were ruined, including the dome over the tomb of Imam Riza, the cupola of the Masjid-i Gauhar Shad and many public buildings. 4000 people were killed. Nishapur was also severely damaged and half the town collapsed with the loss of 1600 lives. Another small town is also said to have been overthrown.[186]

1677 **Lar.** Several settlements and caravanserais on the road north from Lar to Bunaru, particularly at Pa-yi Kutal, were destroyed by two earthquakes which also affected Biriz and Lar itself.[187]

1678 February 3 **Lahijan.** On 10 Dhu' l-Hijja 1088 an earthquake in Lahijan, followed by many aftershocks, ruined all mosques, particularly the *masjid-i jami'* with its minarets, sanctuaries and mausoleums. Bath-houses, bridges and many houses were also ruined.[188]

1678 **Gunabad.** A destructive earthquake in Khurasan destroyed many villages. The town of Gunabad was completely ruined and the casualties were excessive. Only the old *masjid-i jami'* withstood the shock and only one person survived. The town was resettled by survivors from the outlying villages.[189]

1679 June 4 **Erivan.** A severe earthquake in the district of Erivan almost totally destroyed the villages of Gar'ni, Giamrez, Gokht, Dzoragegh and Qanaker, killing 1228 people. Damage extended to Norashen, Karp'i, Erivan and Getargel where many houses were ruined, with casualties. In Erivan part of the fortress, as well as the mosques, their minarets and a few houses, collapsed. In Edjmiadsin only three churches were left undamaged. The monasteries of Havuds T'ar', Geghard, Khor Virap and Dzhrvezh were ruined. In all, about 7600 people were killed.

The earthquake destroyed the bridge on the Razdan river and rockfalls blocked many passes. The Erivan gorge was dammed by landslides and traffic was temporarily interrupted. Elsewhere the ground slumped and in the valleys the ground liquefied, ejecting water and gasses from cracks. The shock was not felt very far but it did cause concern in the region of Ararat. After-shocks continued for more than three months.[190]

1687 **Mazandaran.** A serious earthquake in Mazandaran destroyed many villages and triggered landslides.[191]

1687 April **Mashhad.** A damaging shock occurred in Mashhad.[192]

Figure 3.11. A.D. 1641 (5 February), Dehkhwarqan–Tabriz.

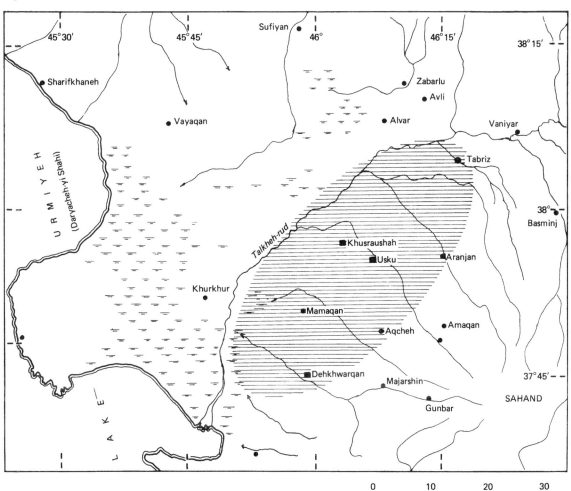

1695 May 11 **Isfara'in.** At dawn on 27 Ramadan 1106 there was a major earthquake in the Isfara'in valley. Many villages in this sparsely populated area were destroyed, with casualties ranging from a hundred to about ten in each of them. The large village of Batay was completely destroyed with 360 casualties, and the loss of many animals. At Banir, at the foothills of the nearby mountain, a landslide overwhelmed the village. In Kuran, a site to the south of the valley, the shock caused some damage and great panic in men and animals. Strong aftershocks continued to be frequent for over a year, until 25 Muharram 1108/24 August 1696, and it was nearly three years before slight shocks had ceased altogether.[193]

1696 April 14 **Chaldiran.** A major earthquake in Armenia completely destroyed the villages in the Chaldiran district. Damage extended to Duavanits where the villages of Vanits were ruined. Near Maku the walls of the monastery of St Taddeus and the newly-built cells collapsed, killing a number of people. The shock ruined many settlements in the upper parts of the district of Van.

Aftershocks continued for a long time. A violent one in May in the region of Maku caused additional damage to the monastery of St Taddeus.[194]

1703 **Qais–Hingam.** A destructive shock affected the islands of Qais and Hingam.[195]

1705 **Basra.** A strong earthquake was felt in the region of Basra.[196]

1715 March 8 **Southeast Van.** A severe earthquake occurred at dawn in the region of Mahmatan, southeast of Van. It destroyed many villages in the Mehmedik plain, killing a considerable number of people (figure 3.13). At Hoshap' the walls of the fort were ruined and at Satmanis the fort towers collapsed. The earthquake caused the collapse of the dome of the church of St Bartholomeus at Deir and the ruin of the fort at Sarai. In Van only one house was ruined and a few people perished.[197]

1717 March 12 **Tabriz.** A little after midnight an earthquake in Tabriz destroyed 4000 houses, killing more than 700 people.[198]

1721 April 26 **Southeast Tabriz.** Early in the morning of

Figure 3.12. A.D. 1648 (31 March), Van–Hayots-dzor.

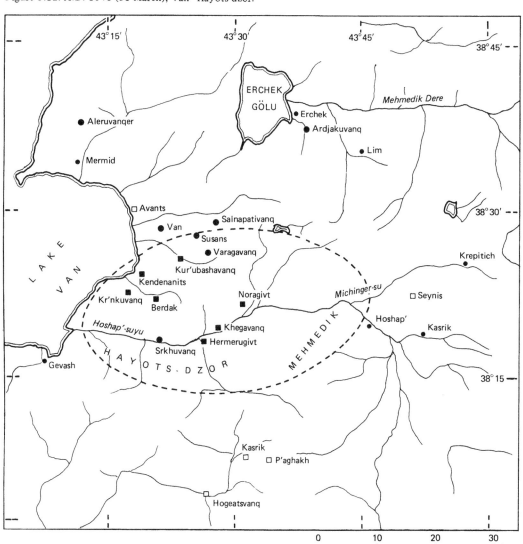

Sunday 28 Jumada II, 1133,[199] a major earthquake shook the region of Tabriz, killing at least 40 000 people.[200] In Tabriz itself the shock ruined about three-quarters of the houses and caused substantial damage to most of the larger buildings which did not, however, collapse.[201] Detailed information about the extent of the affected region outside Tabriz is lacking, but internal evidence suggests that the heaviest destruction, accounting for the large number of casualties, occurred within a zone that extended from near Tabriz to the southeast, through Shibli and beyond Qareh Baba (figure 3.14). The shock triggered many rockfalls and was associated with a fault break that extended for at least fifty kilometres, from Takmeh Dash to near Tabriz. The break through Shibli was still visible in 1809, and parts of the fault trace that seems to be connected with this earthquake can be seen today on the ground.[202] The shock seems also to have been strongly felt in the Qazvin region and was followed by many strong aftershocks.[203]

1755 June 7 **Kashan.** A destructive earthquake in Kashan destroyed 600 houses and more than 1200 people were killed. The caravanserai of the town suffered substantial damage.[204] The shock, which was felt in many Persian towns, was also damaging in Fin, where the irrigation system was ruined. In all, 3000 houses were destroyed and casualties would have been higher had not the

majority of the population been out harvesting cotton,[205] see figure 3.15.

1765 April 23 **Shiraz.** An earthquake in Shiraz caused considerable damage to houses as well as to the structure of the Masjid-i Nau, which was restored four years later.[206]

c. **Makran.** In Ra's Kuchari on the Makran coast, an earth-
1765 quake caused an entire hill to slump into the sea with men and camels on it.[207]

1766 **Lar.** In the district of Lar an earthquake triggered land-slides that overwhelmed a village and its inhabitants. It is possible that Lar itself also suffered some damage.[208]

1769 May 1 **Baghdad.** During a damaging thunderstorm a number of shocks were felt in the city.[209]

1778 December 15 **Kashan.** Just before dawn on Tuesday 25 Dhu' l-Qa'da 1192 there was a destructive earthquake in the Zagros, round the western edge of the *kavir*. The earthquake was strongly felt in Persian Iraq, in the region of Ray and in Qum and Isfahan. Destruction was centred in the districts of Kashan, where more than 8000 people were killed.[210] In Kashan almost all houses were destroyed and the main buildings and fortifications were completely ruined. The damage in the city was so bad that the survivors would have removed elsewhere had not the ruler, Karim Khan-i Zand, organised immediate reconstruction. Damage extended perhaps as far as Sin-Sin to the north and Quhrud to the south, affecting

Figure 3.13. A.D. 1715 (8 March), Southeast Van.

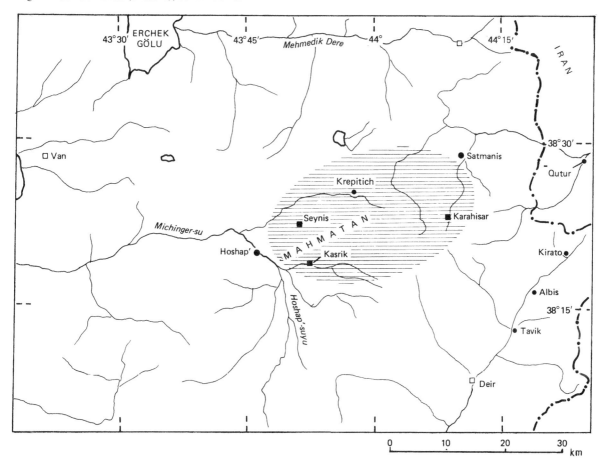

the water supply of the region. The earthquake was followed by an outbreak of cholera which claimed further victims. Aftershocks continued at the rate of two or three a day for the next month (figure 3.15).

The earthquake caused serious damage to the bazaar and the *masjid-i jami'* which, together with other public buildings, were repaired and re-constructed over the following five or six years by 'Abd al-Razzaq Khan.[211] There is no evidence that the Saljuq minarets of Zain al-Din and of the *masjid-i jami'* in Kashan suffered any damage.

1780 January 8 **Tabriz.** Preceded by a strong foreshock, a catastrophic earthquake on the night of Friday to Saturday 29 Dhu 'l-Hijja 1193 to 1 Muharram 1194 (7–8 January 1780)[212] in the region of Tabriz almost totally destroyed the city and devastated about 400 villages, including Marand, Tasuj and Iranaq.[213] In Tabriz itself, all major buildings, weakened by previous shocks, were ruined and all private houses, as well as the fort and walls of the city were totally destroyed, the radius of destruction being variously given as 72 or 120 kilometres from Tabriz.[214] Outside this distance, in Khuy, Salmas, Urmiyeh and Gunayi (?), buildings were damaged but there was no loss of life.[215] Vast numbers of people

perished in this earthquake, estimates ranging to over 200 000. The number was probably somewhere around 50 000.[216] Among these were the son of the ruler of Tabriz, Fazl'ali Beg Dunbuli, along with some 700 retainers who were killed when the palace collapsed.

The earthquake was associated with a fault break at least sixty kilometres long that extended from the vicinity of Shibli in the southeast, to near Marand in the northwest (figure 3.16). Contemporary reports show that faulting extended from northeast of Tabriz, in the foothills of the Surkhab ('Ain 'Ali) mountain, for about forty-five kilometres, heading southeastwards to Shibli. On Surkhab the fault break is described as two metres wide heading for twelve kilometres in a southeasterly direction, while to the northwest of Tabriz the break is described as a fifteen kilometre scarp four to ten metres high, facing southwest and clearly distinguishable by its grey colour, heading in a northwesterly direction.[217] Further to the northwest, in the vicinity of Marand, the ground opened up but closed again.[218]

In the low-lying region to the west of Tabriz, the soil liquefied and mud was ejected from the ground. The shock caused springs and *qanats* to dry up and new streams to flow elsewhere, in some places in such

Figure 3.14. A.D. 1721 (26 April), Southeast Tabriz.

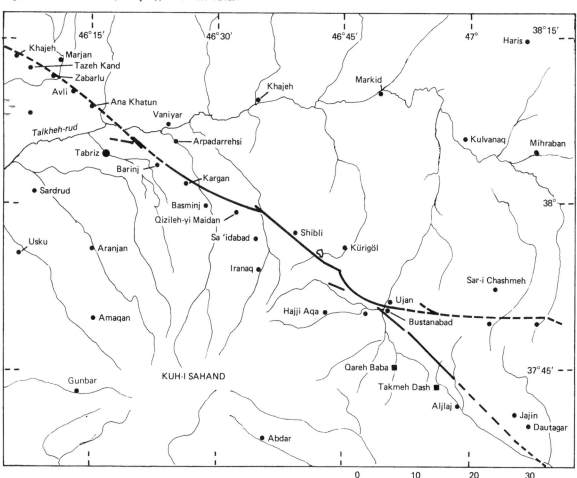

quantities that it threatened to flood the devastated area. The water ceased to flow after about two hours, draining off towards lake Rizaiyeh.[219] About twelve kilometres east of Tabriz the earthquake caused extensive slumping and sliding of a large area of grassland.[220]

The earthquake was felt in Van and as far as Divrigi and Malatya, more than 700 kilometres away. Aftershocks were frequent and continued to be felt at short intervals for three or four years after the earthquake, possibly longer. Among them, strong shocks causing further damage were recorded for 6 Safar 1194 (12 February 1780) and 14 Safar (20 February).[221]

The earthquake was responsible for the destruction of all the historical buildings in Tabriz.[222] Among those which were restored afterwards and still exist today (in greatly altered form) we may note the Friday mosque in the bazaar, work on which was begun by Ahmad Khan Dunbuli immediately after the earthquake. The Sahib al-Amr mosque was restored in 1208/1794 by Ja'far Khan Dunbuli[223] in a complex immediately north of the Mihran-rud, that includes the Thiqat al-Islam mosque. Both these complexes include other buildings named as having been severely affected by the earthquake, for example the Talibiyyeh madraseh adjoining the Friday mosque, which was originally a Safavid foundation

Figure 3.15. A.D. 1778 (15 December), Kashan.

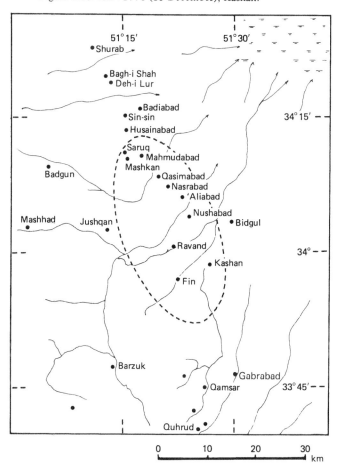

(built in 1087/1676), and the Hasan Padshah mosque and madraseh adjoining the Sahib al-Amr (originally built by Uzun Hasan (d. 882/1478). The Sadiqiyyeh madraseh, also a Safavid foundation, still survives in the bazaar area, and the shrine and madraseh of Sayyid Hamza (d. 714/1314) are in the Surkhab quarter to the north. The mosque of 'Alishah is now reduced to the ruins known as the Arg, and the Ustad–Shagird mosque and the Dal-o-Zal mosque have also survived to the present day.[224] See plate 7 for an illustration of the relative positions of some of these buildings a century before the earthquake.

The fault break associated with the earthquake was throughgoing and of major dimensions. Today, the most southeasterly ground ruptures that can be identified as probably more or less directly related to this event, lie in a narrow zone obliquely across the old road from Tabriz to Ujan, about two and a half kilometres before reaching the Shibli Pass. This zone, mainly tuffs and Quaternary alluvium, can be followed on the ground to the northwest as far as north of Barish (or Barinj), cutting through Miocene marls and branching off to the west towards Tabriz, but not quite reaching the suburbs of the city. From north of Barish, after crossing with the Talkeh-rud, the main branch becomes discontinuous. It continues through Quaternary pediment material and gradually turns into a well-developed scarp as it approaches Avli and Zabarlu, with a throw facing southwest several metres high[217] that either follows the contact between the Upper Red formations and younger, mostly Quaternary deposits or cuts through them. It proceeds up to Chilleh Khaneh Sufla, beyond which the trace becomes again discontinuous. The overall sense of movement of the fault seems to have been normal, with the southwest side downthrown by about six metres. It was not possible to judge the amount or sense of lateral movement along the fault break.[225]

Tabriz was rebuilt gradually on the same site. New houses were built low, without an upper storey, making use of more timber. Even the palace was built with timber bracing and a new system of construction, the *takht-i push*, became widely used.[226]

1780 **Khurasan.** An earthquake caused the deaths of 3000 people in Khurasan in 1194 H.[227]

1783 January 13 **Ararat.** It is alleged that on 13 January and 22 February 1783 the volcano of Ararat became active. This is a doubtful event.[228]

1784 March 1 **Shiraz.** A strong earthquake was felt in Shiraz, without causing damage. To the southeast of the city the shock, which lasted for almost a minute, triggered rock-falls and landslides from the mountains.[229]

1786 October **Marand.** A destructive earthquake occurred in the Zilbir-chai district of Marand. Southeast of Marand and in the region of Sufiyan all the way to Tabriz, villages were ruined and west of the town the shock completely ruined several streets of Khuy. In Tabriz many houses rebuilt after the earthquake of 1780

Figure 3.16. A.D. 1780 (8 January), Tabriz.

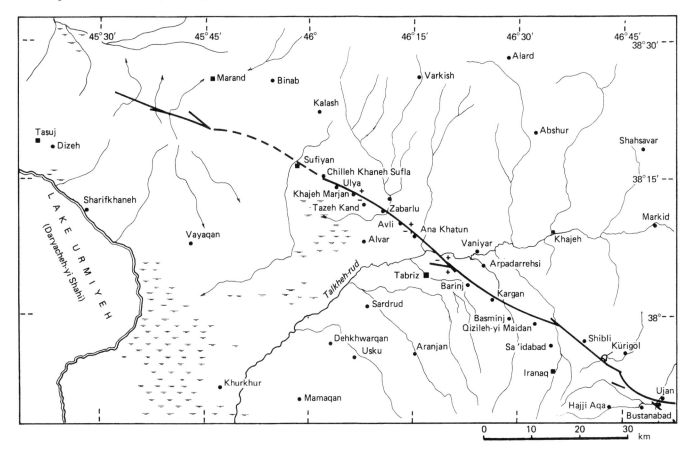

Plate 7. Chardin's drawing of Tabriz in the second half of the seventeenth century. For a description of the various buildings later affected by earthquakes in the city, see Melville (1981). Key: (1) Jahan Shah mosque, (2) 'Alishah mosque (Arg), (3) Ustad Shagird mosque, (4) Chahar Minar, (5) Friday mosque, (6) Sahib-i Zaman, (7) Hasan Padshah (?), (8) Ja'far's castle, (9) Sham-i Ghazan.

collapsed. The shock was widely felt as far as Edjmiadsin and Erivan.[230]

1802 **Sulaimaniyyeh.** A violent earthquake occurred in Sulaimaniyyeh.[231]

c. **Sultaniyyeh.** An earthquake damaged the great mosque
1803 of Sultan Ulja'itu in Sultaniyyeh and caused the collapse of the walls.[232]

1804 **Mihr.** Many houses were destroyed and others damaged at Mihr. The shock, which was felt in Sabzavar, triggered rockfalls from nearby mountains.[233]

1805 **Harhaz.** A destructive shock in Mazandaran ruined a number of villages and caused damage in Barfurush (Babul) and Damavand. The earthquake ruined the *masjid-i jami'* in Barfurush.[234]

1807 July 11 **Tasuj.** A destructive shock at Tasuj almost totally ruined the town, its bazaar and mosques. Damaged extended to the region of Salmas but the shock was not felt very far.[235]

1808 June 26 **Rashm.** A large earthquake in the north-central area of the Dasht-i Kavir devastated the sparsely populated region of Rashm. Preceded by a strong foreshock, the earthquake ruined many settlements along the borders of Mazandaran towards Qum and Sabzavar, but caused very few casualties.[236] This event marked the beginning of a long series of damaging earthquakes in the Alburz.

1808 December 16 **Taliqan.** At the end of Shawwal 1223 a destructive shock in western Mazandaran and Taliqan destroyed many villages. In Qazvin a number of houses collapsed and almost all public buildings, including the 'Abbasid mosque, were badly cracked. In Tehran the shocks caused panic and the inhabitants left their houses and camped in the open. At Tajrish the Imamzadeh Qasim was damaged and the shock was strongly felt in Rasht. Continuing aftershocks felt in Tehran added to the panic.[237]

1809 **Amul.** A destructive earthquake occurred in the districts of Shirgah, Ganj-i Rud and Julab, between the lower reaches of the Harhaz and Talar rivers. In Amul the bridge on the Harhaz was shattered and many houses collapsed, including the remains of the Masjid-i Shah 'Abbas, part of the *masjid-i jami'* and the cupola of the Gunbad-i Shams-i Tabarsi. Other gunbads were also ruined, as well as the bazaar, which was built of timberwork. The shock destroyed the Barfurush bridge on the Babul river and caused extensive damage to Babul. In Sari many of the larger houses were shattered and the Gunbad-i Salm-va Tur was ruined. The Imamzadeh Ibrahim near the Barfurush gate was also destroyed. Damage extended to Ashraf (Bihshahr) where the Safiabad villa was ruined. The shock caused widespread liquefaction in the river valleys and rockfalls in the mountains.[238]

c. **Ghulaman.** The old town of Mashhad-i Ghulaman was
1810 destroyed by an earthquake.[239]

1812 **Shiraz.** Shiraz suffered considerable damage from an earthquake which partly threw down the Bazaar-i Karim

Khan Vakil and shattered the walls of the city, which in collapsing nearly filled up the ditch. The shock also damaged the Imamzadeh-yi Shah Mir 'Ali ibn Hamza, but apparently caused no casualties in the district. Persepolis may have been damaged by the earthquake.[240]

c. **Julfa.** An earthquake caused extensive rockfalls in
1812 Julfa.[241]

1815 June **Damavand.** A strong earthquake was felt in Damavand. At Ab-i Garm it caused a spring of cold water to dry up.[242]

1819 January **Tabriz.** A long series of shocks in Tabriz ruined many houses.[243]

1824 June 2 **Kazirun–Shahpur.** A severe earthquake shook the *dihistans* of Kamarij, Shahpur and Kazirun.[244] The shock destroyed many villages along the Shahpur valley, from Kamarij to Ardashir, as well as in the valley of Kazirun. It also triggered rockfalls that entirely filled up the mountain pass of Tang-i Dukhtar between Kamarij and Kazirun. In Kazirun itself, many houses built of stone with two storeys collapsed killing about 150 people. The whole village of Diris was totally destroyed and Kamarij was also ruined and its caravanserai collapsed. Damage extended as far as Burazjan where the caravanserai was thrown down but not beyond this place or beyond Dasht-i Arjan. The shock was felt in Bushire and Shiraz and it was followed by aftershocks for almost a week (figure 3.17).[245]

1824 June 25 **Northwest Shiraz.** At dawn, on 27 Shawwal 1239, a destructive shock occurred in the district of Shiraz. In the city itself, all houses were damaged and a few collapsed. The eastern city walls and nearly all its towers fell down, and the rest of the fortifications were

Figure 3.17. A.D. 1824 (2 June), Kazirun–Shahpur.

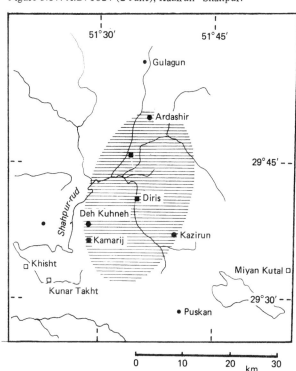

damaged. Of the public buildings, those constructed during the period of Karim Khan-i Zand, such as the bazaar and the Masjid-i Vakil, suffered minor damage. The dome and walls of the Imamzadeh Shah Chiragh, the Madraseh Khan, the Imamzadehs of Shah Mir Hamza and Sa'id Ahmad, as well as part of the palace, collapsed together with several minarets. There is no evidence that damage extended beyond Shiraz, except to the district of Guyum in the northwest where a number of villages, including Kilistan and Qalat-i Guyum, were ruined and a few hundred people were killed, some of the survivors removing to Shiraz. Damage extended as far as Shul (figure 3.18). The shock was felt strongly at Bushire and Imamzadeh Isma'il, and as far as Yazdikh-wast, about 190 kilometres from the meizoseismal region.[246] The earthquake caused a permanent rise in the underground water table in the region of Shiraz, and it was followed by aftershocks for six months. A shock on 28 August caused additional damage near Shiraz.[247]

1825 **Harhaz.** A destructive earthquake in the Harhaz Valley ruined many villages, causing the deaths of a large number of people. Damage extended as far as Jaj-rud, Damavand, Amul and Sari. In the epicentral region almost all bridges and galleries on the Harhaz road were destroyed, particularly those between Kuhrud and Bul Qalam, rendering the road totally impassable for two years.[248] There is some very tenuous evidence that in this locality the shock was associated with ground deformations possibly of tectonic origin.[249]

1825 October **Shiraz.** A strong shock ruined a number of buildings in Shiraz.[250]

Figure 3.18. A.D. 1824 (25 June), Northwest Shiraz.

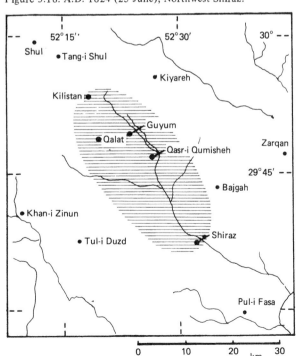

c. **Zurbatiya.** A damaging earthquake in eastern Iraq ruined
1827 Zurbatiya and Badra. The shock was strongly felt in Baghdad.[251]

1830 March 27 **Damavand–Shamiranat.** On the morning of 2 Shawwal 1245 a major earthquake in southern Mazandaran almost totally destroyed the districts of Shamiranat and Damavand, east of Tehran.[252] About 70 villages lying eastwards of the Jaj-rud, along the routes via Damavand to Simnan and Damghan, were ruined and more than 500 people were killed in Damavand alone.[253] Damage extended to Jaj-rud, where the caravanserai was shattered and in Tehran many old houses collapsed killing about 30 people. Not a single house in the capital escaped damage and part of the palace, together with many adjoining houses and part of the bazaar, were thrown down. The Arg, the Great Audience Hall, a number of mansions, as well as the old British Embassy building, were badly damaged and garden walls were levelled with the ground.[254] The loss of property in Tehran was estimated at half a million tumans. The shock caused some damage to a number of public buildings in Amul, Sari and Damghan and triggered rockfalls that blocked the passes on the Harhaz and Talar-rud roads to the north. The earthquake was felt as far as Baku and was followed by violent aftershocks that caused additional damage in the Shamiranat region and great panic in Tehran, where a large proportion of the population camped in tents. The royal court also encamped in the open courts of the Arg. The aftershock of 6 April totally destroyed the old caravanserai at Jaj-rud.[255]

c. **Quchan.** A destructive shock ruined Quchan and caused
1833 some damage in Shirvan.[256]

1834 **Pambukh.** An earthquake destroyed many villages in the region of Pambukh, particularly along the Masun valley. Damage extended to Bagavan where the church of Surb Hovhannes was badly cracked; damage also extended to Arzab.[257]

1837 June **Salmas.** A damaging shock occurred in Salmas. In Tabriz many people left their homes and took refuge in tents.[258]

1838 **Nasratabad.** Some time in 1838 a destructive earthquake occurred in Sistan along the eastern limits of the Dasht-i Lut. Damage in this largely desert region of eastern Persia extended from Chihil Dukhtaran in the north to Gurgaz in the south, a distance of about 150 kilometres (figure 3.19), as well as to Durahi. Between Nasratabad and Gurgaz, as well as between Qal'eh Gurg and Haidarabad, there was extensive faulting. To the west, low-lying areas liquefied to the extent that for years afterwards caravan routes became unsafe. All villages within seventy-five kilometres were ruined but only a few people were killed. The shock ruined the Mil-i Nadiri northeast of Shurgaz and it was followed by aftershocks for almost two years.[259]

1840 July 2 **Maku–Ararat.** Late in the afternoon of 20 June (Old Style)/2 July 1840, a catastrophic earthquake hit

Figure 3.19. A.D. 1838, Nasratabad.

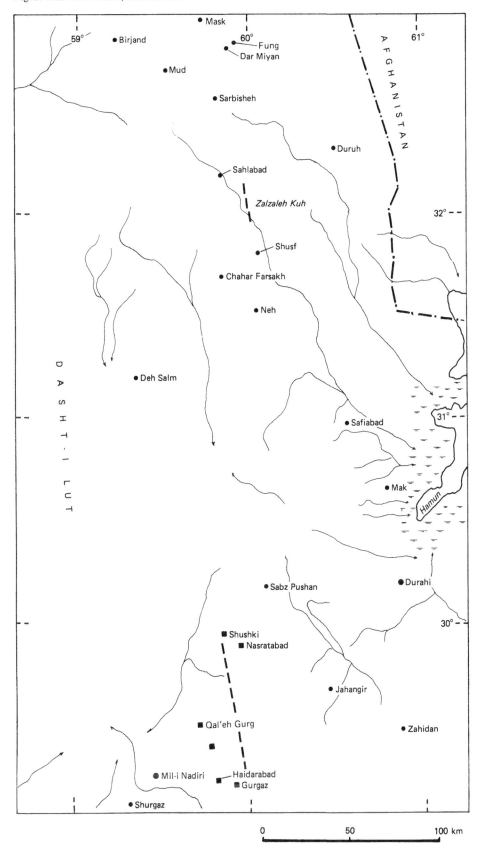

the isolated region between the upper Euphrates and mount Ararat, affecting a large area in Russia, Turkey and Iran.[260] The main concentration of damage was in the districts of Avajik, Pambukh and Gailatu, where almost all villages were destroyed with great loss of life.[261] In Dogubayazit all houses, including the castle, were ruined with the exception of the mosque. Kazl-göl and Masun were also destroyed (figure 3.20). In the region of the upper Euphrates many settlements were severely damaged, and in Bazargan and Maku few houses survived intact.[262]

Further away in Kagizman, Kulp and Igdir, as well as in the plain of the Araxes, from the district of Sharur to Nakhichevan and Urdubad, the shock was less severe but it did cause widespread damage to those villages situated in the low-lying parts of the plain. Here, the ground is marshy and as a result of the earthquake numerous areas to the west of the Araxes between the Qareh-su and the Arpa-chai rivers were intensely fractured. Liquefied sand was ejected from fissures in the ground and on all spots where the banks of the Araxes and Qareh-su are high. Landslides and slumping of the ground added to the damage. Landslides triggered by aftershocks over-whelmed the village of Qareh Khajilu and elsewhere the flow of the Araxes was temporarily dammed, the river over-topping its banks and flooding the surroundings. In the low-lying parts of the Sharur and Nakhichevan districts in Russian territory, the earthquake and its after-shocks ruined 7821 houses, 24 churches and mosques, 107 water mills and killed 49 people, injuring about 30.[263]

The earthquake triggered a colossal slide from above the snow-line of the northeastern side of Mount Ararat. A mass of shattered rock, ice and snow moved down the mountain so fast that a violent air blast was sent out in front of it. Before it was arrested by a natural dam about 900 metres above the valley floor, the slide overwhelmed Arguri, the only village in Ararat, killing all its inhabitants (about 1000 in all) and destroying the small monastery of St James (St Jacob), three kilometres above, burying all the monks and destroying also the holy well of St James.[264] At nine in the morning of 24 June (Old Style)/6 July, the natural dam burst and within minutes the slide debris spread out into the plain below in a twelve kilometre-wide front, destroying Aralik and three other villages twenty kilometres away.

Figure 3.20. A.D. 1840 (2 July), Maku–Ararat.

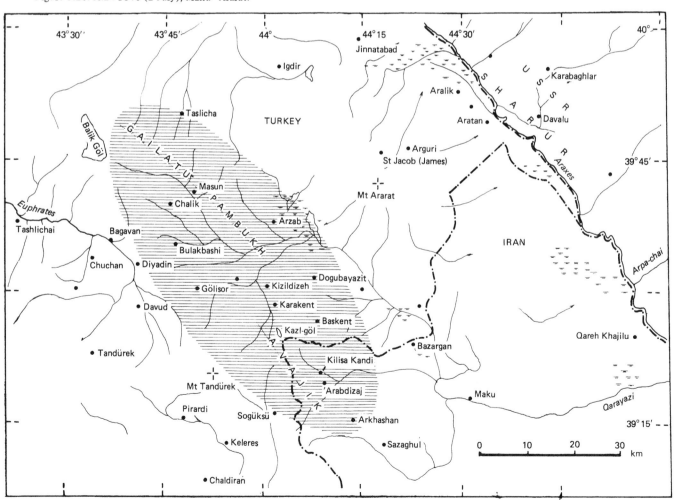

It blocked the Qareh-su river which was forced to change its course.[265]

Rockfalls and slides were reported from other parts of the meizoseismal region, notably from Pambukh and Chingal, where they killed people and herds of sheep. As a result of the earthquake, the flow in many of the streams and springs in the upper Euphrates and in the Sharur district increased, while in the region of Nakhichevan many springs dried up for some time. It is doubtful, however, if the shocks had any permanent effect on the water supply of the region.

The earthquake caused minor damage at Edjmiadsin, Erivan and Gar'ni and it was strongly felt in Alexandropol (Leninakan), Tiflis, in the Karabagh, at Shusha, in Tabriz and in Van. The shock was also felt at Linkuran, in the Talish district on the coast of the Caspian, as well as on the eastern shores of the Black Sea.[266] Aftershocks continued for some time, causing additional damage, particularly in the plain of the Araxes and around Maku and Kazl-göl. They did not stop until early in 1841.

The earthquake ruined the large church at Kilisa Kandi near 'Arabdizaj, and probably damaged the seventh-century church of Surb Hovhannes (Surb Karapet) near Bagavan (Uch Kilisa).[267] It also damaged the monastery of St Thomas at Agulis, the cupola of the monastery at Geghard, the monastery, the church and the mosque at Eghegnadzor, as well as the roof and eastern wall of the monastery at Tatev.

The spectacular landslide from Mount Ararat on which Noah's ark is said to have alighted, the destruction of the monastery of St James and of other churches in the plain of the Araxes (along the most frequented trade route to the East) excited widespread interest and sympathy among European scientists of the time. But this was rather on account of the nature of their locality than because of the special violence of the earthquake there. This, and the exhaustive field survey of the damage carried out by the Russian authorities in their own territory, in which only forty-nine people were killed, induced contemporary and later authors to place the epicentral region of the earthquake on the plain of the Araxes, in the absence of any published information from the Turkish and Persian sectors. Abich (1882) and Filadelfin (1860) place it north of Nakhichevan. Ergin *et al.* (1967) give an epicentre east of Kagizman and another in 1841, at Dogubayazit. The *Atlas* (1977) places the event at Arguri, and Tchalenko (1977) gives it as south of Davalu on the Araxes.[268]

1843 April 18 **Khuy**. A destructive shock in the district of Khuy in 1259 H. killed about 1000 people. In Khuy and the nearby Tajehkand not a single house was left intact and many collapsed. Damage extended to the north of the town as far as Maku and to the south up to Taj al-Din. The shock was strongly felt in Tabriz but not beyond the Araxes. Aftershocks continued to be felt throughout the region and in Tabriz up to 5 December 1843.[269]

1844 May 12 **Quhrud–Kashan**. Preceded by a foreshock in the afternoon of 12 May 1844, a destructive earthquake occurred in the *dihistans* of Jaushaqan and Quhrud, particularly affecting the region between Chaqadeh, Kamu, Kuskan and Chugan. Chaqadeh was totally destroyed and only 3 out of its 103 inhabitants survived. In Qamsar and Quhrud all houses were destroyed or damaged and even garden walls were levelled with the ground. The old Caravanserai at Quhrud collapsed killing a number of people (figure 3.21).[270] Despite the warning of the foreshock, the total loss of life was estimated at 1500.[271] The shock triggered landslides but caused no damage to the two masonry dams downstream of Qamsar and Quhrud,[272] nor did it cause serious damage to Kashan where the minaret of the *masjid-i jami'* was made to lean 2.1 metres between top and base.[273] The earthquake was also felt in Isfahan, without causing any damage except to a small part of the *masjid-i jami'*.[274] Strong aftershocks continued for two weeks.[275]

1844 May 13 **Miyaneh–Garmrud**. In the evening of Monday 13 May an earthquake in Eastern Azarbaijan destroyed a considerable part of the districts of Sarab and Garmrud. In Miyaneh many villages were completely destroyed with great loss of life, and the town itself was half ruined. There are no data about the extent of damage in

Figure 3.21. A.D. 1844 (13 May), Quhrud–Kashan.

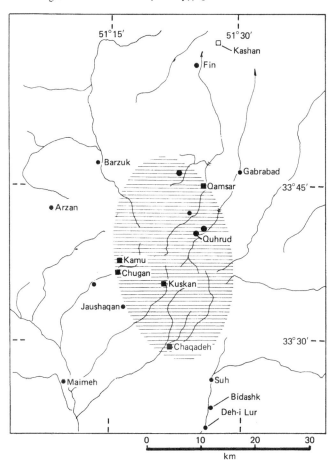

the district of Sarab and to the east, in the Sangavar valley. In the Garmrud and to the southeast in the Kaghaz Kunan districts, however, damage was serious. The affected region covers a large area and damage was reported from Aghkand, Yangikand and Armaghan-khaneh.[276] The shock was strongly felt in Tabriz and in Rasht. In Linkuran slow ground movements persisted for a few minutes.[277]

1851 June **Quchan—Ma'dan.** A destructive earthquake occurred in the region of Quchan (Khabushan) and the Sar Vilayat district of Nishapur, affecting a large area. In Quchan a quarter of the houses collapsed together with the dome of the Imamzadeh of Sultan Ibrahim and the *haram* was ruined. In the town, 160 people were either killed or wounded. The earthquake destroyed many small villages in the adjacent district of Sar Vilayat and as far as Burzinun, killing about 2000 people. It is probable that damage extended to Bar and to the turquoise mines of Ma'dan. The shock was strongly felt in Mashhad, where it lasted a long time and killed one person, and also in Turbat-i Haidariyyeh. In January 1852 an extremely violent aftershock did much damage in Quchan.[278]

1853 May 5 **Shiraz.** At dawn on 4 May/25 Rajab 1269, a series of foreshocks began to cause progressive damage in Shiraz.[279] The first foreshock caused panic and the second, half an hour later, ruined the Gaudehraban quarter. A mosque, a portion of the bazaar, and about seventy houses collapsed killing a number of people, while all houses in the city were damaged. In places the ground liquefied and *qanats* caved in. Shocks continued throughout the day and the following morning, at about a quarter of an hour before noon, a third violent foreshock caused additional damage, and one of the minarets of the Masjid-i 'Abbas collapsed. Then at noon on 5 May the main shock almost totally destroyed the city. Of the more important public buildings, the Shah Chiragh, the Madraseh Khan, the Armenian Church,[280] the Masjid-i 'Abbas and its remaining minaret, the remnants of the city walls as well as several hundred houses, all collapsed, killing many people in the city itself and in the villages around Shiraz. The Masjid-i Nau, with the exception of its large west-facing porch, the adjacent vestibule and *shabistan tarik*, was ruined.[281] Within a radius of twelve kilometres of Shiraz all man-made structures were ruined.[282] The shock triggered rockfalls on the road to Ardakan and caused widespread liquefaction of the ground to the southeast of the city. Six hours later a violent aftershock added to the destruction and loss of life. In all, about 9000 people were killed[283] and the damage to the city was enormous. The government granted a remission of taxes for five years.[284] Aftershocks did not persist for very long.[285]

1856 October 4 **Tabriz.** This was the largest of a series of small shocks felt in the region of Tabriz, causing no damage. Nevertheless, this event, which triggered a 'Cacciatore' type of seismometer in Tabriz, was studied in great detail by N. Khanikoff.[286] His earthquake map contains the first attempt to draw isoseismal lines for an earthquake in Persia, leading to the first determination of the macroseismic epicentre by means of such lines (plate 8).[287]

1862 December 21 **Shiraz.** Preceded by two very strong fore-shocks, an earthquake on the morning of 28 Jumada II, 1279 caused extensive damage in the region of Shiraz. The shock cracked practically every house in the city and caused some damage to the Masjid-i Vakil. An old mosque collapsed, and part of the walls of the city fell down. The ground motions were very intense. People were thrown to the ground and water tanks burst open. In the region of Qareh Bagh, southwest in Islamlu and in Shahpurjan, all the settlements were destroyed and in places the ground slumped and the mountain was 'cleft in twain'. There is no evidence that damage extended beyond Shiraz to the north and east of the city.[288]

1863 December 30 **Hir—Ardabil.** A destructive earthquake occurred in the *shahristan* of Ardabil, particularly affecting the *dihistan* of Hir. At Niyaraq and Kirt some 500 people were killed and Dalilar was totally ruined, with casualties. Half of Hir was destroyed with the loss of 108 lives, and Aralu, as well as Naushahr, were ruined. In all, about 1000 people and many hundreds of animals were killed. In Ardabil there was no serious damage but all houses sustained cracks, and in Linkuran, 80 kilometres from the epicentre, hanging lamps were caused to swing for many seconds by as much as sixty centimetres. Between Hir and Bulgavar the shock triggered many landslides and in the valley the ground slumped.[289] The shock was strongly felt in Prishib and Tabriz (figure 3.22).

On 2 January 1864 a strong aftershock caused additional damage. At Bulgavar it triggered a landslide that added to the damage and caused casualties.[290] Aftershocks continued for some time.[291]

1864 January 17 **Kirman.** On the night of 7 Sha'ban 1280 a destructive earthquake in Chatrud and in the settlements to the northeast of the plain killed many people and animals. The shock caused considerable damage in Kirman where the *ivan* of the Jami' Muzaffar collapsed and the walls of the Qubbeh-yi Sabz were damaged.[292]

1871 December 23 **North Quchan.** On the night of 9 Shawwal 1288 an earthquake devastated the region north of Quchan (figure 3.23). In the Atrak valley half of the town of Quchan, including the remnants of its walls, mosques, madrasehs and the dome of the Imamzadeh Sultan Ibrahim, were ruined. Few people were killed as there had been slight shocks and rumblings for some time beforehand.[293] To the northwest, Ja'farabad and Isfijir, as well as another eight villages in the valley between these settlements, were totally destroyed with casualties. In the mountains further to the north, Ab-Suvaran and all villages, including Darbadam, together with the forts that guarded mountain passes on the roads to Gaudan and Chunli as far as Shamkhal, were completely obliterated and a large number of people were killed.[294] Damage extended to the district of Chinar and to the Incheh Pa'in valley. The shock was felt in Mashhad and as far as Tehran.

On 6 January 1872 an equally strong aftershock completed the destruction, killing a large number of people.[293] Aftershocks continued for almost four years.[295]

1875 May **Kuhbanan**. A violent earthquake in the district of Kuhbanan destroyed the village and fort of Jur as well as the settlements of Tukhrajeh. It is said that before the earthquake many game animals came down from the mountains and entered Jur. The villagers chased them out and this diversion saved them when the earthquake struck. The shock caused the springs at Tukhrajeh to dry up and damaged the settlements of Rashk. The village of Wasit was also ruined and the shock was strongly felt in Kirman and in its dependencies.[296]

1879 March 22 **Buzqush–Garmrud**. Preceded by a foreshock at dawn a few minutes earlier, an earthquake devastated the southwest part of the district of Ardabil and the region of Garmrud, which had suffered the same fate in 1844. Between Saqqizchi and Munaq on the southeast

slopes of Kuh-i Buzqush and along the Garm-rud, all villages were totally destroyed, in most cases leaving no survivors. In the region of Tark, Dizaj and Yangijeh rock-falls and landslides added to the destruction, killing hundreds of people. In all, more than twenty villages were totally destroyed, fifty-four suffered heavy damage, and at least 2000 people and 4000 animals were killed. In Tark the Tash-masjid of Khwajeh Ka'us Tarki, which was ruined in 1844 and rebuilt, was damaged and its minarets fell down. In Miyaneh and Ardabil several houses collapsed killing a few people and in Linkuran the shock caused some damage. The shock was felt as far as Tehran, Shusha and Alexandropol, and it was followed by a succession of damaging aftershocks that continued for two weeks, the sequence ending eight months later (figure 3.24).[297]

No ground deformations of tectonic origin attributable to this earthquake or features in the alluvium produced by geologically late movements along the Garmrud were found. It is uncertain whether an exposure and a very short segment of a Quaternary fault found north of Sariqamish[298] was associated with this earthquake or with an earlier event.

Plate 8. Earliest known isoseismal map for an earthquake in Persia, 1856.

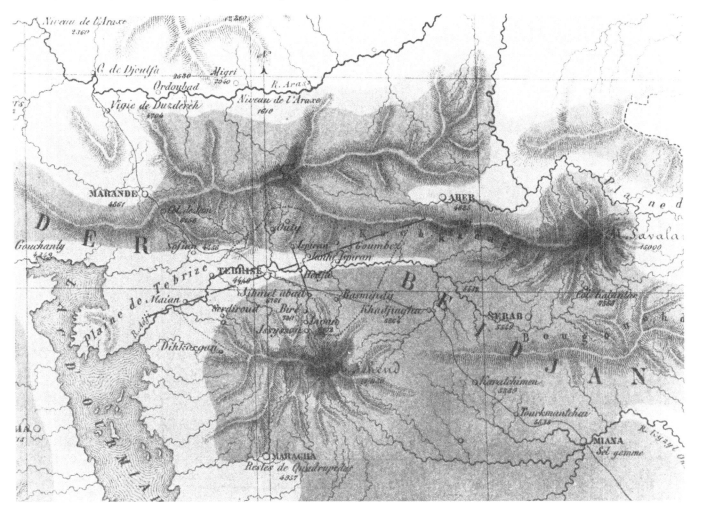

1880 July 4 **Garrus—Takht-i Sulaiman.** A damaging earthquake in the district of Garrus, west of Zinjan, ruined a number of villages, killing about sixty people (figure 3.25). The shock triggered landslides and rockfalls from the mountains, and in one place a new sulphur spring began to flow. Damage extended to Chiraghtappeh and settlements on the plateau of Takht-i Sulaiman where, however, the shock did not affect the discharge of water from the lake on the plateau. There was also some damage at Anguran and in other villages to the south of Takht-i Sulaiman, where almost all houses were ruined by the main shock and by an aftershock on 5 July. The shock was felt at Kavand and it was followed by aftershocks until September.[299]

1890 July 11 **Tash—Shahrud.** A destructive earthquake at dawn ruined a large area in the sparsely populated regions of Kuh-i Shangi and Shahvar between Astarabad (Gurgan) and Shahrud. In the village of Tash only one house was left standing and 140 out of its 200 inhabitants were killed. Shahkuh-i Bala and Pa'in, Mujin and other settlements as far as Purdilu in the 'Aliabad district were destroyed with casualties. Damage extended over a large area as far as Astarabad, Shahrud, Surmeh and Kalateh (figure 3.26).[300] In Astarabad almost all the houses were damaged to the extent that they were evacuated, and in Kalateh and Shahrud many houses

were ruined. Landslides blocked mountain passes, particularly at Shahkuh, and east of Tash rock masses were shattered.[301] Hundreds of people and many flocks of sheep perished. The shock was felt over a large area, mainly to the northwest, as far as Baku. It was also felt at Duzlyolum to the northeast but not beyond Damavand and Mayamay.[302] The shock caused waves in the Caspian Sea that were reported from all along its southern shores between Anzali and Ashur Ada. Aftershocks continued for at least five months.[303]

1893 November 17 **South Quchan.** Preceded by a strong shock on 20 October, an earthquake in the evening of the 8 Jumada I, 1311 devastated Quchan, particularly affecting the populous upper Atrak valley and the sparsely inhabited district of Sar Vilayat (figure 3.27). The town of Quchan, already ruined in 1871 and subsequently rebuilt, was utterly destroyed. All public buildings, including the telegraph office, the bazaar with its 1100 shops, part of the mosque and its minarets, the madraseh, caravanserai, and the palace of the Ilkhan, collapsed.[304] All houses and the remnants of the walls of the town were destroyed and at least 5000 people were killed. The dome of the Imamzadeh of Sultan Ibrahim and a few houses rebuilt with timber bracing after the 1871 earthquake, were the only structures left standing. In the town and its surroundings *qanat* galleries caved in, pro-

Figure 3.22. A.D. 1863 (30 December), Hir—Ardabil.

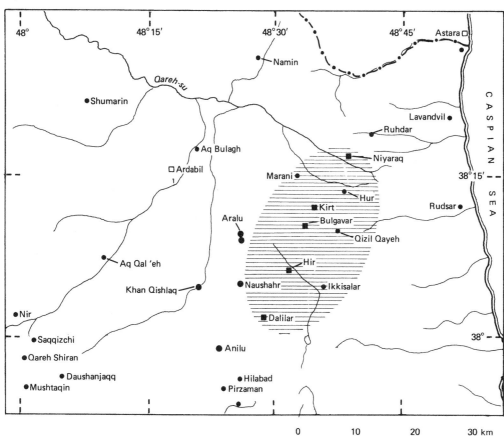

ducing crevices in the ground five to eight metres deep that added to the destruction.[305] In the Atrak valley all the villages between Yazdanabad, Kalbalasi, Kalukhi and Jartudeh were almost totally destroyed, with an estimated 10 000 casualties and the loss of more than 30 000 animals.[306] Damage was equally serious in the mountainous region of Kuh-i Muhammad Beg and in the *dihistan* of Sar Vilayat, as far south as Chakaneh Ulya and Saqi Beg. The number of people killed in this region is not known but the damage was so heavy that a few villages were abandoned. Also, there is some evidence to show that the direct routes to Sabzavar via Chakaneh Sufla and particularly the summer track via Bar had to be abandoned for a year or so.[307]

As a result of the shock large areas in the Atrak valley were intensely fractured and the banks of the Atrak, as well as water-worn ravines and gullies in loess, collapsed. The areas where earthquake fracturing was noticed are irregularly scattered through the Atrak valley, but slumping of loess deposits was most spectacular in the vicinity of Yusufabad and near Utrabad, along the banks and floor of the small valley of the southwest-trending Garmab stream,[308] where slumping and a large landslide in loess produced long fractures running discontinuously for three kilometres, in places gaping open to a depth of more than one metre.[309]

The shock was felt all over Turkmenistan and as far as Tehran 640 kilometres away from the epicentral region. It was followed by aftershocks that continued for almost six months. On 18 November, at night, an aftershock caused serious damage at Katlar and Gabrabad where the mosque collapsed. At dawn on the 19th an even stronger shock, which caused concern in Ashkhabad, triggered rockfalls from all over the east flanks of the Kuh-i Muhammad Beg and caused great damage to Katlar, Kalukhi and Gabrabad. On 12 January 1894, in the evening, another aftershock caused the collapse of ruins in Quchan with loss of life.[310]

Figure 3.23. A.D. 1871 (23 December), North Quchan.

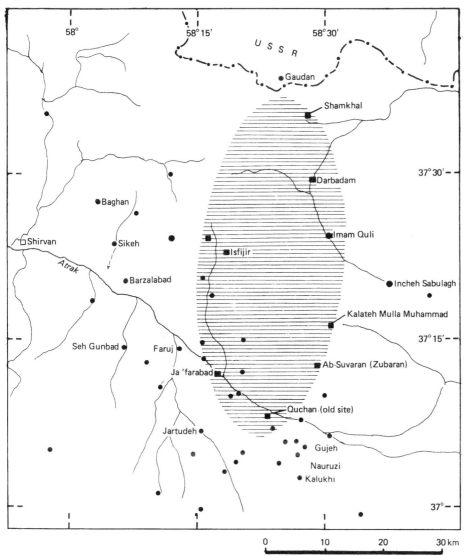

Figure 3.24. A.D. 1879 (22 March), Buzqush–Garmrud.

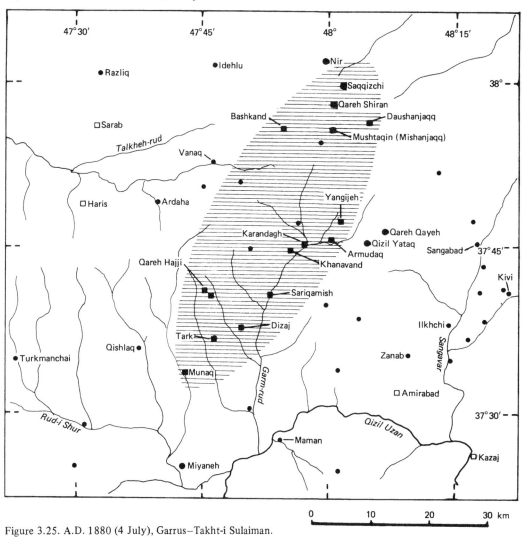

Figure 3.25. A.D. 1880 (4 July), Garrus–Takht-i Sulaiman.

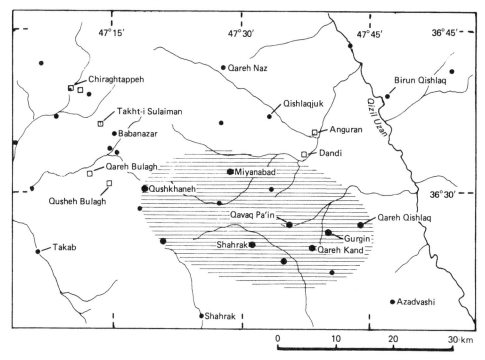

The earthquake was recorded by primitive, un-damped seismographs at Nikolaev and Pavia, 3800 kilometres from Quchan, and at Ashkhabad by a seismoscope.[311]

1894 February 26 **Shiraz.** A destructive shock in the *dihistan* of Kurbal almost totally destroyed the villages of Gari, Mansurabad and Kharameh, killing many people. In Shiraz the shock was felt very strongly, causing the collapse of some old walls.[312]

1895 January 17 **Quchan.** Shortly before noon another earthquake totally destroyed Quchan and a number of villages in the upper Atrak valley, killing about 1000 people. The earthquake, which was followed by incessant shocks until the next morning, utterly ruined the Imamzadeh of Sultan Ibrahim (which had been repaired after the earthquake of 1893), killing one of its inmates. Public baths, the bazaar, the Customs House, the Governor's residence, and the post and telegraph offices were destroyed with casualties.[313] With the exception of a few small shanties, almost all the houses, even those rebuilt after 1893 with timber and light roofs, were ruined. Because of the lightness of the building materials, however, and probably because of the foreshock that warned people, the loss of life in Quchan was much less than it would otherwise have been.[314] In all, about 770 people, including many from neighbouring villages who happened to be in town, perished.[315]

The earthquake destroyed five villages in the Atrak valley and another four in the Kuh-i Muhammad Beg, all of them already seriously affected by the earthquake of 1893.[316] In all, about thirty villages were destroyed or damaged with the loss of a few hundred lives (figure 3.28). The shock again ruined the *qanats* in Quchan, and the telegraph wires were brought down for several kilometres. Slumping of the ground and sliding of river banks extended for more than five kilometres along the Atrak. Rockfalls from the mountains were reported from the regions of Kalukhi and Katlar. There is no evidence that the earthquake was associated with faulting.

The earthquake was felt as far as Shahrud and Birjand and it was recorded faintly at Wilhelmshaven.[317] It was followed by a relatively short sequence of aftershocks which lasted about one month, the strongest of them causing some damage at Ja'farabad on 22 January.[318]

1896 January 4 **Khalkhal—Sangabad.** On the night of 2 January 1896 a destructive foreshock[319] in the *shahristan* of Khalkhal completely ruined Sangabad and almost all the villages in the upper reaches of the Sangavar-chai as far as Pirzaman and Hilabad. In Sangabad 300 people lost their lives. Damage extended mainly to the north of Sangabad. In Ardabil a few houses were damaged and cracks appeared in the government building in the citadel. The foreshock was felt as far as

Figure 3.26. A.D. 1890 (11 July), Tash—Shahrud.

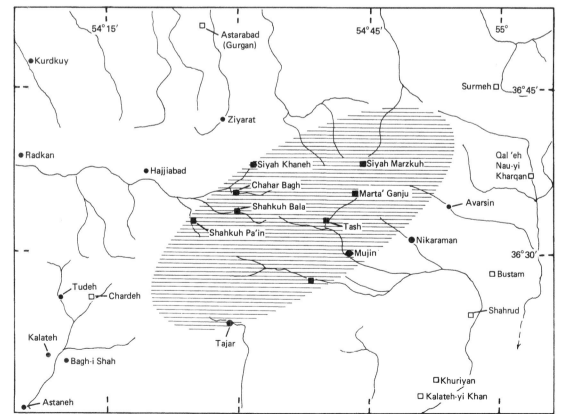

Tabriz, Qazvin and Linkuran, and it was recorded in Strasbourg, Kharkovo and Nikolaev.[320]

Two nights later the main shock totally destroyed not only the villages affected by the foreshock, but also the region to the south up to the confluence of the Sangavar with the Arpa-chai. Kivi was totally destroyed and 800 people lost their lives. Many other villages as far as Sukrabad, Lumbar, Hilabad and Ilkhchi were ruined and 1100 people were killed. A large number of cattle and sheep also perished. Rockfalls and landslides added to the damage, and in the plains the ground slumped.[321] Figure 3.29 shows the extent of the meizoseismal area due to the main shock and its aftershock. The shock was widely felt and it was noticed for its long duration. Long-period ground movements caused minor damage in

Ardabil, Linkuran and in Tabriz where the 'Ababafi factory was ruined. The shock was felt in Qazvin, Tehran and along the southwest coast of the Caspian. It was recorded in Strasbourg and Nikolaev.[320] Aftershocks, some of them damaging, continued for nine months. The aftershock of 14 January was particularly severe and caused additional damage, a number of deaths, and slumping of the ground.[322]

1897 May 27 **Kirman**. An earthquake on 25 Dhu' l-Hijja 1314 almost totally destroyed Chatrud and the settlements of Sarasiyab, together with most of the water mills in the district. The water supply was cut off and in Kirman most people fled to the countryside. A few people were killed in Kirman and several public and private buildings were damaged, some of them beyond repair. The dome

Figure 3.27. A.D. 1893 (17 November), South Quchan.

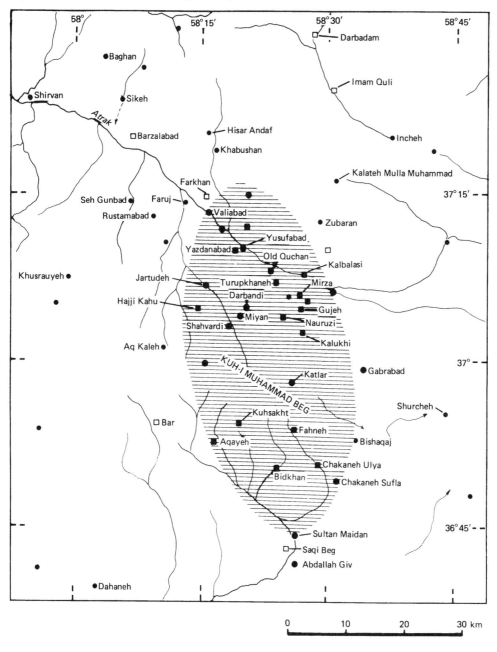

of the Qubbeh-yi Sabz, already supported by a half-ruined structure, collapsed injuring those inside and killing their animals. Also, the building of Bagh-i Nasiriyeh and the roof of the Ta'ziyeh pavillion were damaged. After the earthquake a few of the villages around Chatrud were abandoned, the survivors re-settling in Kirman.[323]

1900 February 24 **Khuy**. An earthquake at dawn near Khuy ruined the villages of Amir Beg, Imam Kandi, Quruq, Shirin Kandi, and the region of Sufla Kuh, where several people and many animals were killed (figure 3.30). In Khuy all the houses and shops were damaged and the remains of the Fuqani Dagh collapsed and the wall of the fort was ruined. The shock was very strongly felt throughout the district of Salmas and it was perceptible in Urdubad. Aftershocks continued for many months, the strongest occurring on 14 March. It caused widespread damage and forced the people in Khuy to abandon their homes, which had become uninhabitable, and live in tents.[324]

1903 March 22 **Durukhsh (?)**. It is probable that Durukhsh in

Figure 3.28. A.D. 1895 (17 January), Quchan.

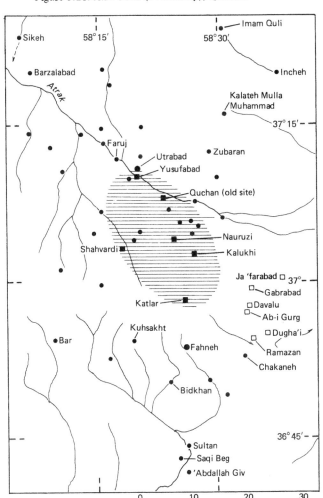

Khurasan was destroyed by an earthquake (refer to figure 3.9).[325]

1903 September 25 **Turshiz**. On 3 Rajab 1321 a severe earthquake caused extensive damage in the region of Turshiz (Kashmar) to the west of Turbat-i Haidariyyeh in Khurasan. The earthquake killed 350 people and destroyed, among other buildings, the carpet factories of the district, particularly those at Kundur and Kashmar. In Turshiz, the largest settlement in the region, damage was extensive, particularly in the southern parts of the town where almost all the houses were destroyed and 80 to 100 people lost their lives. In the northern suburbs no houses collapsed but almost all of them were damaged. Part of the bazaar also fell down, and the *ivan* of the mosque collapsed but not its minarets, which withstood the shock with minor damage. Turshiz was not as badly damaged as other outlying settlements to the west of the town, where the destruction was complete and more than 150 people were killed (figure 3.31). The earthquake caused local ground slumping, particularly north of Nasrabad, and rockfalls occurred in the hills north of Turshiz. A temporary change in the flow of water from springs and wells was also noticed. There was no damage to the Minar-i Kishmar, a thirteenth-century brick masonry tower 13.4 metres in diameter and 18 metres high. Nor did the earthquake have any effect on the Minar-i Firuzabad, another minaret of the thirteenth century, 7 metres in diameter and 18 metres high, situated just outside the epicentral area.

From figure 3.31 the connection of the Turshiz earthquake with the Duruneh fault zone is of course only too obvious. However, there is no evidence whatever that this earthquake was associated with movements of this major tectonic structure. The meizoseismal area of the Turshiz earthquake not only lies to the south of the zone, but also many sites (some of them very old, such as the Minar-i Kishmar, located literally in the zone or to the north of it) suffered no damage.

The shock was felt in Shahrud and Turud, and it was perceptible in Dustabad but not in Mashhad. Aftershocks were felt for over two months, causing further damage in the area.[326]

1909 January 23 **Silakhur**. A catastrophic earthquake on the morning of 1 Muharram 1327 devastated the Silakhur valley, southeast of Burujird in the Zagros.[327] Damage was particularly heavy not only in the densely populated valley of the Silakhur, but also further to the southeast in the mountain settlements as far as Arjanak (figure 3.32). In all, 128 villages were affected, of which 64 were totally destroyed with a loss of life estimated between 6000 and 8000.

The earthquake was associated with faulting which extended for a distance of forty-five kilometres from near Kalanganeh to the south of Saravand. Ground deformations beyond this, further to the southeast towards Arjanak, suggest that faulting perhaps extended beyond this point. Field evidence and local information

Figure 3.29. A.D. 1896 (4 January), Khalkhal–Sangabad.

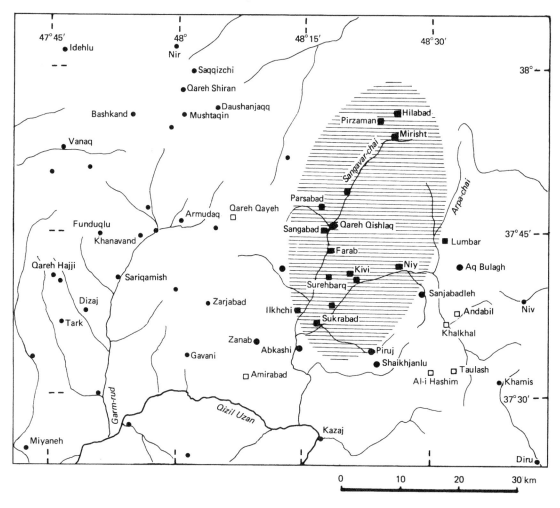

Figure 3.30. A.D. 1900 (24 February), Khuy.

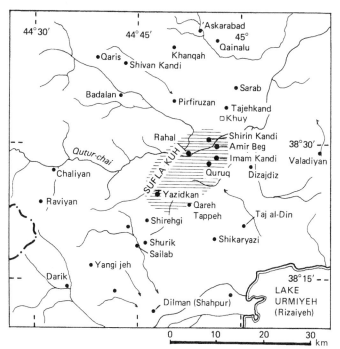

indicate that the average vertical displacement along the fault break was between one and two metres, with the northeast side downthrown. There is no evidence of strike-slip motion. In late Quaternary alluvium, such as northwest of Durud (Bahrain), the fault break is heavily eroded but it can be seen very clearly on the ground and on aerial photographs (plates 9 and 10).

The shock was strongly felt in Burujird, where water was thrown out of ponds, as well as in Kirmanshah and Hamadan,[328] and it was perceptible as far as Kharput in Turkey, Georgia and eastern Iraq. The earthquake was followed by a long sequence of aftershocks that lasted at least five months.[329] Most of the stronger shocks occurred in the southeast and northwest parts of the meizoseismal area, adding to the damage caused by the main shock.[330]

1911 April 18 **Ravar.** A destructive earthquake at night in the *dihistan* of Ravar, preceded by a strong foreshock, killed about 700 people. The small villages of Abdirjan, Maki and Deh Lakarkuh in the sparsely populated area east of Ravar were totally destroyed with many casualties. Almost all the houses in Ravar and its adjoining settlements were ruined. In the town itself, which had a population of 6000 at the time of the earthquake, several of the carpet-weaving factories and the *masjid-i jami'* collapsed, killing fifty people. Ravar remained in ruins for a long time, its public buildings only being rebuilt thirty years later.

The main shock and its aftershocks triggered many rockfalls from the northeast face of Lakarkuh, and it is very probable that the shock was associated with faulting west of Abdirjan, extending for a few kilometres in a south-southeasterly direction. However, when parts of this region were visited some years ago there was no discernable evidence of very recent faulting, except that the alignment pointed out to us as having been activated in 1911 happens to belong to a thrust zone bearing 160°E.

Minor damage extended to a number of villages, shown in figure 3.33, and the shock was strongly felt in Kirman, Deh Zu'iyeh and Kuhbanan. It was also felt in Birjand, Nasratabad and Duzdab. Aftershocks continued for the next three months, many of them being felt as far as Kirman. They persisted with long intermissions for some years.[331]

1923 May 25 **Kaj Darakht.** Preceded by a local but damaging foreshock at 'Abbasabad, an earthquake at dawn on 5 Khurdad 1302 completely destroyed five villages and severely damaged another twenty in the populous area of Kaj Darakht, with the loss of 770 lives. The earthquake caused some temporary changes in the flow of the underground water channels supplying the villages, a number of which were abandoned after the earthquake (plate 11). Damage extended mainly to the east of the meizoseismal area, where another ninety people were killed, forty of them in the region of Turbat-i

Figure 3.31. A.D. 1903 (25 September), Turshiz.

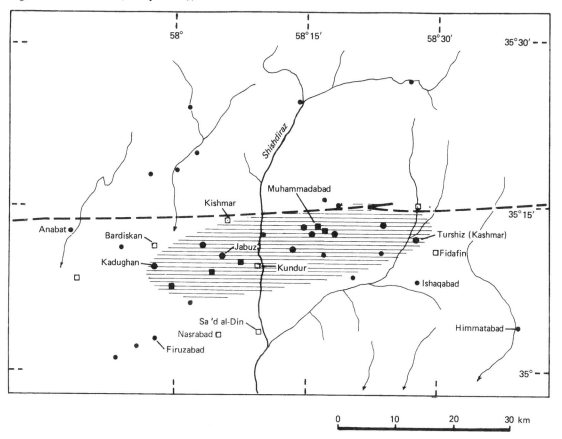

Haidariyyeh, a town with a population of about 10 000 (figure 3.34). In Turbat itself, one-fifth of the local houses were slightly damaged, and a few collapsed killing seven or eight people. The bazaar, built in 1860, as well as the seventeenth-century shrine of Qutb al-Din Haidar, suffered no damage.[332] The large adjacent *jami'*, which bears an inscription dated 1040/1630, suffered only a long vertical crack through the brickwork of the left wing of its *ivan*.[333]

There is no evidence of ground deformations of tectonic origin despite the proximity of the Duruneh fault zone. The shock was felt at Ribat-i Safid and Turbat-i Jam and it was perceptible in Mashhad. After-shocks continued to be felt strongly in Turbat-i Haidariyyeh for more than three months, causing further damage but no loss of life. The aftershock sequence

ended with a shock at 'Abbasabad where a few houses collapsed and one person was killed.[334]

1923 September 17 **North Bujnurd.** On the morning of 25 Shahrivar 1302 there was a destructive earthquake in the villages along the Atrak river, northwest of Bujnurd (figure 3.35). The area chiefly affected was on the north-east slopes of Akhir Dagh and on the southwest flanks of Kuh-i Bab Bulnar, where ten villages were destroyed and another twenty-two severely damaged. Casualties are given as 157 dead and 146 wounded in this area, where there was considerable loss of property. In Bujnurd there was damage to public buildings and some destruction of houses, but no loss of life. Damage extended to Shirvan, Gifan and Darband. In the region of Qal'eh Jaqq the shock triggered extensive landslides and elsewhere there was evidence of ground deformations.[335] The shock was

Figure 3.32. A.D. 1909 (23 January), Silakhur.

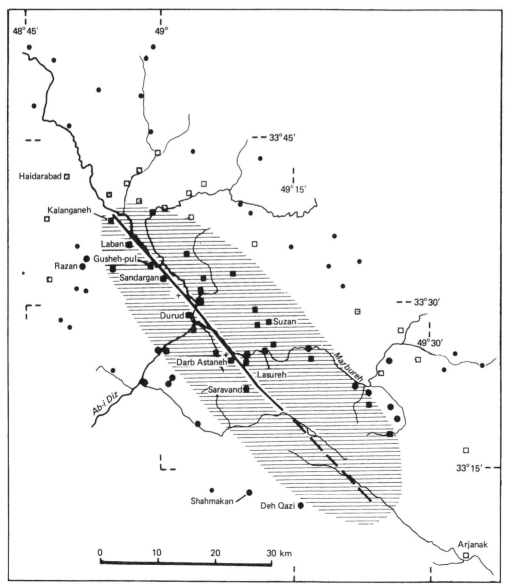

Plate 9. Aerial view of the Silakhur fault: (*a*) Segment northwest of Durud in the region of Sandargan. North is to the right; (*b*) Segment of fault passing through the switch yard of Durud town (D).

(*a*)

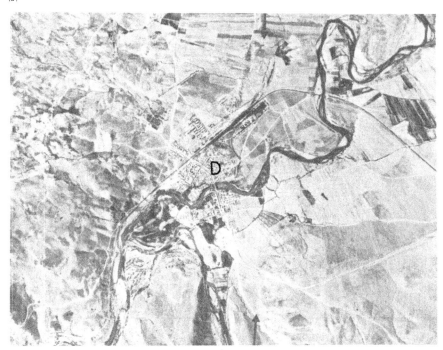

(*b*)

Plate 10. Fault break of Sandargan associated with the 1909 earthquake in Silakhur. Photograph taken by the Russian mission, members of which are seen in the background.

Figure 3.33. A.D. 1911 (18 April), Ravar.

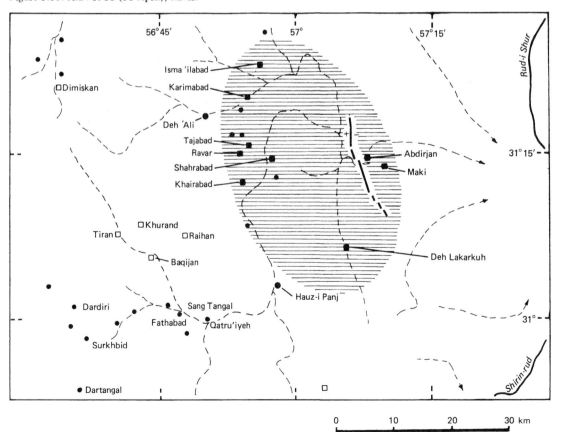

Plate 11. Kaj Darakht earthquake of 1923: remains of the tower of old Shadmihr.

Figure 3.34. A.D. 1923 (25 May), Kaj Darakht.

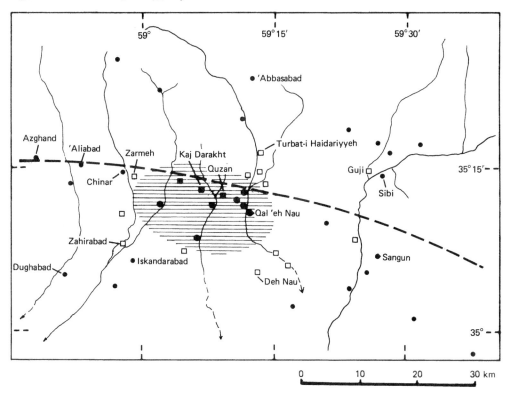

strongly felt in Ashkhabad and it was perceptible in Mashhad and Kazanszhik. The earthquake was followed by aftershocks lasting at least a month, causing further damage in Bujnurd, where no dwelling was considered safe for occupation.[336]

1923 September 22 **Lalehzar**. Late at night on 31 Shahrivar 1302, a destructive earthquake in the province of Kirman ruined parts of the districts of Sirjan and Baft. The *dihistan* of Gughar and the villages of Khatib, Qal'eh 'Askar and Lalehzar were destroyed and more than 200 people were killed. In Lalehzar alone sixty people lost their lives. Within a large area (figure 3.36) almost all houses were ruined, including the forts at Chinalu and Chaman Rang on the Kuh-i Ahurak, and many of the smaller settlements were abandoned after the earthquake. The shock triggered rockfalls that blocked the pass west of Gughar that leads to Sirjan.

Minor but widespread damage extended to Nigar, Rayin and as far as Kirman ninety kilometres away, where a number of public buildings, including the telegraph office, were badly cracked. Private dwellings outside the Vakil Gate as well as the Uakik clock tower were damaged and two people were injured. In Sirjan minor damage was equally widespread, and *qanats* collapsed reducing the water supply of the town. At Rafsanjan, 120 kilometres away, most of the houses were fissured. The shock was felt in Bam and Anar and it was

perceptible in Bafq, Yazd, Niriz and Bandar 'Abbas. A strong aftershock seven hours later caused additional damage, and was also felt in Kirman.[337]

1927 July 22 **Dasht-i Kavir.** On the night of 31 Tir 1306 north-central Persia was shaken by an earthquake which was felt within an area of 280 000 square kilometres. It caused no casualties but it was particularly strong in the region of Khuvar, where it caused some damage. A few houses and part of the interior of the fort at Aradan were ruined. In Simnan the top of the minaret of the *masjid-i jami'* and an older tower in the town were ruined. In Ivan-i Kay one of the towers of the castle was fissured. In Firuzkuh a few houses were cracked, and in Yazd water was thrown out of ponds. The shock was felt in Isfahan, Khunsar, Kashan, Tehran, Sangsar, Ashraf, Bandar Gaz, Damghan, Turud, Khur and Yazd (figure 3.37).

The location of the macroseismic epicentre of this earthquake, which must be sought in the Dasht-i Kavir, is problematic. A field trip to the regions southeast of Varamin as far as Hisar Quli and the vicinity of Siyah Kuh, as well as south of Aradan, proved fruitless,[338] the region being almost totally uninhabited with very few sedentary people.

The main shock was followed by a number of very strong aftershocks which were widely felt.[339]

1929 May 1 **Kopet Dagh.** During the evening a catastrophic

Figure 3.35. A.D. 1923 (17 September), North Bujnurd.

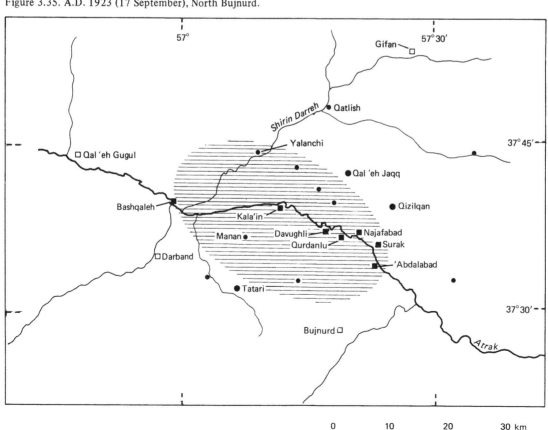

earthquake shook northeastern Persia and the Soviet Turkmen. The shock devastated the *dihistans* of Gifan, Ribat, Zaidar, Qushkhaneh, Jiristan, Amiranlu, Aughaz and Baghan, in which eighty-eight villages were destroyed with the loss of more than 3200 lives. In all, about 300 villages were affected, in which 1100 people were injured and thousands of animals perished. The shock was particularly severe in the northwestern part of the meizoseismal area between Gifan and Zaidar where more than half of the total loss of life occurred (figure 3.38). In Gifan, with the exception of an old Imamzadeh which is still standing, everything collapsed killing 300 people out of a population of 1500. The town was rebuilt on a different site lower down by the river. In Ribat only the *darvazeh* remained standing and out of 500 people more than 330 were killed. The village was later rebuilt on a different site. In Zaidar all but one of the 300 inhabitants were killed. Kurkulab and Garmab in the USSR were completely destroyed with loss of life. The villages of Shinarghi, Sarani, Aq Qal'eh, Savali, Nazar Muhammad, Kugli, Beg, Birzu and Sikeh were almost totally destroyed with great loss of life. To the northeast of the epicentral region the earthquake caused extensive damage over a large area. In Qatlish, Darband, Firyuza and Bajgiran almost all dwellings were damaged, and in Shirvan, with the exception of a few public buildings, all dwellings were shattered and seven people were killed. Some damage also occurred in Quchan, Bujnurd and

Ashkhabad. The shock caused serious damage to the *qanats* of Sikeh, Allahabad and Baghan in the Shirvan region and to those at Bagh. It also affected the spring water supply in many villages.

The earthquake was associated with movements along a post-Pliocene fault for fifty-five kilometres, from Sikeh to Bagh. At Sikeh the scarp of the fault break is eroded but still visible, standing about one metre high with the southwest side downthrown. It can be followed through the village to the east-northeast of Baghan, up to its crossing with the Rud-i Karganli, bearing 345°. At this point it steps to the left and clearly continued to the northwest, bearing 325° to 335°, passing to the southwest of Zakiranlu, to the northeast of Birzu and southwest of Chalikanlu, Palkanlu and Nazar Muhammad. It follows low, heavily eroded ridges on the side of the hills, and crosses them through saddles without any clear indication of lateral motion of the two sides. Nowhere does the scarp exceed one metre in height. Further to the northwest, between Nazar Muhammad and northeast of Aq Qal'eh, local information suggested that in places the throw was more than two metres. However, on examination these large displacements proved to be due to landsliding, the sliding utilising the trace as an upper boundary and modifying it. Landslide topography is widespread (plate 12). Local people recalled large cracks running in all directions, particularly towards Kakili.[340] Beyond Aq Qal'eh the

Figure 3.36. A.D. 1923 (22 September), Lalehzar.

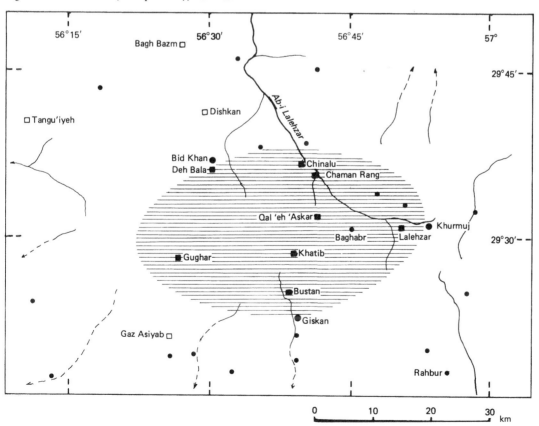

scarp passes to the northeast of Taftazan and Bagh but it disappears half-way to Zaidar, opposite Shinarghi, on the northeast side of the new road from Qulhak Pa'in. At this point another system of ground ruptures takes over, that bears 290°. Villagers remember ground fractures in strands running for 100 to 200 metres discontinuously

Figure 3.37. A.D. 1927 (22 July), Dasht-i Kavir.

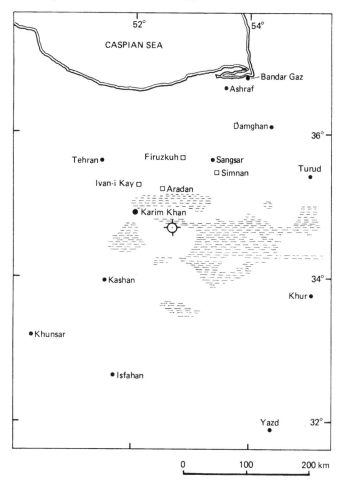

from about one kilometre south of Zaidar to Ja'farabad. These features today are heavily eroded but where indicated to us they resembled a series of ponding ridges interspersed with numerous sink-holes, reminiscent of eroded pressure features. There is no evidence of faulting to the north of this alignment. Other geologic effects, such as landslides, rockfalls, slumping of river banks, and changes in the ground water and stream flow, however, abound particularly in the valleys of Gifan, Ribat, and from what has been reported, in Garmab. Thus, the available field evidence suggests that the fault break associated with the Kopet Dagh earthquake runs for seventy kilometres from Sikeh to Ja'farabad. It has an arcuate trace and the overall movement seems to be oblique thrust. No conclusive evidence of strike-slip motion could be found.

The earthquake was felt as far as the Amu-Darya, in the Kara Kum, in places adjacent to the Aral Sea, in Mazandaran, and all along the south and southeast coast of the Caspian. It was followed by a long sequence of aftershocks, many of which caused further destruction and loss of life. Garmab and Kurkulab were totally ruined on 3 May and on 13 May Atabeg and other settlements along the Kirgan Ulya were ruined. On 13 July Qal'eh Faruj, Sinakli and another ten villages which had previously been damaged were completely destroyed and twenty-one people were killed. Quchan and Shirvan suffered additional damage and only a few wooden houses remained habitable. Aftershocks were still being reported in January 1930.[341]

1929 July 15 **Izeh—Andika**. At noon a damaging earthquake shook the southeast provinces of Khuzistan and Bakhtiyari. The villages of Qal'eh Zaras, Taqa, Andika and the settlements between these villages were ruined and many people were killed (figure 3.39).[342] On the southeast face of Kuh-i Landeh the earthquake triggered a landslide and in places rockfalls that produced a discontinuous scarp about one kilometre long. Aftershocks and

Plate 12. Landslide triggered by the Kopet Dagh earthquake of 1929, utilising the fault scarp as an upper boundary.

Figure 3.38. A.D. 1929 (1 May), Kopet Dagh.

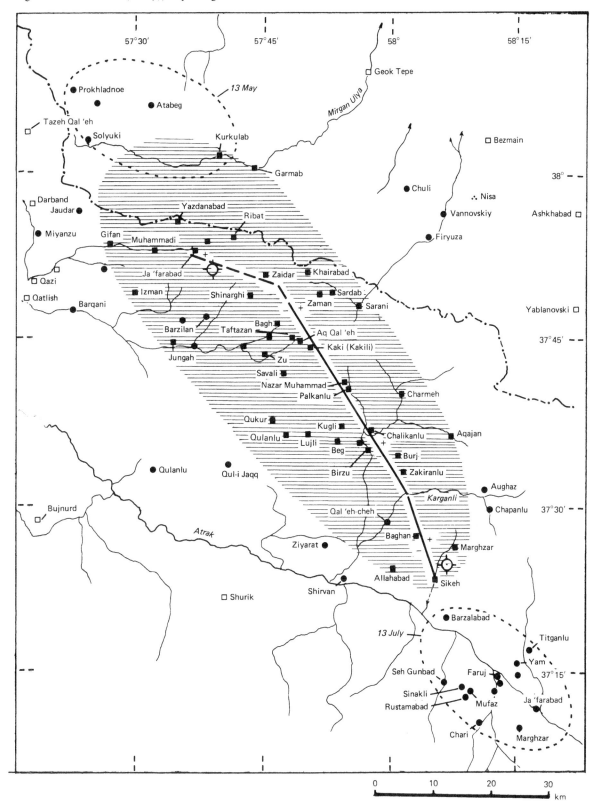

also new springs of water that emerged at the foot of the scarp after the earthquake caused additional movements of the slide mass.[343] Local rockfalls blocked the road east of Bard Qumchi.[344] The shock caused minor damage to the plant and private property at Masjid-i Sulaiman as well as at Malamir and it was strongly felt at Haft Gil and in the Bakhtiyari district. It was perceptible in Baghdad.

1930 May 7 (6) **Salmas.** Preceded by a damaging foreshock fifteen hours earlier, an earthquake devastated the *shahristan* of Shahpur in the early hours of the morning (local time). The foreshock damaged eight villages including Dilman, killing twenty-five people. It had the effect of warning the majority of the inhabitants of the area where the main shock followed, and they slept out of doors that night, thus saving their lives.[345] The main shock totally destroyed Dilman, see plate 4*b* (afterwards rebuilt as Shahpur) and about sixty villages in the Salmas plain and the bordering mountain regions, killing about 2500 people (figure 3.40). Damage extended from the Salmas plain to the *dihistan* of Qutur and the upper reaches of the Zab in Turkey. In the region of Salmas about fifty Armenian and Assyrian churches were destroyed or damaged beyond repair. Ten of those destroyed dated from the eleventh to the fifteenth centuries, and another eleven from the sixteenth to the eighteenth centuries. At Kuhneh Shahr (rebuilt as Tazeh Shahr) the mediaeval tower of Mir-i Khatum and the *masjid-i jami'* were also destroyed.[346] As a matter of

fact, no early historical monument exists any more in the Salmas valley.

The earthquake was associated with faulting, which is still visible and can be followed on the ground discontinuously for about sixteen kilometres from northwest of Shurgil to the vicinity of Kuhneh Shahr. For most of its course, which bears about 300°, it is possible to judge the sense of lateral motion (which is dextral), but not the amount of movement except at two points where it can be measured on right-lateral offsets of 1.0 and 4.0 metres. Between Shurgil and its crossing with the Zula-chai, the northeast side of the fault break is downthrown. The vertical offset is variable and in places attains apparent throws of 4.0 to 6.0 metres. However, for the whole length of the rupture the average throw does not exceed 1.2 metres. Beyond Kuhneh Shahr the trace is no longer visible. Nevertheless, local information suggests that it ran for another six to twelve kilometres in the same direction, along the southwest bank of the Dau-shivan river. Northwest of Darik another fault break in rock and alluvium can be seen running for about three kilometres, bearing 50° to 60°, with the northwest side downthrown by no more than about one metre. There is no evidence of lateral motion, nor is there any conclusive evidence that this feature extended to the northeast much beyond its crossing with the Dau-shivan. To the southwest, however, local information supports the assumption that it extended towards Hablaran, but this may refer to non-tectonic

Figure 3.39. A.D. 1929 (15 July), Izeh–Andika.

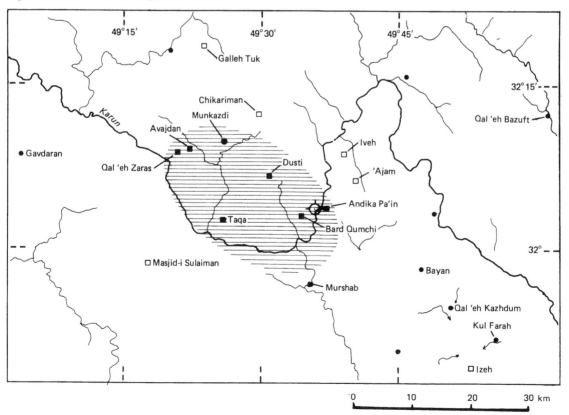

features, such as rockfalls, which did occur at the confluence of the Rud-i Aqkarzeh and Rud-i Rushadeh. Careful searching and enquiry revealed no ruptures in any other parts of the epicentral region. Considering, therefore, the rapid erosion and degrading of the fault break in the last forty-five years, it is reasonable to assume that the break extended for about thirty kilometres with about two metres of average throw and lateral offset.

The shock affected the flow of the streams and springs in the area but it is doubtful if this had any permanent effect on the recharge areas or on the permeability of the aquifers. As a result of the earthquake, the ground water table rose temporarily and flooded low-lying regions in the valley, but soon subsided to its old level. Landsliding in the Salmas valley and rockfalls in the mountains as far north as Qutur added to the damage. Incipient slides were also noticed in two archaeological mounds or 'tells', at Drishik Tepe and Haftavan Tepe. Recent excavations in the latter site

disclosed internal deformations of the 'tell', found also elsewhere,[347] which may be attributed to incipient sliding caused by the 1930 event or by an earlier earthquake.

The earthquake caused minor damage in Marand, Khuy and Urmiyeh (Rizaiyeh) and it was strongly felt in Tabriz where the electric lines were cut off and water sloshed out of reservoirs. The shock was felt as far as Leninakan and Tiflis to the north and Kirkuk and Baghdad to the south.

The aftershock sequence lasted for about three and a half months. The largest aftershock, on 8 May, caused widespread damage over a large area to the northeast of the meizoseismal region. It almost totally destroyed the villages of Givaran, Mir 'Umar, Raviyan, and Chaliyan to the south of Qutur, which had already been ruined by the main shock. It also caused serious damage to the southeast as far as Shikaryazi in a region that had not been seriously affected by the main shock.[348]

Figure 3.40. A.D. 1930 (7(6) May), Salmas.

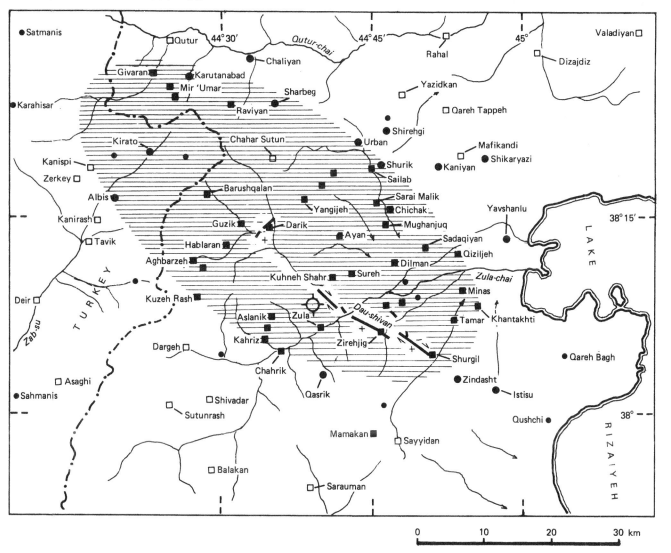

1930 October 2 **Ah–Mubarakabad.** Late in the afternoon an earthquake in Mazandaran ruined the small district of Ah and damaged a number of neighbouring villages (figure 3.41). In Ahmadabad a few people lost their lives, and in the nearby spa of Chashmeh 'Ala the bottling factory was damaged beyond repair. The shock did not cause any change in the flow of the hot springs. Damage and loss of life were heavier at Mubarakabad and in the district of Ah, where rockfalls blocked streams and roads, particularly to the northeast of the former locality, where the *qanat* was also blocked. In the Ira-rud valley, the shock ruined a few houses in Ardineh and caused the collapse of walls at Sang Darvazeh. The main shock and its strong aftershock of 7 October caused additional damage in the Ira-rud valley and triggered landslides from the north banks of the stream between Javard and Ira.[349]

The meizoseismal area of the shock aligns with the Musha–Fasham fault zone which did not show any evidence of re-activation.[350] Minor damage extended to Musha, Ira and Damavand, where a number of public buildings were rendered uninhabitable. The shock was felt strongly at Karaj, Tehran, Garmsar and Firuzkuh.[351]

1933 November 28 **North Buhabad.** During the afternoon an earthquake totally destroyed the small cluster of desert villages northwest of Buhabad, situated on the western fringe of the clay flat (*daqq*) of Muhammad Rafi' (figure 3.42).[352] The village of 'Aliabad Mulla 'Ali Riza was

totally destroyed with casualties, and it was abandoned (plate 13).[353] In and around the village, where at the time of the earthquake the perched water table was about five metres below ground level, the shock caused widespread liquefaction and the ejection of mud from mud volcanoes. Rahimabad was totally destroyed but without fatalities, and Khairabad was partly ruined with casualties. The settlements of 'Aliabad and Deh Muhammad Rafi' were also destroyed with casualties. Damage was less severe in Ja'farabad, Chilleh Khaneh, Karimabad, Jannatabad and Ahmadabad.[354] There were no casualties in Buhabad but almost all the houses were damaged.[355]

The shock caused extensive slumping of the ground and ejection of silty sand from cracks along a narrow zone that extended from south of 'Aliabad Mulla to Deh Muhammad. These features seem to be connected with the west limits of the clay flat (*daqq*) and with areas of high water table. In one place ground fractures follow a disused *qanat* causing an apparent throw of one metre down to the west, with diagonal cracks suggesting right-lateral motion. Ground ruptures were virtually absent elsewhere, except at a site about six kilometres southwest of 'Aliabad Mulla, where, according to local information, the shocks 'caused the rocks to open up'. The site is on conglomerates and sandstones, above a series of cones produced by a ravine that debouches from the west into the plain. It does corre-

Figure 3.41. A.D. 1930 (2 October), Ah–Mubarakabad.

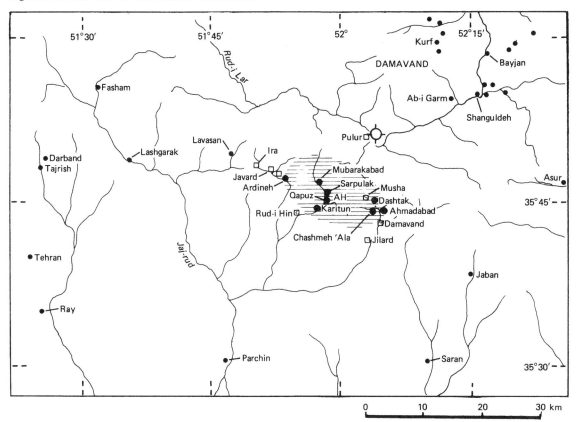

spond with a thrust zone that dips to the southwest but no evidence of recent movements could be found except open joints and cracks, running along the zone discontinuously for about five kilometres.[356]

The earthquake caused slight damage in Anar and Bafq and it was strongly felt in Rafsanjan, Yazd, Taft, Ardakan, Mahdiabad and Kirman. Aftershocks con-

Figure 3.42. A.D. 1933 (28 November), North Buhabad.

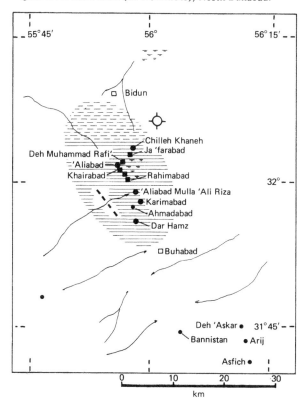

tinued for three months, the shock of 12 December causing more damage in Buhabad than the main shock.

1935 March 5 **Talar-rud.** Early in the afternoon a damaging earthquake east of the Talar-rud in Mazandaran ruined eight and damaged nineteen villages in the isolated and sparsely populated districts of Banaft and Dau-dangeh, killing about sixty people (figures 3.43 and 3.44).[357] Much of the damage was concentrated in the *dihistans* of Rast-u Pay and Khanqah, but because most houses were of timber construction the number of casualties was rather small. In the Talar-rud valley, between Dau-ab and Surkhabad, the railway line which was then under construction suffered considerable but localised damage. The portal of tunnel 10 was blocked by rock-falls and the roof of the tunnel collapsed killing a number of workers. Rockfalls also blocked the roads out of the valley at Dau-ab and 'Abbasabad killing forty-six labourers and destroying the houses of the construction gangs at Kuhpayeh and near Paru. The earthquake caused widespread but minor damage throughout Dau-dangeh and along the road to Firuzkuh as far as Shahi, where a few old houses, the roof of the textile factory and its chimney stack collapsed.[358] The shock was felt at large distances, mainly to the northwest, at Shahsavar, as well as at Gurgan and all along the southern coast of the Caspian.

1935 April 11 **Kusut–Mazandaran.** A few hours after midnight on 23 Farvardin 1314 (local time) a destructive earthquake and a damaging sequence of aftershocks[359] in the districts of Gulijan and Chahar-dangeh in Mazandaran destroyed twenty-six villages and ruined or damaged beyond repair another eighty (figures 3.44 and 3.45).[360] In spite of the fact that most houses in this part of Persia were built of timber with thatched roofs,

Plate 13. Ruins of the settlement of 'Aliabad Mulla 'Ali Riza, near Buhabad, abandoned after the earthquake of 1933.

many of them collapsed killing about 480 people. At
Kusut and Varand the destruction was followed by fire,
while at Gardashi, Qadikula and Churat it was followed
by landslides triggered by violent aftershocks. Throughout the meizoseismal region and at some distance from
it, landslides blocked roads, passes and streams. Near
Dausalleh the Tijan-rud was dammed up by a half-
kilometre long slip from its northern banks, while
between Varand, 'Alamdar and Kusut the mountain
slopes slumped on both sides of the river. Scarps of
slides triggered by the earthquake are still visible along
the north banks of the Tijan-rud and up the Qadikula
forest, as well as between Jinasim and Shit on the
Zalim-rud, where tilted and leaning trees, some of them
at least eighty years old, suggest an overall unstable
topography.[361] Even on relatively flat terrain the shock
and its aftershocks caused extensive slumping and
sliding, as for example between Talukula and north of
Kunim, as well as west of Amuri and between Aryam
and Sankur, the scarps of the slides being still visible on
the ground.[362] Landslides and the rains that followed
the earthquake made the region impassable. The first
relief team sent from Sari immediately after the earthquake found it impossible to proceed beyond Varand up
the Tijan-rud and returned to base on 17 April.[363] There
is no evidence that any of these ground deformations
are of tectonic origin.

Outside the meizoseismal region damage was widespread but minor. At Shahi many public buildings,
including the cotton mill, the sugar factory and the textile factory damaged by the 5 March earthquake,
sustained additional damage and an elevated water tank

Figure 3.43. A.D. 1935 (5 March), Talar-rud.

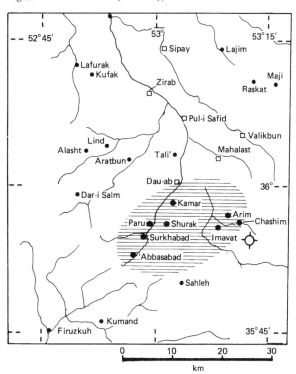

and two chimney stacks collapsed. In Sari the walls of
many houses, including the Governor's, cracked and
water sloshed out of ponds. The railway line to Nika
slumped at its sixth kilometre. In Amul a few old houses
collapsed and in Ashraf the Bagh-i Qasr suffered minor
damage. Also in Babul the Baghshah-i Shahi was damaged
and a few houses cracked.[364]

The earthquake was felt over a relatively large
area, mainly in an east–west direction, from Shirvan to
Rasht and from Gazan Quli in the USSR to Simnan. It
was followed by a twenty-four hour sequence of violent
aftershocks and by a period of minor shocks that lasted
about six months, mainly recorded by the USSR network.[365]

The earthquake caused absolutely no damage to
the early eleventh-century tower, the Imamzadeh
'Abdallah, at Lajim (plate 14),[366] and to timber framed
houses in the village. However, it did destroy all dwellings built of adobe, but without casualties, and caused
slumping of the ground some distance below the tower.
This was partly attributed to the heavy rains that followed the earthquake.

Also near Raskat, a somewhat smaller tower of
the same period (plate 15)[367] built on rock, suffered
insignificant damage. Pre-existing cracks in the brickwork of its northern face opened up more, while in the
nearby villages of Raskat Sufla and Raskat Ulya a number of adobe houses collapsed.

1941 February 16 **Muhammadabad.** Early at night an earthquake caused heavy damage in the sparsely inhabited
region northwest of Birjand. Muhammadabad and its old
caravanserai were totally destroyed and out of its 920
inhabitants, 680 were killed. The isolated settlements of
Nurah, Qaran, Kamiran and the villages of Quminjan and
Chahak were ruined. Damage extended to Qaisar,
Tighdar and Afriz, where houses collapsed without
casualties. The fort of Afriz and the Imamzadeh Zaid ibn
Imam Musa were also damaged and houses cracked as
far as Khur and Sarayan (figure 3.46).

Field evidence suggests that the earthquake was
associated with surface faulting that extended from
about four kilometres south of Quminjan for about eight
to ten kilometres to the south. Local people verified
some of the details of surface faulting still visible,
though confined within a narrow zone that follows the
contact between flysch in the east and Eocene volcanics
in the west, bearing N-180°-E. They consist of a series of
crushed and weathered zones and of open cracks in rock,
the latter arranged *en echelon*, running discontinuously a
few tens of metres, mainly in volcanics. They suggest
thrusting from the east with a considerable component
of right-lateral motion. Some of the wider cracks have
been eroded into sinkholes and others have in part been
filled up with silt. It was not possible to judge the
amount or sense of vertical movement but our informants, who still remembered many of the details very
well, claim that the west side was downthrown by about

Figure 3.44. Map showing area covered during successive field trips to establish the location of Firrim and the effects of various Mazandaran earthquakes. The map has not been redrawn or corrected for spelling, but it is included for comparison with existing (1975) topographic maps which give a misleading picture of watersheds.

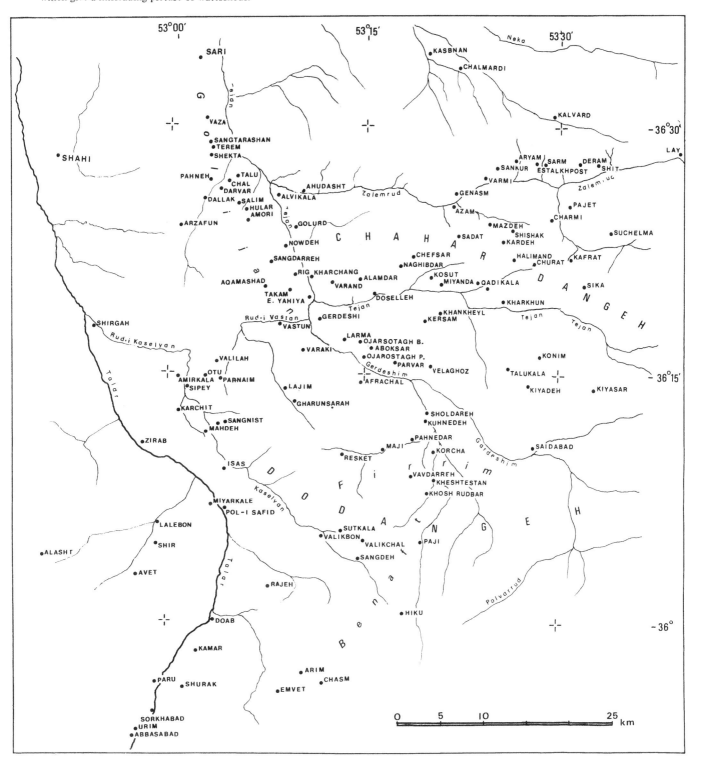

50 centimetres. This zone of ruptures follows the western foothills of Chaliyu. It then passes to the east of Kuh-i Kamar Qal'eh and again to the west of Kuh-i Kamiran and dies out at its crossing with the Rud-i Abshur. It is of interest that to the south of this point a recent scarp in playa and clay pans (*daqq*) can be followed for another twenty-five kilometres to the south (plate 16). However, according to our informants, no ground deformations except collapsed 'shutts' and slumping of the 'daqq' were noticed in this part of the epicentral region, nor do they remember displacements along this scarp where it crosses the track from Muhammadabad to Chahak.

Another series of fractures was noticed running for a few kilometres from north of Hauz Qal'eh Kuhneh (Khuni) to the vicinity of the spring of Turshab, striking N-80°-E. The scarp indicated to us follows the contact between the pediment and the clay flat of the Daqq-i Muhammadabad, showing a throw to the south of less than one metre.[368] These fractures, which are attributed by the local people to the 1941 earthquake,[369] are of interest because they resemble a fault trace, but they were probably of slumping origin of the clay flats, similar to those observed in the Dasht-i Biyaz valley in 1968.

The earthquake caused widespread liquefaction of the clay pans to the east and south of Muhammadabad where the water table at the time of the earthquake was

Figure 3.45. A.D. 1935 (11 April), Kusut–Mazandaran.

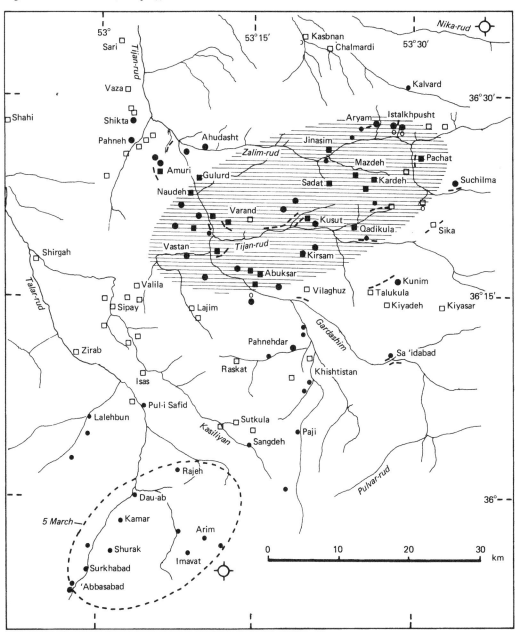

Plate 14. The Imamzadeh of Lajim in Mazandaran.

Plate 15. The tower of Raskat.

Figure 3.46. A.D. 1941 (16 February), Muhammadabad.

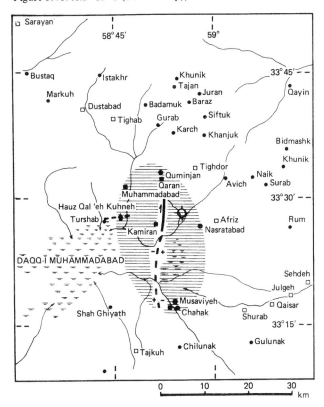

Plate 16. Aerial view of the Muhammadabad fault.

very near the surface. Between Muhammadabad and Tighdar and south of Qal'eh Kuhneh, the ground slumped and mud was ejected from cracks.

Muhammadabad was rebuilt about 500 metres to the north of its old site and many of the settlements that were ruined were abandoned. The shock was strongly felt in Birjand and Qayin and it was perceptible in Turbat-i Haidariyyeh.

1945 November 27 **Makran.** In the early hours of 28 November (local time) the Makran coast of Pakistan was shaken by one of the largest earthquakes of this century ($M = 8.1$). Fortunately the region affected, between Karachi and the Persian borders, is singularly sparsely populated with only three small coastal towns, i.e. Ormara, Pasni and Gwadur, which are devoid of any engineered structure, most buildings consisting of mat huts and one-storey adobe houses.[370]

Pasni is an open roadstead and port of about 2000 inhabitants, built on a sandbank connecting the headland of Zarrin with the mainland. The earthquake destroyed eighty per cent of the houses, killing about forty-five people. The telegraph building and the few better built official buildings were rendered unusable. A part of the town was involved in a submarine slide which submerged a zone along the shore so that the coast today is about 100 metres inland.[371] Ormara (see figure 3.47) is a smaller port on sand where sixty per cent of its houses, mainly of adobe construction, collapsed killing about twenty people. The shock caused many landslides and rockfalls along the steep bluffs south of the town. In places the ground slumped and it was flooded by a sudden rise of the underground water table. Between Ormara and Pasni, a distance of about 120 kilometres, there are very few settlements of grass mat huts, which suffered little or no damage. Further to the west Sur (Sor) was damaged and in the port of Gwadur some houses were ruined without casualties. However, to the north of Gwadur, over the hills in the Akara—Jawar region, the ground was broken up and huge cracks ran along the north-facing slopes of the ranges of hills for a

Figure 3.47. A.D. 1945 (27 November), Makran.

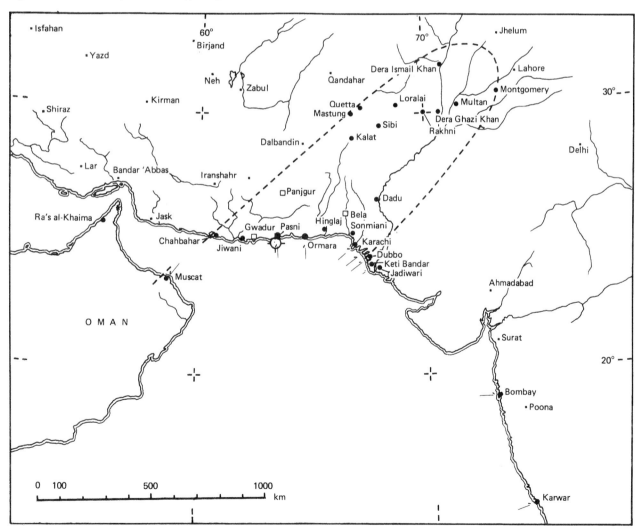

considerable distance. The shock caused no damage at Jiwani or further west along the coast at Chahbahar, where the ruins of the old fort located about two kilometres from Tiz, to the north, and constructed or at least renewed in the early 1870s, were not affected by the shock; nor was its stone masonry ten-metre high tower affected. As a matter of fact, along the Makran coast west of Jiwani and east of Sonmiani, the earthquake was not felt very strongly and it was not perceptible beyond Karachi and Dadu. The earthquake was also felt as a slight shock at Muscat. In contrast, to the northeast of Pasni it was felt as far as Montgomery and Dera Ismail Khan in the Punjab, more than 1000 kilometres away. Damage was reported from Panjgur, Bela and from the Barkhan Tehsil of the Loralai Agency, where casualties were also reported.[372] There is no evidence that the shock caused undue concern in Persian territory.[373]

The total number of people killed in this earthquake does not seem to have exceeded 300. This figure includes those drowned by the seismic sea-wave (*tsunami*) that accompanied the event and which added significantly to the overall damage caused by the main shock. At least three waves followed the earthquake, of which the first (noticed shortly after the shock) did not come very far inland. The other two followed 90 to 120 minutes later, around 5.00 a.m., and swept over the one-storey houses at Pasni and Ormara, causing great damage and reaching heights of five to ten metres on shore. At Jiwani the waves sank a *dhow* and at Gwadur drowned three people. Elsewhere on the Makran coast they destroyed sailing vessels, littered the land with debris and silted up streams. At Karachi, 360 kilometres away, the waves had a height on land of about 1.5 metres, but they persisted for such a long time that they caused damage to the harbour works and loss of life around Keti Bandar on the coast of the Indus delta. There, during the recession of the waves and the rapid drawdown of the water that followed during the strong ebbing of the sea, low-lying hills collapsed and spread out, totally destroying a number of fishing villages between Dubbo and Jadiwari, causing many casualties. In the region of Bombay, more than 1100 kilometres away from Pasni, waves reached heights of up to two metres, causing some loss of life. At Karwar, about 1500 kilometres away, waves flooded creeks and inlets. At Muscat the shock was followed by a very high tide and at Mahe in the Seychelles, 3400 kilometres away, wave heights reached about thirty centimetres. There is no information about the effects of the waves in the Persian Gulf,[374] but in the Arabian Sea at least one *dhow* on its way from Muscat to Karachi was sunk with casualties.[375]

Following the earthquake, four large mud volcanoes rose near the shore eight to thirty metres above water, seven to thirteen metres deep and emitting gas, but they were soon eroded by the sea.[376] Also inland, the mud volcanoes near Hinglaj were re-activated and a large volume of gas ignited, causing an eruption the glow of which could be seen from a great distance. Mud volcanoes along the Makran, both inland and offshore, do occur under normal conditions and they should not be considered as evidence of the severity of the shock.[377]

The earthquake damaged the trans-oceanic cable between India and Great Britain, which broke in eight places off the Makran coast, presumably due to submarine landslides triggered by the shock.[378] The land lines between Pasni and Karachi also broke down.

There were substantial ground failures at Pasni and Ormara. About five kilometres to the north of the latter, the ground slumped, forming tensile fractures sixty centimetres wide running in an east—west direction with the south side downthrown by more than 1.5 metres. Though there are previous reports (Sondhi 1947) of uplift at Pasni, tectonic changes in elevation were established only in the Ormara area, where the land rose about two metres. Older raised beaches and marine terraces were observed along the whole Makran coast, from Karachi to Jask, and radiocarbon and uranium—thorium dates on shells from these beaches indicate the beaches were elevated during the last 10 000 years, attesting to numerous past earthquakes comparable to the 1945 event.

Aftershocks continued for some time. On 2 January 1946 Pasni was again damaged, and on 5 August 1947 another shock in the evening ruined many houses at Pasni. In the Kulanch the ground slumped, ejecting water from cracks. Rockfalls were noticed from the mountains and landslides occurred along the coast to the west of Pasni. The shock was widely felt.

1947 September 23 **Dustabad.** In the morning a destructive earthquake in the sparsely populated region of Daulatabad, southeast of Firdaus (Tun), demolished a number of villages killing about 400 people. Dustabad was totally destroyed and 170 people were killed. Half of Muhammadabad, rebuilt after the earthquake of 1941, collapsed and heavy damage with casualties extended to the *dihistans* of Sarayan, Charmeh and Badamuk. To the west and southwest of Daulatabad and for tens of kilometres, the area was in 1947 totally uninhabited, while to the south there is nothing more than the clay flats of Daqq-i Muhammadabad and the Shikasteh-yi Muhammadabad. The only man-made structures there (a few rubble-stone houses and *ab-anbars*) were levelled with the ground (figure 3.48).

The earthquake was associated with a complex system of faulting. Ground fractures reported by local people, their trace still visible in a number of localities, can be followed for about twenty kilometres in a N-350°-E direction, from west of Kuh-i Qirmiz, cutting across the west slope of Kuh-i Chargraqsh and extending to the north as far as the east flank of Chang-i Kulagh. Beyond this point the trace is no longer visible on the ground, but according to local information it extended across the pediment between Dustabad and the settlement of Istakhr. From Kuh-i Qirmiz to Kuh-i Chargraqsh the zone of fractures is in late Eocene volcanics (plates 17 and 18). In places it consists of *en echelon* cracks, some of them open and eroded into sinkholes (plate 19) bearing N-25°-E, and elsewhere of crushed zones with the west side downthrown by thirty to eighty centimetres. Two exposures in gullies suggest right-lateral movement of at least one metre.

In the plain, between Dustabad and Sarayan and also between Tighab and Tighdar, the ground slumped and *qanats* were blocked. A series of scarps between Badamuk and Gurab striking N-140°-E, mostly in conglomerates, as well as between Turshab and Hauz Qal‘eh Kuhneh south of Muhammadabad, were attributed to this earthquake by the local people.

The shock was widely felt, causing panic in Qayin, Bushruyeh and Birjand, and it was perceptible in Ravar and Mashhad. It was followed by numerous aftershocks that lasted for about six months, causing additional damage, particularly in the region of Charmeh and at Sarayan.[379]

1948 July 5 **Gauk.** Late in the afternoon of 14 Tir 1327, a strong earthquake was felt in the province of Kirman. Published details of the damage it caused are lacking, but local information indicates that the few settlements in the sparsely populated *dihistan* of Gauk were ruined. It is alleged that the earthquake originated in the Kuh-i Dau Shah mountains north of Gauk and triggered rockfalls from the mountains to the northeast of Malik and Tirkan. The shock was strongly felt as far as Bam and Kirman (*Ittila‘at* 1327, no. 6684). At Sikunj, the shock

Figure 3.48. A.D. 1947 (23 September), Dustabad.

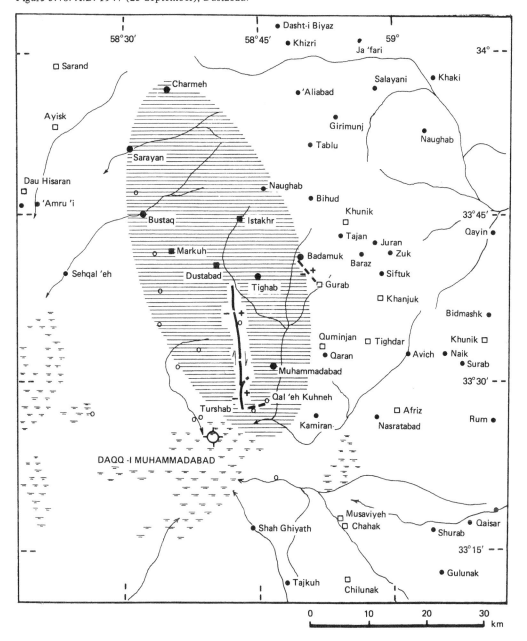

Plate 17. Aerial view of the Dustabad ground ruptures associated with the 1947 earthquake.

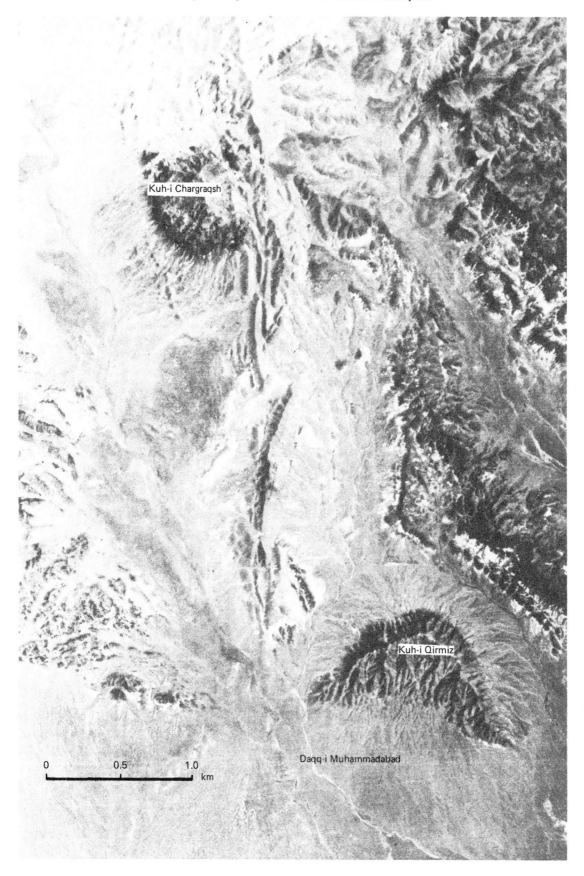

was particularly severe, causing panic and some damage, while at Jupar the *qanat* yield increased considerably for some time after the event (Beckett 1953: 53).

The importance of this earthquake lies in the fact that it occurred on the southeast extension of the region affected by earlier, nineteenth-century events. An earthquake in 1877, particularly, ruined the villages of Sirj, Hasanabad and Hashtadan, and caused the hot springs around Ab-i Garm to cease flowing. Damage extended to settlements mainly south of Ab-i Garm, where the shock caused ground cracks and rock failures (Houtum Schindler 1881*b*: 355). There is no evidence that the earthquake was felt very far.

1957 July 2 **Bandpay—Mazandaran**. At dawn on 11 Tir 1336

a destructive earthquake ruined the mountainous region of Bandpay, which is situated to the north of the drainage divide of the Alburz mountains.[380] In the meizoseismal region, which lies between the upper courses of the Harhaz and Talar rivers and comprises the *dihistans* of Bandpay, Beh and Dala Rustaq and Chalav, about 120 villages were totally destroyed with an estimated 1500 casualties (figure 3.49).[381] The heaviest destruction caused by the earthquake occurred between Nandal, Sangichal, Chaliyasar, Nasal, Andavar and Pardimeh, while rockfalls and landslides added to the destruction particularly at Burun, Varzaneh, Shanguldeh, Nal and Dinan. In the upper reaches of the Sajarud and Shir Qal'eh, streams were dammed up and passes were

Plate 18. Ground view of the Dustabad fault trace; Kuh-i Chargraqsh in the background.

Plate 19. Weathered features of the Dustabad ground ruptures northwest of Kuh-i Qirmiz.

blocked. A massive rockfall blocked the Harhaz-rud near 'Aliabad, creating a twenty-metre high dam and a reservoir about one kilometre long. The damage sustained by the few engineering structures in the region, which can be found only along the Harhaz road, was minor. The abutments of the eighty-metre long masonry spandrel arched bridge at Bayjan settled and its deck cracked. The lining of the road tunnel between Kuhrud and Bayjan was badly cracked and the haunch sheared. There was also some damage to the short tunnel near the Nur confluence and to the abutments of the wooden bridge at 'Aliabad.[382]

There is no evidence that the earthquake was associated with surface faulting,[383] and no changes in the fumarolic activity of the Damavand volcano were noticed.

Outside the meizoseismal region shown in figure 3.49, damage was widespread but not serious. It decreased more rapidly with distance to the northwest and southeast than to the northeast and southwest. It extended as far as Pul-i Safid, Shirgah and to the Kasiliyan river where at Utu a few houses collapsed.[384] In the opposite direction at Pulur and Fasham a few roofs caved in and in the southern parts of Tehran a few houses were damaged.[385] To the southeast a few houses were damaged as far as Firuzkuh. At Shahi and Sari, in

the opposite direction, a few modern houses were damaged and at Babul the cross-bracing of an elevated water tank snapped.

The shock was felt from Khurasan and Shahrud to Hashtpar in Azarbaijan and as far south as Kashan. About 25 million dollars worth of damage was done and many sites and settlements in the region were totally abandoned.[386]

A small Imamzadeh at Vaneh and the Imamzadeh Hashim east of Ab 'Ali were seriously damaged.[387]

1962 April 1 **Musaviyeh**. At dawn a damaging earthquake ruined a number of villages north of Birjand on the extension of the region devastated by the earthquake of 16 February 1941. The lower part of Musaviyeh was totally destroyed and the rest of the village was ruined. Chahak, Chilunak, Tajkuh and Nuj were also ruined with the loss of a few lives and a large number of animals. Damage was confined within a small area (figure 3.50), but springs of water and liquefaction phenomena were reported from as far as Muhammadabad and Shah Ghiyath. There is no evidence that the shock was associated with ground deformations of tectonic origin along the Rud-i Abshur scarp (plate 16) or elsewhere. The earthquake was strongly felt at Birjand, Khur, Khusf and as far as Firdaus.[388]

1962 September 1 **Buyin Zahra**. Late at night on 10 Shahrivar

Figure 3.49. A.D. 1957 (2 July), Bandpay–Mazandaran.

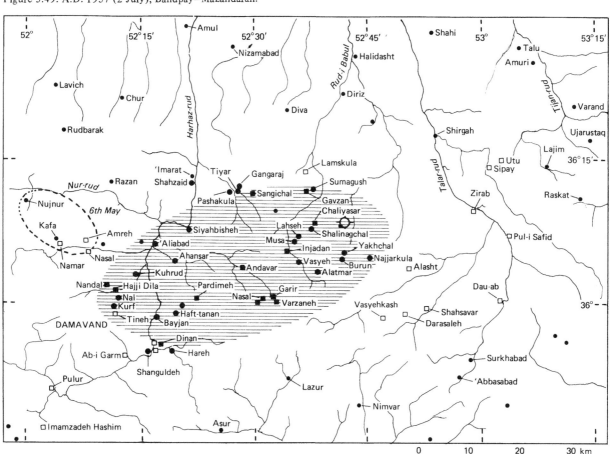

Figure 3.50. A.D. 1962 (1 April), Musaviyeh.

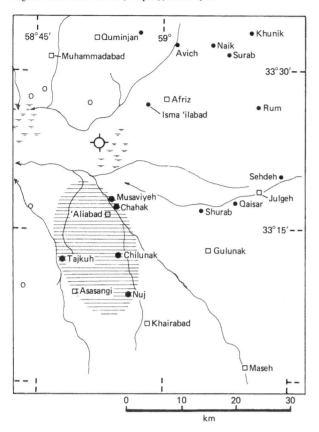

1341, a catastrophic earthquake in the densely populated region south of Qazvin totally destroyed 91 villages, killing 12 200 people and injuring 2800. In all, over 300 villages were damaged or destroyed, 180 of them with loss of life. Among the places worst affected were Buyin Zahra, Danisfan (Danisfahan), Arasanj, Rudak and Ildarchin (figure 3.51).[389]

The earthquake was associated with faulting that runs discontinuously from west of Ipak to near Ildarchin, a distance of about eighty kilometres, striking 103°E (figure 3.51).[390] The observed ruptures, which do not follow any major throughgoing tectonic feature, are mainly in Eocene volcanics and they form a wide zone within which one or two main fractures show the largest displacements. On average there are a few tens of centimetres of throw to the north and a left-lateral strike-slip half as large. These factures (shown in figure 3.51) do not follow the small pre-existing faults in the volcanics, which strike mostly east-southeast, but rather cut across them, forming a new and complex system of ruptures at a sharp angle. For instance, the alignment A−A in plate 20 shows a geological fault in Eocene volcanics near Ahangiran which passes between the ruins of the village and its new site, west of Ipak.[391] No evidence of dislocation was found after the earthquake, and ground ruptures along most of this ten-kilometre long feature were virtually absent. Ground ruptures, however, did develop within a broad zone to the north of this fault, the main fractures (shown in plates 21 and 22) exhibiting a throw to the north and northwest of about forty-

Figure 3.51. A.D. 1962 (1 September), Buyin Zahra.

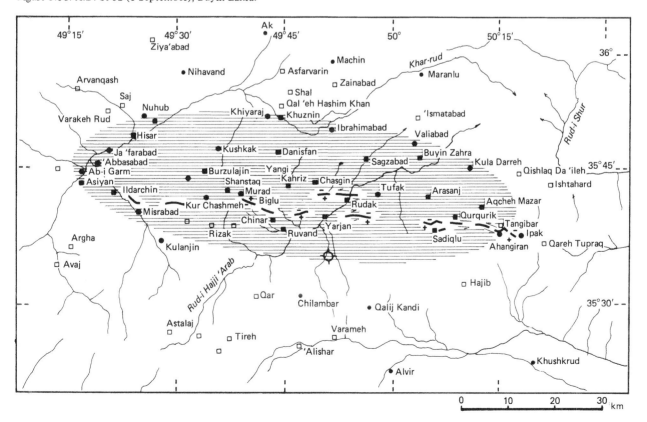

five centimetres and a left-lateral displacement of about ten centimetres. A similar pattern of fractures was noticed elsewhere at Qurqurik, Chinar and west of Kur Chashmeh, where the largest and most extensive ruptures developed between or across the small pre-existing faults in the volcanics rather than along them, forming a wide zone of dislocation. This complex system of faulting is typical of the eighty-kilometre long rupture zone, the existence of which had not been known before the 1962 earthquake. It was produced by a large thrust component with some left-lateral strike-slip on a fault striking about 100°E.[392] The average thrown and strike-slip displacements taken across the width and along the whole length of the zone were found to be 140 and 60 centimetres respectively.[393]

Within the meizoseismal region, the variation in damage and hence of Intensity showed inexplicable anomalies. On a statistical basis damage was much more closely related to the foundation conditions and type of construction than proximity to the fault zone. For instance, Tangibar, a few hundred metres away from the main rupture, suffered practically no damage. In contrast, Ahangiran was totally destroyed (plate 20). In fact

in the immediate vicinity of the fault zone damage was not as serious as at some distance away from it.

The earthquake triggered numerous local rockfalls and caused slumping, particularly along the Hajji 'Arab river and in the Buyin Zahra plain. The shock caused changes in the underground water supplies and in the flow of spring water. Permanent changes in the underground channels, however, only occurred in the areas adjacent and to the north of the fault zone.

The shock was widely felt. It caused minor damage in Qazvin and as far as Tehran, particularly in the southern districts of the city, where about fifty houses were seriously damaged and a number of public buildings including the railway station developed plaster cracks. Waves in ponds and reservoirs occurred throughout the area up to 170 kilometres away, and irregularities in the water level in the Caspian were noted. The earthquake was felt in Azarbaijan and it was strong enough to frighten the population in Tabriz, Bihbahan and Yazd.

Aftershocks continued for about two months in great numbers but with little strength and not sufficiently to significantly increase the damage caused by the main shock.[394]

Plate 20. Aerial view of the central segment of the Buyin Zahra fault break; small arrows show curved path of trace between Tangibar and Ahangiran.

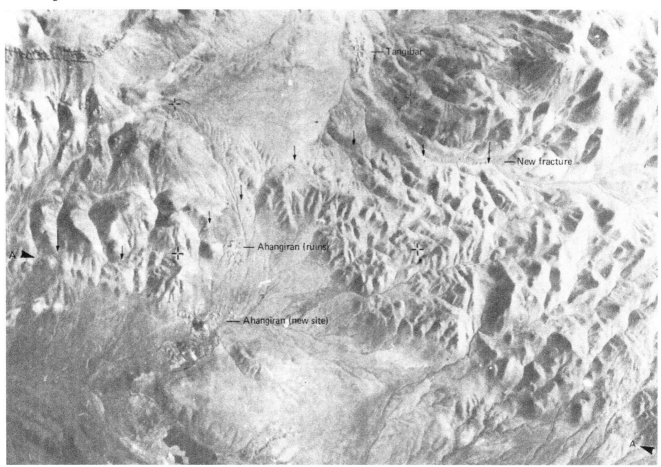

There were no major engineering structures or properly constructed houses, nor were there any important historical monuments in the meizoseismal region. A number of late Safavid–early Qajar structures in the vicinity, such as the Husainiyeh and the Imamzadeh Shah Sulaiman at Ishtahard, the Imamzadeh at Palangabad, the caravanserai at Hajib, and the twelfth-century Imamzadeh near Takistan, suffered minor cracks in the brickwork of their walls and lower parts of the domes. The Masun Khani bridge on the Khar-rud, about fifteen kilometres north of Buyin Zahra, was damaged by the sliding of its south abutment. There was no damage to the remaining two arches of the bridge on the Rud-i Shur, thirty-five kilometres east of Ishtahard. In Qazvin, no damage to the historical buildings was noticed except plaster cracks. There was some damage, however, to the dome of the Imamzadeh-yi Abazar, east of Qazvin.

1968 August 31 **Dasht-i Biyaz.** In the early afternoon of 9 Shahrivar 1347 a catastrophic earthquake hit northeastern Khurasan, affecting the region northwest of Qayin. The main concentration of damage was in the Nimbluk valley, where over 2500 people perished, or about a quarter of the total number of casualties, estimated at about 10 000. In Dasht-i Biyaz all houses collapsed completely, killing 1230 out of its 1670 inhabitants.

Other large villages which were most seriously affected were Khizri, Miyam, Buskabad, Binavaj and Charmeh (figure 3.52).[395] The only engineered structures in the meizoseismal region, consisting of a few steel-framed elevated water tanks and two-storey houses, were not destroyed.[396] Damage decreased rapidly away from this region. To the north, in Gunabad and Bidukht, as well as in Firdaus and Qayin to the west and southeast respectively, the earthquake was far less severe. The exception was Kakhk, where apart from a few properly built houses, the town collapsed with the loss of 1379 lives.

The earthquake was associated with about eighty kilometres of fresh faulting within a zone that has been active during historical times (figure 3.52). When the numerous smaller fractures are taken into account, the main fault break can be seen to lie within a shear zone, one to three kilometres wide, which trends almost east–west from beyond Zigan, across the northern side of the Nimbluk valley, whence numerous branches lead off towards Tak-i Zu and Mus'abi. The overall fault movement was left-lateral, showing maximum relative horizontal offsets of 4.5 metres (plate 23), and an average displacement of about 2.0 metres. Vertical offsets varied along the break. Measured on principal fractures, displacements showed the north side up on the east and west and down in the middle (between chainage 2W and

Plate 21. Detail of fault break northeast of Ahangiran.

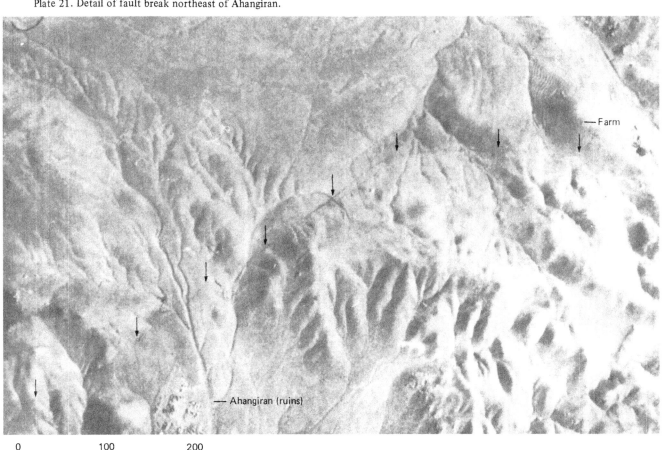

Farm

Ahangiran (ruins)

0   100   200
|_____|_____| metres

8E, figure 3.53), with maximum throws of 2.1 metres and average of about 0.7. However, when measured across the whole width of the fracture zone the overall vertical movement shows that the north side was up-lifted.[397]

The break follows old faults with features of Quaternary and Recent activity. Ridges on the side of hills, offset streams and *qanats*, scarps in alluvium and eroded saddles mark the trace of the fault break.[398] In rock, the trace passes through zones of fault breccia and mylonite, and shows generally smaller apparent displacements. Slickensides and marked striations can be seen on exposed fault surfaces (plate 24). The displacement of *qanats* indicates that the fault zone of the 1968 earthquake has been active during historical times (plate 25). It is of interest to note that in the Nimbluk valley *qanats* are double or triple where they cross the fault zone and that the general alignment of the older shafts shows left-lateral offsets across the fault zone of as much as ten metres. It appears that the most recent lines of shafts and adits across the zone is in fact a detour round that portion of the original *qanat* across the zone which was destroyed by past fault movements. In figure 3.53 the old *qanat* that runs along the fault zone suggests that earlier fault movements had dammed the aquifer towards the valley, thus creating an underground reservoir of water that was tapped by the construction of an east–west *qanat* branch.[399]

In the immediate vicinity of the fault break, the

Plate 22. Weathered scarp of the Buyin Zahra earthquake photographed thirteen years after the event.

Figure 3.52. A.D. 1968 (31 August), Dasht-i Biyaz.

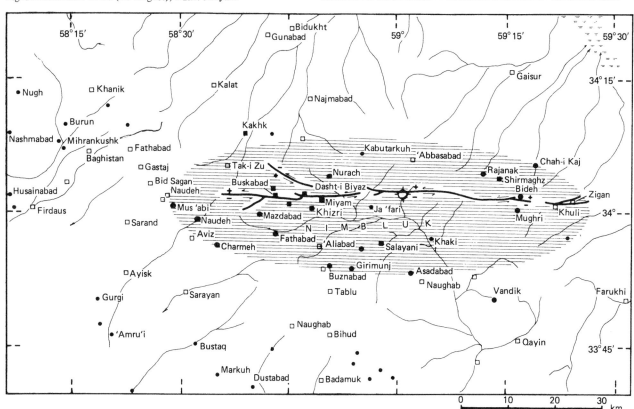

damage caused to houses by shaking was found to be similar to that produced five to ten kilometres away, and that proximity to the fault within such distances was not a criterion for heavy damage in such a large magnitude event. It is also difficult to explain why fault scarps in weathered rock or alluvium (in places standing vertical over two metres high[400]), and also adobe walls straddling the break and displaced up to one metre, were not shaken down. It is probable that much of the observed fault displacements occurred through rapid creep rather than transiently.

Extensive ground deformations of non-tectonic origin, mud volcanoes, and large-scale slumping of the ground, were found to the south of the fault zone in the Nimbluk valley, and they were responsible for much of the destruction in this part of the meizoseismal region. Over large areas the valley floor settled by as much as thirty centimetres, due to the densification of loose subsurface deposits. The flow of water in the *qanats* which were not damaged increased temporarily, the earthquake having no permanent effect on the underground and surface water flow.[401] A few *qanats* blocked by the fault movements were soon repaired and a disused section of the *qanat* at Miyam was used to install a strain-meter. Left-lateral fault creep of several milli-

metres per annum continued for some years, creep movements still being monitored instrumentally.[402]

About twenty-one hours after the main shock a violent earthquake seventy kilometres west of Dasht-i Biyaz almost totally destroyed Firdaus, which had been only slightly affected by the main shock. Aftershocks continued to be reported for some time. In the first six weeks nine strong tremors were experienced, the most damaging occurring in the Nimbluk valley on 11 September.[403]

The main shock was felt in Yazd, Simnan, Bujnurd and throughout the Badghis district in Afghanistan, i.e. within a radius of more than 600 kilometres. It caused slight damage as far as Tayabad, half-way between Mashhad and Herat, and caused water to splash out of ponds at Herat and Bust–Lashkaragh in Afghanistan, 570 kilometres from the epicentral region.[404]

By the end of 1971 the largest villages to the northwest of the Nimbluk plain were rebuilt. New Dasht-i Biyaz, which was re-named Shahabad, was built to the southeast of its old site, closer to the fault break, while villages in the plain were re-grouped into two agricultural co-operatives.[405]

1968 September 1 **Firdaus.** At eleven o'clock the following morning, about twenty-one hours after the Dasht-i Biyaz

Plate 23. Aerial view of the main rupture associated with the Dasht-i Biyaz earthquake, northeast of Miyam.

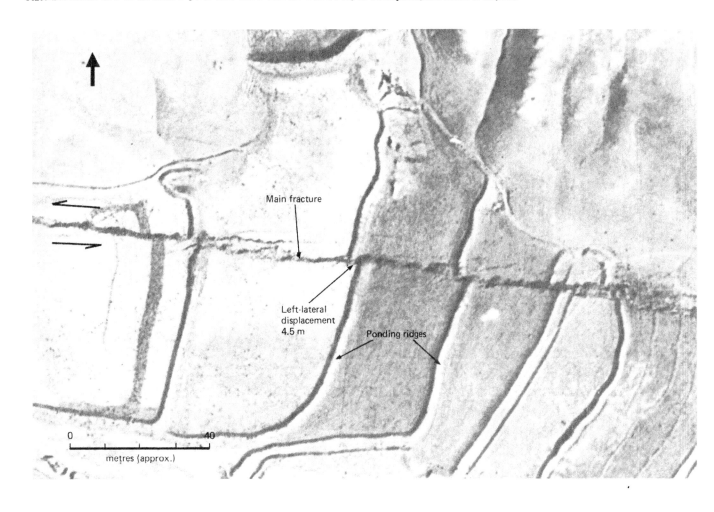

Figure 3.53. Detailed mapping of the central part of the fault break associated with the Dasht-i Biyaz earthquake of 1968. (1) Principal displacement zone, (2) small fractures, (3) pre-earthquake fault lineaments, (4) *qanat* lines, (5) mud volcanoes, (6) isolated mud houses, undamaged, (7) isolated mud houses, damaged, (8) isolated mud houses, destroyed, (9) wells, (10) relative displacements in centimetres. *U* up, *D* down.

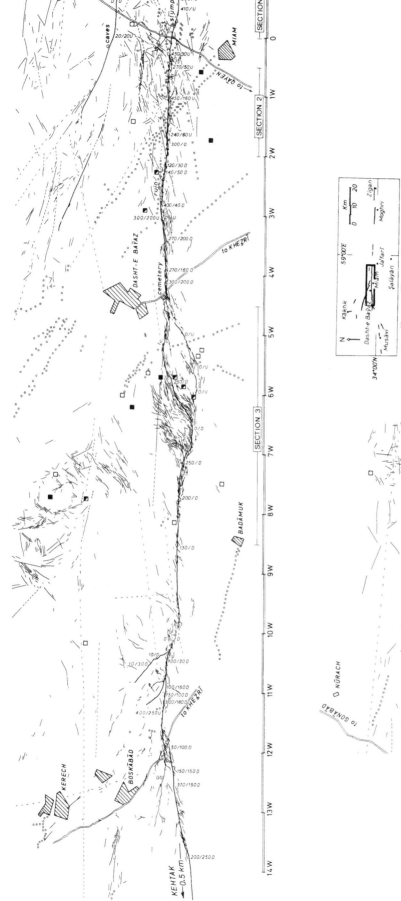

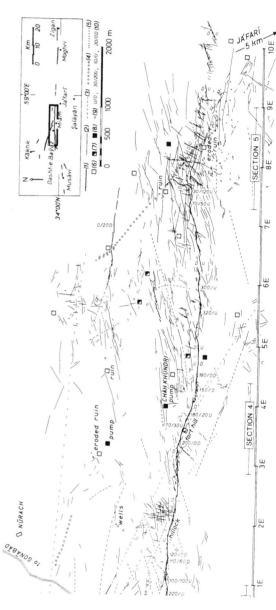

earthquake, a violent shock caused serious damage in the district of Firdaus, about seventy kilometres west of Dasht-i Biyaz. The shock almost totally destroyed the town of Firdaus and ruined a number of villages (figure 3.54) which had been only slightly affected by the first shock.[406] Damage extended mainly to the north of Firdaus where about 150 people lost their lives. To the south, and for about twenty-five kilometres from the town, the region is almost totally uninhabited. Nevertheless, the shock of 1 September was strong enough to cause damage without casualties to the outlying villages of Sarand, Ayisk and 'Amru'i and block a number of *qanats* belonging to these villages. In Firdaus, almost all adobe houses were demolished and about 750 people were killed.[407] However, better built houses, as well as steel-framed constructions and a forty-metre high chimney stack made of good brickwork, suffered no structural damage, nor was there any damage to a number of steel-framed elevated water tanks.

The earthquake produced minor discontinuous fractures and cracks in the pediment east of Firdaus, principally the result of lurching, showing no consistent direction or displacement. The most continuous fractures and small landslide cracks were noticed near Ishratabad, between Khartud and Rahmatabad. These features follow the strike (N-160°-E) of the Plio-Miocene marls which dip against the hill slopes to the east-northeast and run discontinuously for about four kilometres to near Khatimabad. The tensional nature of the cracks and their location along bedding plains lead to the inference that they are primarily due to movement of weathered marls down the slope during the shaking rather than to faulting.[408] Damage to nearby settlements during both earthquakes was relatively small. The shock caused no permanent changes in the flow of *qanats*, and it was felt as far as Sabzavar, Mashhad and Tabas.

1972 April 10 **Qir–Karzin**. Throughout March 1972 a swarm of shocks in the districts of Hingam and western Karzin had forced the people to live outside their homes. Early on the morning of 21 Farvardin 1351 a destructive earthquake shook central Fars, almost totally destroying the *dihistans* of Qir, Karzin and Afzar.[409] The earthquake was preceded by a number of foreshocks, which at Hingam and Sarbisheh were sufficiently strong to forewarn the inhabitants. The shock nevertheless killed 5010 and injured 1710 people in 50 villages, a loss of life that amounts to about 20% of the population of the region.[410] The town of Qir, the only large settlement and centre of exchange for the Khamaseh and Qashqai tribes, was totally destroyed and almost three-quarters of its 5000 inhabitants perished. Other villages and winter camps were totally destroyed (figure 3.55).[411] Almost all adobe and engineered houses, particularly those built by the Government in recent years, collapsed.[412] By contrast, the only properly built structures in the region, i.e. a 10-span, 120-metre long reinforced concrete bridge across the Rud-i Mand, and two elevated steel tanks, survived the shock with insignificant damage.

Plate 24. Slickensides on exposed fault surface, looking north.

Plate 25. (*a*) Aerial view of fault break between Dasht-i Biyaz and Miyam; (*b*) Fault break east of location shown on 25*a*; note the association of *qanats* (underground irrigation channels) with the fault zone.

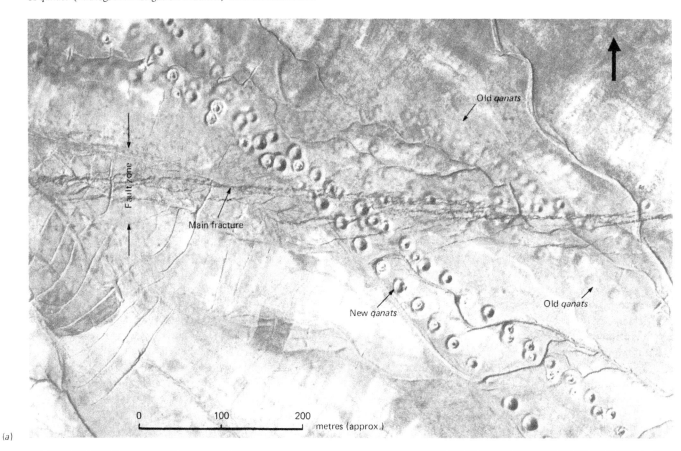

(*a*)

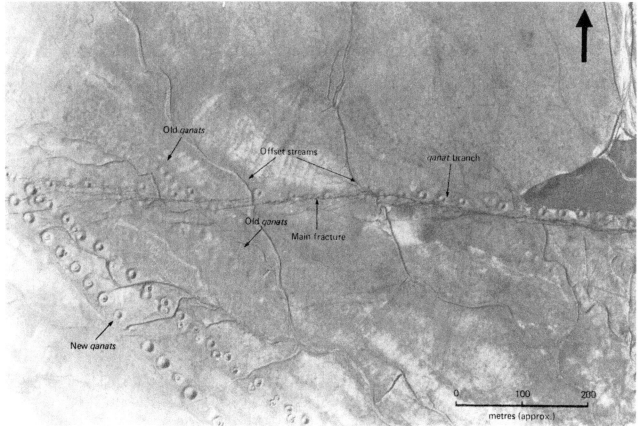

(*b*)

The earthquake was associated with minor and discontinuous ground deformations. The fractures found south of Hasanabad and east of Tang-i Ruyin are due to slumping and lurching. All other fractures, south of 'Aliabad ('Abbad), northeast of Barikhun and between Mubarakeh and Manganuyeh, seem to be of secondary tectonic origin, i.e. the result of strain re-adjustment on pre-existing tectonic lineaments. These fractures were found in zones up to 160 metres wide and many hundreds of metres long, striking between N-135°-E and N-165°-E in recent alluvium or in gravel fans without any evidence of vertical displacements. Individual cracks were invariably oriented in an *en echelon* pattern, indicating right-lateral displacements of a few centimetres and they may or may not reflect features of the causative fault.[413] However, this type of secondary faulting is not unusual, particularly in complex structural systems that lead to a complex surface response.

The earthquake caused considerable damage to the underground water channels in the area, and altered the flow in those that did not collapse. Although *qanats* are not very widely used in this part of Fars, about five kilometres of channels were ruined and 180 wells collapsed. The shock also caused temporary springs to flow in the eastern part of the Qir valley, north of Tang-i Ruyin. In a few low-lying areas near the Rud-i Mand the ground liquefied and in places the banks of the river slumped. In the mountains north of Qir, several rockfalls and landslides blocked the road towards Firuzabad. In the Hingam and Sarbisheh region to the west of Qir, the

ground motion was so strong that cobbles, stones and small boulders embedded in flat ground were dislodged or thrown from their sockets.

In Firuzabad, Jahrum and Fasa, about sixty kilometres from Qir, the earthquake caused minor damage and injured a number of people. In Shiraz, 135 kilometres away, a few adobe houses cracked and window panes of the upper floors of the twelve-storey high Cyrus Hotel were broken. In Shiraz a Wilmot seismoscope recorded a double amplitude of 1.5 centimetres.[414] In Lar, 170 kilometres away, the shock caused panic. It was felt as far as Bandar 'Abbas, Kirman and Yazd. Aftershocks were felt in the region for over a year after the main event, some of them of comparatively low magnitude and causing heavy but localised damage.[415]

In the years that followed the affected region was partly rebuilt and new roads were constructed.

The earthquake caused no apparent damage to the palace of Ardashir near Firuzabad or to the tenth-century *masjid-i jami'* in Niriz.[416]

1977 December 19 **Gisk—Kirman**. Preceded by two strong foreshocks during the night an earthquake destroyed a few villages in the district of Zarand, killing 665 people (figure 3.56).[417] The shock destroyed the villages of Dartangal, Gisk and Sarbagh and originated on the Kuhbanan fault (plate 26) which runs just behind these villages, eight kilometres northeast of Zarand.[418] Within this part of the fault zone ground displacements were small and discontinuous and they run for a distance of not more than ten kilometres, striking N-140°-E. Fault movements in this zone were mainly right-lateral with an average displacement of about ten centimetres measured across the zone. The average vertical displacement was about seven centimetres with the valley side downthrown. The ground deformations observed suggest an overall intense compression from the northeast with an apparent right-lateral component. An interesting result of this earthquake is the elongation of the shape of the lower intensity (I = V+) isoseismals in a direction perpendicular to the small causative fracture.[419]

Within the meizoseismal region there are no important engineering structures, nor are there any properly built houses. In Zarand, a number of modern structures suffered minor damage. The earliest construction in the town, the eighteenth-century *masjid-i jami'* built on the site of a tenth-century mosque and the remains of a Saljuq minaret, suffered no damage.

1978 September 16 **Tabas**. Late in the evening an earthquake devastated the region round Tabas.[420] There were no foreshocks, but some unusual premonitory occurrences were reported, such as abnormal animal behaviour and the appearance of a glow over the Shuturi Kuh. The shock totally destroyed thirty villages, killing 18 220 people, of whom 80% perished in Tabas itself. Damage extended over a large area in this sparsely populated part of Iran, affecting eighty-five villages in all (figure 3.57). Damage was less serious to the few well built dwellings,

Figure 3.54. A.D. 1968 (1 September), Firdaus.

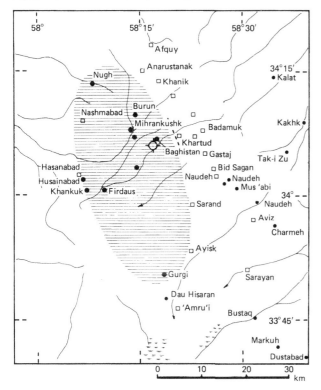

while elevated water tanks made of steel sustained little or no injury.

Ground movements in Tabas were very violent. A strong-motion accelerograph recorded peak horizontal accelerations of just over 80% g.[421] Fortunately, there were few damaging aftershocks, the largest hardly exceeding magnitude five. Shallow thrust faulting was observed within a zone many kilometres away, running discontinuously for a distance variously reported as between fifty-five and eighty-five kilometres. Relative displacements have been reported only for very few points on these fault breaks,[422] but the average slip should not have exceeded 1.7 metres. The fault breaks were associated with pre-existing surface expressions that suggested Recent movements, but which could not have been classified as active without additional evidence; the proximity of the major Quaternary Naiband fault, which runs for 450 kilometres, overshadowed these secondary, but important features.

The shock was widely felt and was perceptible in Tehran, 610 kilometres from the epicentre. It ruined the twelfth-century Madraseh-yi Dau Minar and shattered the fortress of Abu 'l-Hasan Gilaki in Tabas. A lunar

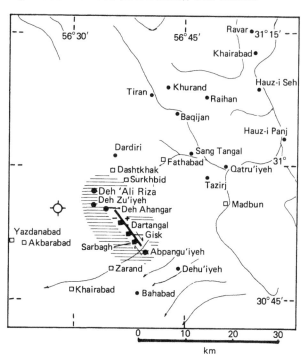
Figure 3.56. A.D. 1977 (19 December), Gisk–Kirman.

Figure 3.55. A.D. 1972 (10 April), Qir–Karzin.
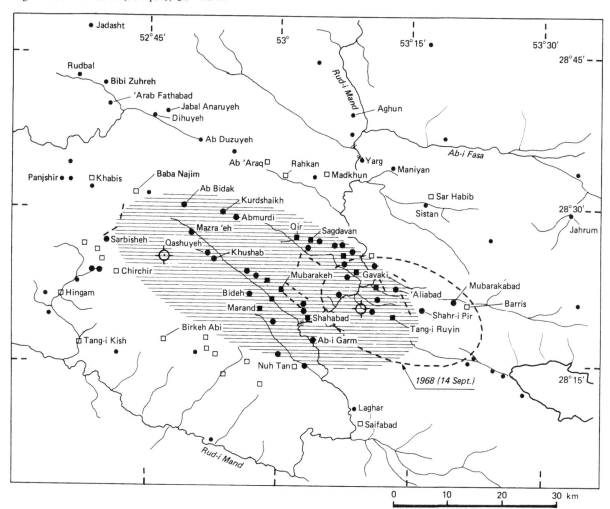

Plate 26. Aerial view of the Gisk earthquake of 1977.

eclipse that followed the same night and the tense political situation at the time caused serious rescue problems.

1979 November 14 **Karizan–Khwaf.** Preceded by a damaging earthquake earlier in the year (16 January),[423] This earthquake destroyed the villages of Karizan, Istind and Buhnabad and caused extensive damage in the Zirkuh region. In all, about 240 people were killed. It destroyed the *qanats* of Karizan and caused damage and casualties as far as Khwaf. The earthquake was associated with twenty kilometres of faulting along pre-existing Quaternary faults that extended from Istind to the north of Karizan, showing right-lateral displacements and throws to the west of about one metre respectively (figure 3.7).[424]

1979 November 27 **Khuli–Buniabad.** This earthquake occurred on the eastern extension of the 1968 Dasht-i Biyaz earthquake. It affected a very sparsely populated region, and the area damaged by the shock of 14 November 1979, killing only a few people but causing widespread damage over a large area. The earthquake was associated with a fault break along the extension of the rupture of 1968, from Chah-i Zandar, west of Khuli, to Buniabad in the east, a distance of about sixty kilometres. Along this section of the break the vertical offset amounts to 2.5 metres with the southern block downthrown, although in places the sense of the displacement is reversed. The trace bears N-80°-E and shows consistently left-lateral strike-slip displacements from one to four metres. Near Mihrabad, three kilometres north of Buniabad, the fault break backs up to the southwest and runs for ten kilometres before it joins the rupture of the 14 November shock, showing a right-lateral motion (figure 3.7).

3.4 Collected data on individual topics

The following sections bring together isolated details, from the same macroseismic sources of information, about particular earthquake effects or associated subjects.

Figure 3.57. A.D. 1978 (16 September), Tabas.

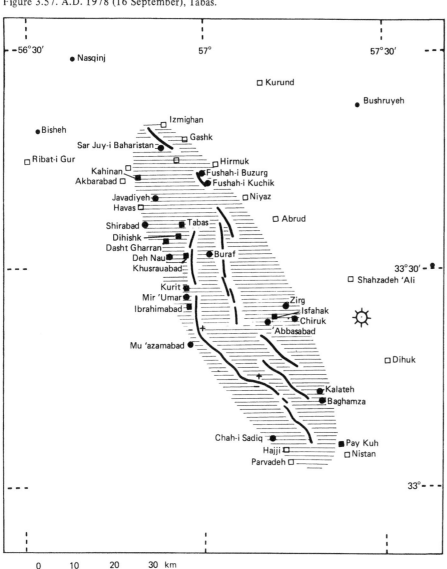

3.4.1 Faulting

Evidence of faulting associated with historical events is rarely clear and in most cases inconclusive. For some of the earlier earthquakes of this century also, evidence is poor and occasionally insufficient. Table 3.1 at the end of this chapter summarises all events reported to have been associated with faulting, some of them already discussed in the preceding section.

The cases of faulting in 856, 1336, 1648, 1838 and 1879 are based on very tenuous evidence, but they should not be dismissed without further field study. Faulting attributed to the earthquakes of 18 April 1911, 15 July 1917, 27 July 1946, 12 February 1953, and 7 November 1976 is also supported by meagre and inconclusive evidence, suggesting that the observed features were perhaps associated with local re-adjustments caused by movements in deeper structures. They may otherwise represent secondary deformations in the upper part of the rupture zone, as in the case of the ground deformations observed after the earthquake of 10 April 1972. Quite often exposures and short segments of Quaternary faults are found in the epicentral regions of the earlier events, showing geologically late movements. Whether these movements can be associated with a particular historical earthquake or with an altogether earlier event is uncertain in the absence of contemporary evidence. Quite often, however, contemporary evidence helps to disprove alleged faulting, as for instance with the Mishan break which was wrongly identified with the earthquake of 2 July 1972 ($M = 5.3$).[425]

The ground deformations associated with the earthquakes of 13 December 1957 and 16 August 1958 are also of questionable primary tectonic origin. A recent field study of these features suggests that although they align with what seem to be features of very recent tectonic origin, the local information about their re-activation in recent years is at best conflicting and of little reliability.[426]

It is also perhaps interesting that no known faulting associated with historical and recent earthquakes occurred on a major Quaternary fault (figure 5.11). Almost all cases of faulting are connected with relatively minor fault zones of recent age, mainly in the northern and eastern parts of Persia.

3.4.2 Tsunamis

Tsunamis are large water waves set up by earthquakes through the sudden deformation or tilting of the Earth's crust under water, large-scale subaqueous slides or sub-aerial landslides into deep waters. Tsunamis or seismic sea-waves deserve study because of the loss of life and damage to property that they may cause along coastal areas and also because of the indirect evidence that they may provide for submarine crustal deformations.

Instances of tsunamis in Persia are relatively few throughout the period. The earliest case appears to be associated with the 1008 earthquake at Siraf in the Persian Gulf. According to al-'Umari 'houses were destroyed and about 10 000 people were killed; the sea produced a wave which sank a number of ships from which no-one survived.' Al-'Umari quite clearly attributes the sinking of the ships to a tsunami generated by the shock, but there is no evidence that the sea-wave flooded the coast. Moreover, it should be mentioned that it may simply have been storms that were responsible. High winds affected the Tigris valley and southern Persia that year and some sources refer to the sinkings without mentioning the earthquake (see § 3.3). It is perhaps of interest that thirty years earlier Siraf was affected by a more destructive earthquake, and although there is no mention of a seismic sea-wave associated with this event, it has been suggested that as a result of the shock, a large part of the water-front sank below the sea, making the harbour impracticable. Le Strange (1905: 259) suggests that the harbour had silted up, but there is no river there sufficient to cause this. It might be supposed that the land was uplifted, but again there is no evidence that this happened and historical sources anyway give no indication of an event of such significance.

The evidence for the tsunami associated with the 1608 earthquake in the Mazandaran is more conclusive. The shock set up sea-waves in the Caspian that caused grave concern but relatively little damage. This was a major earthquake, comparable to that of 958, which seems to have affected the level of the Caspian Sea (see § 3.3).

Large sea-waves in the Caspian, causing no serious damage, have been associated with the earthquakes of 1890 in Tash and in the Mughan on 4 December 1910. In the first instance the wave was noticed all along the southern coast of the Caspian, from Ashur-Ada to Anzali and in the second instance only at Astara. Abnormal fluctuations of the sea level of the Caspian attributed to earthquakes were noticed on 26 April 1868 and 26 April 1960, mainly at Baku and Linkuran, where the sea level oscillated within the hour by fifty to more than a hundred centimetres.[427]

The only reasonably well documented tsunami occurred on 27 November 1945 in the Makran, in conjunction with the large earthquake of the same date (see § 3.3). Preceded by two smaller waves the main tsunami flooded the Makran coast about one to one and a half hours after the shock, at Ormara and Pasni, reaching heights inland of four to five metres. The wave swept the whole shore of the Arabian Sea, causing casualties along the Pakistani coast of the Makran as far away as Bombay and producing noticeable effects up to Karwar and the Seychelles at a distance of more than 3000 kilometres. Off the Makran coast the trans-oceanic cable between India and Great Britain broke in eight places, presumably due to submarine slides triggered by the shock. Even though the wave came at low tide and probably after the coast of Ormara was elevated by two metres, the damage it caused was very serious.

Large waves on the coast of Makran were known in the past to have caused considerable damage and peculiar effects. Late in December 1897 hundreds of tons of fish were driven on shore at Gwadur, allegedly by a storm which had increased the activity of gas which poisoned the sea during that period. The dead fish cast inland formed a bog through which a track had to be beaten down (Sykes 1902b: 128). This, and other instances of extensive flooding of the coast as far as Las Bela in Pakistan, are being investigated.

3.4.3 Prediction

Earthquake prediction was a serious preoccupation of the early soothsayer, astrologer or prophet, and there are many recorded instances in Persian history of destructive earthquakes having been forecast. The earliest omens found are in Assyrian letters from the Sargonid Dynasty, probably of the eighth century B.C. (Waterman 1929) and in the Sibylline Oracles.

However, later predictions in Persia do not seem to have been very well received, and even allegedly accurate forecasts were almost invariably ignored. For instance, the earthquake of 1042 in Tabriz was predicted by the astrologer Abu Tahir Shirazi who tried in vain to persuade the people to leave the city. Earthquakes had happened in this region sufficiently frequently to be likely to occur again, yet the main reaction to the prediction was one of apathy. The earthquake did occur and about 40 000 people perished (Mustaufi, *Nuzhat*: 75). In his later poem on the event, Qatran nevertheless feels that auguries and prognostications are absurd. The earthquake of 1549 in Kuhistan was also predicted by the Qadi of the district who tried unsuccessfully to convince people to stay out in the open that particular night. They refused to listen and only the Qadi stayed out with his family, but finding the night very cold they returned to their house where they soon perished along with 3000 people in the district (Rumlu: 342). Apparently, the Qadi himself was not too sure about his prediction. Astrologers had more success in predicting the earthquake in the summer of 1593 in Lar, and persuading the inhabitants to leave the city (Natanzi: 529), probably with the help of the occurrence of foreshocks.

Indifference to predictions, probably as a result of the fatalistic outlook of the people, is noticed by many foreign travellers who point out that in Tabriz, for instance, shocks are so frequent that the inhabitants think very little of them. Even when a destructive earthquake was predicted, no-one was unduly concerned: 'So singular is the combined effect of habit, of hope and of attachment to their place of birth' says Kotzebue that 'they show no symptoms of alarm'.[428] Earthquake predictions, usually made by astrologers, mullas, Armenian priests and dervishes, become more numerous in the aftermath of a destructive earthquake, near the time of an eclipse, or during a period of political unrest. These predictions invariably failed and occasionally those responsible were arrested.

Unusual animal behaviour prior to earthquakes is often mentioned, but there are relatively few cases where such behaviour was significantly unusual or it was recognised as premonitory. The most interesting case is that of the earthquake of 1875 at Jur in Kirman. We are told that before the earthquake, large numbers of game came down from the mountains and entered the village. The villagers chased the animals with rifles and sticks and the women and children too went to watch the sport. The earthquake then struck, causing great destruction, but because of this diversion no-one was killed (*Iran*: 1292, no. 255).

The case of the 1868 earthquake at Hisar Quli in the Darya-yi Namak is interesting because it is the earliest attempt found to link abnormal animal behaviour before an earthquake with the phenomenon itself, as well as with its effect on communications by telegraph. An anonymous Persian scientist in *Ruznameh Daulati*: 1285.6.7, says

> On 10 Jumada I, great difficulty and poor reception were experienced in communications everywhere by the telegraph wires, from which it follows that when there is going to be an earthquake, the strength of the electricity in the Earth becomes greater the day before, with the result that communication by telegraph is difficult . . . On the night in question, before the motion in the ground and near to the earthquake, animals and birds cried out and a stinking smell like sulphur was noticed by most of the people, resulting in a feeling of nausea. And since the membranes of animals are more sensitive than those of men, it is possible that this smell was the cause of the cry that the animals gave out. For there are fumes confined inside the globe of the Earth, which, because of an earthquake, emerge from gaps and cracks in the ground.

By 1910 one of the telegraph operators at Kirman claimed a discovery, involving copper wire which picked up the Earth's movement and gave six seconds' warning of the imminence of an earthquake. In an article in *Iran-i nau*: 1327, no. 55, the inventor describes how he first noticed the warning effects of the 1897 earthquake on the telegraph line.

Earthquake prediction is also increasingly preoccupying the modern scientist, with results that hardly differ from those of the early times. The prediction made after the Buyin Zahra earthquake of 1962 (*Kaihan*: 1341.6.13) and more recent events had little constructive effect on the people, who either seemed strangely reluctant to believe that an earthquake would occur, regardless of whether it was an accredited astrologer or a seismologist who told them so, or took full advantage of the warning to serve their vested interests. Our field studies show that with earthquakes in Persia, it is not the prediction and warning problems that are important, but mainly the social and economic implications of forecasting such disasters. False alarms and inaccurate timing create more problems than already exist. The case of a prediction causing more damage than an earthquake, described by Garza & Lomnitz (1979) for Mexico, is a typical example of science seriously disrupting economic normality in developing areas, witnessed on a small scale in some recent forecasts made in that part of the world, or by astrologers in earlier periods. For the prediction in 1156 of a catastrophic earthquake to occur in the Middle East on the 14 September 1186, but which never happened, and for an account of its repercussions, see Michel (iv, 730–1).

3.4.4 Social implications

The aftermath of earthquakes in Persia follows a pattern typical in many other parts of the world. The sudden crippling of the local economy often leads to population movements, emigration, increase in taxation and crises in human affairs. After a short period of enthusiasm for ambitious reconstruction plans, interest in the application of various schemes, particularly for small towns and villages, begins to die out as the available funds are found to be insufficient (most of them having already been squandered), and as the problems which reconstruction poses become more involved and less exciting.

The usual pattern is that as the interest of the authorities gradually begins to decrease, many sites (particularly the smaller ones) are not rebuilt and the larger ones turn into slum areas, their inhabitants beginning to migrate, leaving the less active members of their community behind. The human and political aspects of an earthquake disaster, not only in Persia, but also elsewhere, were and still are rarely made known by contemporary investigators who are not always free to express their views on relevant, sensitive matters (*Disasters* 1978: i, 85; ii, 265). For instance, after the earthquake of 1824 in Shiraz, the Government did not allow the inhabitants to seek shelter in the surrounding villages with the result that the mortality due to exposure, sickness and probably aftershocks, rose to 200 daily. This measure was said to have been suggested by the Kalantar, who was afraid of being left alone in his jurisdiction should the people be allowed to leave (*IO* 9 83). The local government was so weak that in the next earthquake disaster at Shiraz in 1853, it could not prevent plundering of the survivors by bands of robbers inside and outside the city. Nevertheless, the disaster in no way lessened the government's oppression of Fars (*FO* 248 153; 60 179; Wills 1894).

Often the misuse of funds allocated for reconstruction led to the abandonment of villages and to crises in local affairs. For instance, the official view after the Kaj Darakht earthquake of 1923 was that there was little need for a relief fund for the villagers, as the few survivors inheriting the estates of their deceased relatives would be better off than they were before (*FO* 371 9035). Attempts by officials to administer funds justly were often frustrated. After the Kopet Dagh earthquake of 1929, the Governor-General of Khurasan, hearing of the appalling corruption and misuse of funds, took over their administration himself. Two days later he had to hand over his governorship because of his too honest and close supervision (*FO* 371 13797).

On a number of occasions an earthquake would trigger local unrest, as for instance after the Mazandaran earthquake of 1935, when in Sari and Babul the event was interpreted as divine punishment for the Government's banning the Muharram religious processions. This led to anti-Government demonstrations that ended in bloodshed in Babul after the army moved in (Divanbegi 1969).

The idea of centralising small villages into large agricultural units after a destructive earthquake has been tried by the Government a number of times. The case for centralisation is usually made on regional economic grounds, and in those terms it is nearly always a very strong one. However, the adverse effects on the individual and on certain social groups have never been discussed. A rather indiscriminate centralisation after the Dasht-i Biyaz earthquake of 1968 had an adverse effect on the social and economic life of the survivors of a number of settlements, who resisted re-settling away from their villages for some time. The same was noticed after the

Buyin Zahra earthquake of 1962, the newly-built village of Rudak remaining uninhabited for many years. Also, after the 1972 earthquake of Qir, officials feared that the villagers, who were mostly former nomads who had been given land in the 1961 reform programme, would take the opportunity to return to their old way of life. As a matter of fact, a number of them who left after the earthquake for their usual summer camps did not return in the autumn. This frustrated plans of the Government to re-group the destroyed villages into large agricultural co-operatives as originally planned.

As with prediction, where damage to property can be comparable with that caused by the earthquake that did not occur, improper handling of the aftermath of an earthquake in a developing area can cause more serious and long-lasting damage to the region's social fabric than if nature were allowed to take its natural course in local rehabilitation.

Despite the undoubted impact of a destructive earthquake on a regional level, the effects are almost invariably short-term. It is tempting to regard a major earthquake as an event of some historical significance, responsible for the collapse of a civilisation, culture, or economy and the disappearance of a whole city. While this may well have been the case in the earliest days of man's settlement in and around the Iranian plateau, it would not be accurate to maintain this for the period that we have investigated.

Sometimes an earthquake may have beneficial side effects, such as the creation of an oasis where there was formerly desert (as in 819), or the similar rise in the water table at Shiraz in 1824. Even without such compensations, physical destruction seems to have been made good remarkably quickly. This is largely explained by the fact that building materials commonly used in Persia allow not only easy destruction, but also rapid reconstruction. New cities arose on the ruins of the old, and we often find them described a few years later as flourishing once more, populous and commercially active. An earthquake is merely one of the many hazards of climate and environment that may cause dilapidation of buildings and decimation of inhabitants in Persian communities, who meet such visitations with a grim determination. The absence of long-term effects of earthquakes in Persia is thus to a large extent a tribute to the resilience of Islamic society. Only disasters of the first magnitude, such as the Mongol invasions of the mid-thirteenth century and Timur's campaigns a century and a half later, seem to have inflicted lasting damage (Bosworth 1977*b*: xvi, 85), and nowhere does a historian describe the destruction of a city by an earthquake with the same horrified shock as those injuries inflicted by his fellow men.

Nevertheless, earthquakes may be seen as contributory factors in the decline of certain areas. The eclipse of Qumis in the ninth century, of Siraf in the eleventh century, and of Nishapur after the twelfth century may all be seen partly in these terms. Even if not to the extent that has been claimed, the earthquake at Ani in 1320 clearly had a lasting effect on its prosperity. Earthquakes in Firrim in the thirteenth century, and three apparently disastrous shocks in Sistan by the early ninth century were certainly coincident with, even if not responsible for, a gradual regional decline and erosion of political and economic importance.

    Frequent stories of advantage being taken of an afflicted community for looting, robbery or government intervention reveal human nature in its poorest light. The sack of Ganja

(Kirovabad) in 1139, the extortions of the army occupying Gurgan (Gunbad-i Kavus) in 874, and the Bakhtiyari raids in the Silakhur valley in 1909 reveal some of the insecurities prevalent in society, not by any means restricted to the Middle Ages nor to the Middle East.

Table 3.1. *Ground deformations associated with earthquakes in Persia*

Date	Epicentre N°	E°	M	m	L (km)	Az	R_m (cm)	R_{av} (cm)	Q	Region
856 Dec. 22	36.20	− 54.30	(7.9)		100†				P	Qumis
1336 Oct. 21	34.70	− 59.70	(7.6)		100	155			Pd	Khwaf
1493 Jan. 10	32.96	− 59.76	(7.0)		30	130			M	Mu'minabad
1648 Mar. 31	38.30	− 43.50	(6.5)		3†	70			Pdk	Hayots-dzor
1721 Apr. 26	37.92	− 46.66	(7.7)		50†	125			Pd	Tabriz
1780 Jan. 8	38.21	− 46.01	(7.7)		60†	120	700	250	M	Tabriz
1825	36.10	− 52.60	(6.7)						Pdk	Harhaz
1838	29.60	− 59.90	(7.0)		70	170			P	Nasratabad
1879 Mar. 22	37.78	− 47.90	(6.6)		2	170			Mkk	Buzqush
1909 Jan. 23	33.41	− 49.13	7.4	7.2	45	135	250	100	Gd	Silakhur
1911 Apr. 18	31.23	− 57.03	6.2	6.7	15	155	50	15	Md	Ravar
1917 Jul. 15	33.48	− 45.82	5.6	6.3	2†	140			Mk	Tursaq
1929 May 1	37.72	− 57.81	7.3	7.1	70	140	210	60	G	Kopet Dagh
1930 May 6	38.24	− 44.60	7.2	7.0	30‡	125	640	135	G	Salmas
1933 Nov. 28	32.01	− 55.94	6.2	6.4	5	140			Gdkk	Buhabad
1941 Feb. 16	33.41	− 58.87	6.1	6.4	10	5	50	20	Gd	Muhammadabad
1946 Jul. 27	35.60	− 45.83	5.5		2†	145			Mk	Panjwin
1947 Sep. 23	33.67	− 58.67	6.8	6.4	20	175	130	50	G	Dustabad
1953 Feb. 12	35.39	− 54.88	6.5	6.9	8*	70	140		Mdk	Turud
1957 Dec. 13	34.58	− 47.82	6.7	6.5	20*	135			Gdk	Farsinaj
1958 Aug. 16	34.30	− 48.17	6.6	6.2	20*	130	150	30	Gdk	Firuzabad
1962 Sep. 1	35.71	− 49.81	7.2	6.9	85	103	95	30	Gd	Buyin Zahra
1968 Aug. 31	34.02	− 58.96	7.4	6.0	80	95	510	230	G	Dasht-i Biyaz
1972 Apr. 10	28.38	− 52.98	6.9	6.3	20*	120			Mdkk	Qir
1976 Nov. 7	33.82	− 59.19	6.4	5.8	9	140			Pk	Vandik
1976 Nov. 24	39.12	− 43.92	7.3	6.2	55	110	350	190	G	Chaldiran
1977 Dec. 19	30.90	− 56.61	5.7	5.8	10	140	15	12	G	Gisk
1978 Sep. 16	33.40	− 57.13	7.3	6.7	80*	150	150	150	Md	Tabas
1979 Nov. 14	33.91	− 59.81	6.6	6.0	20	165	140		Md	Karizan
1979 Nov. 27	34.05	− 59.63	7.1	6.1	60‡	80	470		Md	Khuli

The notation used in the table is as follows:

M = Surface-wave magnitude; (M) deduced from macroseismic data.

m = Body-wave magnitude.

L = Length of ground deformations or faulting in kilometres; $(L^\dagger)$, actual length probably longer than shown; (L^*), length of wide zone of deformations consisting of parallel or sub-parallel fractures; $(L^\ddagger)$, arcuate or complex fault break.

Az   = Azimuth of general trend of ground ruptures in degrees.

R_m = Maximum observed relative displacement in centimetres.

R_{av} = Average relative displacement along fault break. Occasionally, the average throw and strike-slip taken across the width of a fault zone was found to be larger than the maximum displacement on individual fractures.

Q = Quality of evidence of faulting or of the nature of ground deformations:

G = good, derived from detailed field studies

M = moderate, based on cursory field surveys of the whole or part of the fracture zone as well as on field reports or incomplete surveys in need of authentication

P = poor, deduced from historical data with little or no field evidence

d = trace discontinuous, in places not accessible or eroded; total length of features deduced from few and widely spaced observations

k = all or most of observed or reported ground deformations probably not of tectonic origin.

4

Instrumental data

4.1 Instrumental epicentres

Instrumental recording of earthquakes in Persia began very late in the last century and continued for some time with instruments, operating mainly in Europe, which by modern standards were very imperfect.

Strictly speaking, the first earthquake in Persia to be recorded by an instrument occurred in Azarbaijan at 19 hours 49 minutes on 4 October 1856. It triggered a Cacciatore-type seismoscope in the observatory of Tabriz, which was established by the Russian Consul N. Khanikoff early in 1855 (Abich 1857, 1858 – see p. 62). It is after 1892, however, that the larger events in Persia began to be recorded more regularly. The Quchan earthquake of 17 November 1893 did trigger a Voznesenski-type seismoscope in Ashkhabad, about 1° away from the epicentre, but it was also registered by a Rebeur-Paschwitz pendulum at Nikolaev and by a Brassart seismoscope in Pavia, 22° and 37° from the epicentre respectively (Rebeur-Paschwitz 1895). The Voznesenski seismoscopes, which were deployed in Turkmenistan and other neighbouring regions by the Russian Geographical Society, recorded a considerable number of local earthquakes, particularly in the Krasnovodsk region (Musketoff 1899). However, it was the large earthquake of 8 July 1895 that was first recorded by practically all the seismographs in operation at the time. At Strasbourg a Rebeur–Ehlert seismograph wrote a maximum amplitude of thirty-two millimetres, and in Ischia periods of twenty-two seconds persisted for some time (Rebeur-Paschwitz 1895). At that time there were only about ten stations equipped with seismographs.

In the years that followed there was a relatively rapid increase in the number of stations equipped with proper recording instruments. Excluding seismoscopes and magnetometers, in 1903 there were sixty-nine seismological stations in all, most of them in Europe operating direct recorders of low magnification and very small damping (Tams 1908*b*). By 1910 the number of seismographic stations had increased to almost 200.

The instrumental location of epicentres in Persia, or rather the first attempts at it, also dates from the turn of the century. For the period before 1914 the British Association for the Advancement of Science (BAAS) published a considerable number of epicentres of the larger shocks for which macroseismic information was used to determine the approximate position and time of origin of the event, whenever it was available. When local observations in epicentral regions were not available, as was the case even with most of the larger earthquakes in Persia, epicentres are in gross error. In fact some of these early events occurred more than 300 kilometres away from the epicentres given by the British Association for the Advancement of Science.

In the following decades the number of seismological observatories increased rapidly. Figure 4.1 shows a chronological conspectus of the seismological observatories in operation throughout the world in the period 1913–77. The information used to construct this figure was obtained from a variety of sources, including the mailing lists of stations to which International Seismological Summary (ISS) bulletins were posted.[1] However, it was found to be impossible to estimate the precise number of active stations in any particular year. Lists of seismological observatories by various authors (e.g. Bullen 1933, Poppe *et al.* 1978), were invariably found on examination to contain a large number of stations which had already ceased operation, or which rarely contributed any useful information for the location of epicentres.

Figure 4.1 also depicts the overall growth of the number of seismological stations throughout the world. Some of these stations ceased to function long ago. Others are of low sensitivity belonging to local networks, while quite a few of them sent data to seismological centres for epicentral locations only

rarely. For instance, in 1913 out of a total of about 200 stations, only 93 supplied the ISS with readings, and of these stations only 73 did so regularly. With time, the proportion of the total number of stations contributing to the ISS and the International Seismological Centre (ISC) decreased. In 1976 out of a total of 3390 stations, 1040 had ceased to operate leaving 2350, of which 940 stations did contribute to the ISC, but only 600 of them regularly. Figure 4.1 shows the variation with time of the number of stations that contributed readings to the ISS/ISC and also of the maximum number of stations used each year by ISS/ISC to estimate any one epicentre.

The maximum number of stations used by ISS/ISC may be taken as a measure of the sensitivity and overall activity of the world-wide network at any particular year, although it depends on a number of factors such as magnitude. It represents not only the number of active stations, but also the efficacy of an international network to make use of the data from individual stations for the calculation of epicentres. From figure 4.1 we notice that in spite of the relatively large number of inactive or defunct stations, even after these deductions, the number of active stations has been increasing on average from about 70 in 1913 to over 600 in 1978. The effect of the two World Wars (which is shown in the figure) was not so much due to the closure or temporary suspension of stations, as to the drastic discontinuing of the regular reporting of data to the ISS and to other central agencies. These effects were of course more serious for certain regions such as Europe and the Middle East than for others such as Central and South America. But although it is true that in many cases station bulletins of the war period were not published and distributed, there are many instances where these bulletins are still available in manuscript form in various archives.

However, in spite of this rapid increase in the number of seismological observatories and improvement of instruments, they remained ill-distributed around Persia, with most of the stations located at distances less than 60° but confined within the northwest quadrant. During the first three decades of this century, stations situated nearer and the few observatories in the southeast quadrant which supply much of the information and control for the siting of smaller shocks, not only operated discontinuously, but were also slow in replacing their old direct recorders with the more advanced electrical seismographs. Most of the near stations in the west and wouthwest sectors (Istanbul, Asmara, Massawa and Dar-es-Salaam)[2] had a very short life and ceased to function before the First World War. The station at Kharput, near Elazig in Turkey, apparently remained in operation between 1906 and 1909 and provided useful information on local earthquakes.

Instruments of that period aimed primarily at the reliable location of earthquake foci, rather than at recording ground motions faithfully. In the mid-1920s radio time signalling improved timekeeping and station bulletins do reflect this interest in reporting arrival times accurately, not only of P- and S-, but also of compound phases. However, the azimuthal distribution and number of stations of the growing world network around Persia remained very poor. With time the net-

Figure 4.1. Chronological conspectus of the seismological stations in the world.

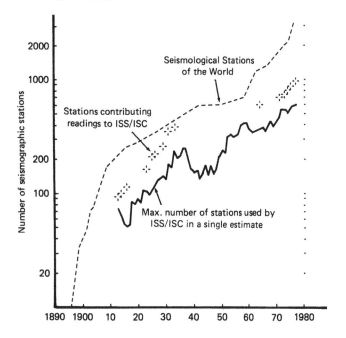

work's sensitivity and accuracy of reporting arrival data did improve markedly, but its overall detection capability was significantly reduced for long periods of time during the unsettled years between 1914 and 1922 and again between 1940 and 1947. During these periods near stations either closed down or continued to function without publishing their data. The following three decades saw a continuing improvement of the abilities of the network of seismographic stations, particularly with the establishment of local stations, figure 4.2.[3]

4.2   Routine determinations of epicentres: ISS period (1918−63)

The determination of epicentres requires the collection and processing of data from as many stations of the world-wide network as possible, and as such it depends entirely on international co-operation. This often makes it a rather protracted and time-consuming operation. Disregarding epicentres reported by BAAS before 1918 which were too crude, as well as epicentral estimates calculated by ISC after 1963 which are reliable on average, we may look into the routine deter-

mination of epicentres during the ISS period from 1918 to 1963.

In principle there are three stages of reporting epicentres which depend on the amount of input data and the speed with which the results are circulated.

Rapid locations are based on relatively few local stations, occasionally incorporating macroseismic observations, and the results are circulated shortly after the event. Since 1952 the Bureau Central International Sismologique (BCIS) has circulated a bulletin giving rapid determinations of events located within a radius of 5000 kilometres from the centre of Europe which includes Persia. Where macroseismic evidence is used to determine the location of the smaller events, the accuracy of the rapid determinations can be quite good.

Preliminary estimates of epicentres, on the other hand, cover the whole world and they are released by the BCIS, the United States Coast & Geodetic Survey (USCGS) or Moscow (MOS) as soon as sufficient information has been accumulated to ensure a reasonable degree of accuracy. Results in the form of Preliminary Determination of Epicentres (PDE) cards are circulated by the USCGS within a few weeks of the date of

Figure 4.2. Distribution of seismic stations in Iran.

occurrence and a bulletin is issued by the BCIS within a few months. Delays in the circulation of preliminary results do occur and these are becoming longer.

Final locations calculated from all suitable data after revision is the definitive stage reserved for the ISS/ISC. The accuracy of ISS/ISC determinations is on average better than that of the preliminary estimates, but this is at the expense of long delays in the publication of the bulletins. In the early days of the ISS the delay was five to six years; in the early 1930s this was reduced to about four years, but partly as a result of the Second World War, the delay increased to twelve years in 1948. Today, the ISC is operating only three years behind current data.

The number of epicentres calculated or revised by the ISS is on average smaller than that of preliminary determinations made by other agencies. For instance, in 1958 the BCIS determined 2523 epicentres as compared with only 623 calculated by the ISS. However, the number of stations used by the BCIS for 80% of its determinations was much smaller than the minimum number of about thirty-five stations used by ISS. For 36% of its determinations the BCIS made use of only six to ten stations (*IUGG Comptes Rendus*, no. 13, pp. 55–77, Strasbourg 1961).

Epicentres were calculated by ISS with techniques that were basically similar to those used today for routine determinations, including the laborious least-squares procedure which was carried out by hand up to 1957.[4] To reduce this enormous amount of work the ISS often 'adopted' old locations for new sets of arrivals from the same neighbourhood without calculation, particularly for successive shocks. ISS made no attempt to work out the most accurate locations, but residuals at all stations were given if a more precise location was needed. From station readings received, the epicentres and origin times of the events were computed together with the focal depths if there was a clear sign that they were appreciable. The arrival data from the best of the early stations was probably very good for the larger events. The problem was that it was mixed up with arrivals from poor stations with poor timing and there was confusion as to what was required to identify the first arrivals from emergent onsets on low gain instruments. Summing up the situation in the late 1940s, R. Stoneley says that what was done by ISS on a large scale was to determine the epicentres with tolerable accuracy, a conventional origin time with comparable accuracy, and (not very precisely) the focal depth (*IUGG Comptes Rendus*, 1947 Meeting, p. 54, Strasbourg). With time the location abilities of the ISS increased and they were further improved by its successor the ISC with a larger network of seismographic stations.

Figure 4.3 shows the annual world-wide variation of the number of epicentres calculated by BAAS and by its successors the ISS and ISC. In the early period 1913–17, BAAS estimated a total of 577 epicentral locations of which 47% were adopted without calculation. In the ISS period of 1918–63 that followed, 24 959 epicentres were reported,[5] 51% of which were adopted or accepted from other agencies without revision. As a matter of fact, during the period 1918–63, 60%

of the ISS determinations for any year (with the exception of 1937, 1939 and 1940) were adopted or they were of quality *R* or *X* (*ISS*: January 1930, pp. 4–5) in the ISS classification of determinations. From the same figure we notice that between 1918 and 1963 the number of epicentres found or revised by ISS increased from 158 in 1918 to 687 in 1961, the annual number of epicentres following very closely the growth of active stations up to the late 1950s and showing minima that correspond to the period of the two World Wars, the decline starting earlier in the mid-1930s. The total number of epicentres reported by ISS follows a similar pattern but with a sudden drop in 1952, from a total of 1429 epicentres to only 414 in the following year. The sudden drop after 1952 was probably the result of the decision taken at the IUGG General Assembly in Helsinki in 1960 that with few exceptions ISS should concentrate on the location of only those earthquakes which appeared to be of magnitude 6 or over. This decision was precipitated by the major activity in the Kurile Islands from 1950 to 1962, which swamped the ISS's manual location procedure. In 1952 the greatly increased observational data supplied to the ISS made it difficult to continue operating without some limitation on the quantity of events to be processed. Nevertheless, as figure 4.3 shows, after a brief reduction in its output, ISS again began to process data from earthquakes of magnitude less than 6 and its output again increased with a huge jump after 1963. This was due to the advent of its successor, the ISC, which brought all the epicentres of the major agencies and their data together for the first time and listed them. Another factor may be the PDE service starting at about this time. From 414 locations by ISS in 1953, ISC shows more than 13 000 listed in 1969. There are two points to note about these figures. First the peak in 1968 and 1969 is most probably due to increased activity; there were three of four very active periods in those years. Secondly, a new level is reached from 1974 onwards which is due to ISC

Figure 4.3. Time variation of number of seismographic stations and of their location efficiency.

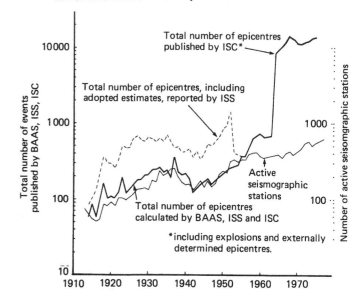

searching for unidentified events from its file of unassociated readings.

During the period 1913–17, BAAS reported 9 epicentres in Persia. The total number reported by ISS for the period 1918–63 is 528. However, of these 239 or 45% were epicentres adopted or accepted from other sources without revision. The question now arises of how accurate are the other 289 calculated epicentres, particularly the earlier ones.

4.3   Accuracy of instrumental epicentres

It is often believed that instrumental epicentres are more accurate than macroseismic epicentres. In regions that are well instrumented with modern seismographs this is of course true in principle. However, this is not the case with Persia and probably with other parts of the world where neither the distribution and number of stations nor the quality of input data are adequate to allow accurate focal determinations, particularly before the early 1960s.

The greatest outstanding problem is the accuracy of the pre-1960 ISS epicentres, not so much internally with respect to the method used for their calculation, but with regard to their actual locations. The internal accuracy of a location may be improved by improving the consistency of the solution even beyond the actual limit of accuracy of the input data. This, however, does not necessarily imply more precision in placing an epicentre, and with stations to one side (as is the case with Persia) high apparent consistency is possible but the precision in locating an epicentre will be low.

Confirmation of the low accuracy of the earlier ISS locations, not only in Persia but also throughout the Middle East, began to emerge gradually as more instrumentally-determined epicentral regions of large shocks were investigated in the field. It was found that they had often been located many tens to hundreds of kilometres away from their true position. Soon it became clear that the problems of instrumental location were common to all of the earlier epicentres calculated by ISS, BCIS, USCGS, and Gutenberg & Richter (1965), as well as to later re-calculations made by the Institute of Geological Sciences (IGS) (1972),[6] Nowroozi (1971, 1976), Alsan *et al.* (1975) and others. It became equally clear that it was perfectly feasible to study earthquakes that happened as early as a century ago in the field and locate their epicentral regions with undeniably better accuracy.

A small sample of glaring cases of mislocations in Persia may be given here in full. Other examples can be found in the case histories in chapter 3, while a considerable number of mislocations of small magnitude events in Persia and of large shocks in its immediate vicinity are left out because of shortage of space.

Until recently, the earthquake of 23 January 1909 was widely known from its instrumental epicentre which was published in the first edition of Gutenberg & Richter's *Seismicity of the Earth* in 1949. Gutenberg and Richter calculated the epicentre of a 7.4 magnitude earthquake at 33°N-53°E which placed this relatively large event on the southwest margin of the *kavir*, between Ardistan and Na'in, right in the middle of

one of the main trade routes and only 130 kilometres from Isfahan.

However, a field trip in that region and a systematic survey of the accounts left by travellers after 1908 was enough to prove that no major or even minor earthquake had occurred in that region during the period 1908–9, and that the epicentral region of the 1909 earthquake should be sought elsewhere.

A particular difficulty with the location of this event was that we could find no Persian sources that mentioned the earthquake, except a brief note in the press (*Iran*: 1324, no. 61). The earthquake occurred at a time when the country, particularly its northwest provinces, was in a state of anarchy and disorder. The Persian revolution had begun in 1906 but it was early in 1909 that disorder and violence spread throughout the country, after the Nationalists failed to persuade Muhammad 'Ali Shah to convene a National Assembly and restore the Constitution. A few weeks before the earthquake, on 5 December 1908, Bakhtiyari forces occupied Isfahan and declared for the Constitution, expelling the Shah's representative. Isfahan's example was soon followed by Hamadan and other towns, and a few weeks after the earthquake Silakhur, a comparatively remote and isolated district in Bakhtiyari country where the earthquake in fact happened, found itself surrounded by Nationalist centres which were in contact with each other and controlled by local provisional governments. In contrast and in spite of the telegraph line between Burujird and the capital, Tehran's contact with those decreasing areas in western Persia, including Silakhur, in which the Shah's authority was maintained, was very poor and irregular. Early in 1909 the state of the roads in western Persia had become insecure, particularly the routes to Hamadan, Kirmanshah and those to the south through Silakhur to Isfahan, to the extent that communication by courier had ceased even for the Russian and British consulates in Kirmanshah.

It is not surprising, therefore, that unofficial news about the earthquake only reached Tehran three weeks after the event by means of consular couriers, while the full extent of the disaster did not become known through the official press until late in March 1909. Even then, the news did not permeate through to the foreign press. The reason for the belated and very brief news of the earthquake in the Tehran press is that after the *coup d'état* of June 1908, all or nearly all newspapers were immediately suppressed, while those which were allowed to continue were issued at irregular quarterly or monthly intervals containing no news of interest.

However, a search in the seismological bulletins of the Russian stations and of Strasbourg revealed that the shock of 23 January was not only widely recorded, but was also felt outside Persia in the Caucasus, in eastern Iraq and as far as Kharput where, from the manuscript notes kept by Sieberg, we learn that it was felt with an Intensity IV by Mr Riggs of the Euphrates College who was in charge of the seismic station. Thus, the news of a large earthquake appeared in the European press before it was reported in the Tehran press, but the whereabouts of the event remained unknown. As a matter of fact the event would probably have passed without notice but for the attention directed to earthquakes at that time, only a month after the disastrous earthquake on 28 December 1908 at Calabria, Italy, which had aroused international concern. Thus, by the end of January recordings of the earthquake of 23 January (see figure 4.4b) had attracted world-wide scientific interest, but there was still no information about the region in which the shock had taken place. On 25 January the Russian press reported the results of instrumental studies of Russian recordings of the event, locating the earthquake within the confines of the Chinese Turkestan, suggesting the occurrence of an event that should have been much more serious than the one which destroyed Calabria. European seismologists disagreed, and the press reported the event in the Persian Gulf or the Indian Ocean a few days later, while seismologists in Leipzig attributed the recordings to a more distant volcanic origin in the Pacific Ocean. This theory they found supported by news of extraordinary sea-waves on the coasts of British Columbia and California. By the end of January, still without any information about the actual location of the earthquake, seismologists agreed on an improved epicentre in western Central Asia, the large amplitudes recorded implying an earthquake as large, if not larger, than the Italian earthquake of 1908. In the meantime, Prince Galitzin, the doyen of Russian seismology, calculated the origin of the earthquake at 36°N-56°E using data cabled to him from the seismological stations of Tiflis, Irkutsk and Pulkovo. This places it southeast of Shahrud, not far from where the earthquake of 25 September 1903 had caused damage. This is reported on 30 January 1909 at a meeting of the Seismological Committee and again on 3 February at a meeting of the Imperial Academy of Sciences in St Petersburg. Prince Galitzin's new location of the earthquake was soon carried by almost all European newspapers, well before the news of the actual location of the earthquake had reached Tehran.

About the same time the Russian consul in Kirmanshah, N. Nilolski (who, like everybody else in the town, had felt the earthquake), decided to send a reconnaissance mission to find the origin of the shock and assess the extent of damage. It is not clear from the correspondence that we have been able to find between Nikolski and his superiors in St Petersburg whether he did this at the instigation of Prince Galitzin, who we know was anxious to verify the epicentre that he had calculated from Russian recordings, or on his own initiative. The Russian mission left Kirmanshah by the end of January in search of the epicentral region, the mission being in the charge of Major Sukhenko of the Kirmanshah consulate (the town still being under the Russian sphere of political and commercial influence). Guided either by local information that should have reached the mission in Kirmanshah before their departure, or by Prince Galitzin's epicentre, the mission proceeded eastwards and reached Burujird ten days later. From this town the mission applied to the government in Tehran for permission to enter Luristan and Bakhtiyari country to study the effects of the earthquake. While the mission was waiting for a reply from Tehran, destitute survivors began flooding to Burujird applying to the governor for help. They reported that

the district of Silakhur was totally destroyed and that the Bakhtiyaris were looting and pillaging the less affected districts of Japalak and Faridun south of Silakhur, and that the roads through the region, particularly those passing through Chahar-Lang, were unsafe. After a long delay at Burujird, the mission was allowed to proceed to Silakhur on 16 February. The unpublished reports of the mission, particularly that contributed by Asadallah Mirza the consular agent at the end of the mission, make fascinating reading, particularly when they are considered in the light of the reaction they created in the British consular agents in Kirmanshah, Ahvaz and Tehran, who were suspicious of the Russian mission penetrating into neutral territory 'ostensibly examining the places devastated by the earthquake'.[7] These reports, which are accompanied by photographs one of which shows the damage and ground deformation at Sandargan (plate 10), together with the confidential diaries and despatches of the British consuls, leave no doubt about the actual location of the earthquake in the Silakhur valley. As a matter of fact, on 1 March Prince Galitzin reported to the Imperial Academy in St Petersburg a revised location of the event with an epicentre in the Burujird region at 33.9°N-48.8°E. At about the same time the actual origin of the earthquake became known in the capital and in Europe, almost certainly due to the cable sent by Nikolski's mission to Tehran. However, it is not until later when the full text of Major Sukhenko's report was received in St Petersburg that Prince Galitzin made his final attempt to locate the event instrumentally with disappointing results at 33.5°N-55.0°E, i.e. near Turud. At a meeting of the Seismological Commission in St Petersburg on 24 April, Dr Shtelling concluded that the macroseismic epicentre should be near 33.3°N-50.2°E. The reason for placing the epicentre to the southeast of Durud is that Major Sukhenko's report suggested damage further to the southeast of Durud in an area that the Russian mission could not visit.

With this information in hand, a series of field trips in the Silakhur area not only confirmed the extent of damage reported, but also established the southeastern limits of the meizoseismal area and the extent of faulting. They establish beyond doubt that the earthquake of 23 January 1909 did occur in the Silakhur valley with a macroseismic epicentre of 33.40°N-49.13°E, and that Gutenberg and Richter's instrumental epicentre at 33.0°N-53.0°E was incorrect, being some 360 kilometres out.

Figure 4.4a shows a facsimile of Gutenberg's worksheet for the relocation of the Silakhur earthquake. For this event, as for almost all other locations of shallow teleseisms, Gutenberg used the simple technique of fitting a sine curve to the distance residuals plotted against distance azimuths (see for instance Richter 1958). To start with he took the location given by BAAS as the preliminary epicentre of the event, i.e. 33°N-50°E, which is not very far off the macroseismic epicentre. Then he chose the earthquake of 22 July 1927 which had been referred by ISS to a nearby location at 34.7°N-54.0°E, so that distances and azimuths for all stations were already calculated for and tabulated in the ISS. From figure 4.4 we notice that with the exception of two stations, Zikawei

Figure 4.4*a*. Gutenberg's worksheet for the Silakhur earthquake of 23 January 1909. (Millikan Memorial Library, California Institute of Technology.) See also table 4.5.

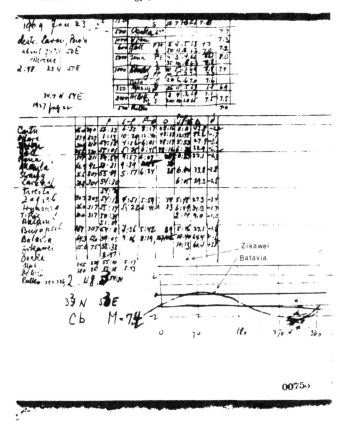

and Batavia with azimuths of 75° and 120° respectively, all the other stations have azimuths clustered within the range 290° to 330°, and a mean distance residual of −2.2, an uneven distribution typical of the pre-1930s. However, the problem here is that in sketching a sine curve through the data-points, Gutenberg leaves out completely the Zikawei residual of +3.8 which lies outside the grid and uses only that of Batavia of +0.1. As a result of this, the preliminary epicentre is shifted to the east by 3° to the location Gutenberg gives as final at the bottom of his worksheet (33°N-53°E) which is of course incorrect. Had he used the Zikawei residual, the shifts in latitude and longitude of the preliminary epicentre would have been very small, of the order of 0.5°, and in the right direction. Of course this kind of oversight is rather rare in Gutenberg's work, which in part justifies dissentients who are prepared to dismiss the macroseismic epicentre in favour of Gutenberg's instrumental location or even accept both. Nowroozi, for instance, suggests that there might have been two large earthquakes on that day, one at Silakhur and another between Ardistan and Na'in which would satisfy Gutenberg's location (Nowroozi 1976: 1271; 1979: 646). But our field studies show no evidence of a destructive earthquake in the Na'in−Ardistan area[8] and station bulletins and seismograms for 23 January 1909 show only one major event, and this occurred in Silakhur (figure 4.4*b*).

A similar problem arose with the earthquake of 18 April 1911, which Gutenberg and Richter place near Buhabad, figure 4.5. This was easily resolved, however, by field studies which proved that in fact the shock occurred near Ravar, 130 kilometres from the instrumental epicentre (see the case histories in chapter 3).

Another example of gross mislocation is that of the

Figure 4.4*b*. Recording of the Silakhur earthquake at Potsdam ($D = 32°$, Wiechert astatic).

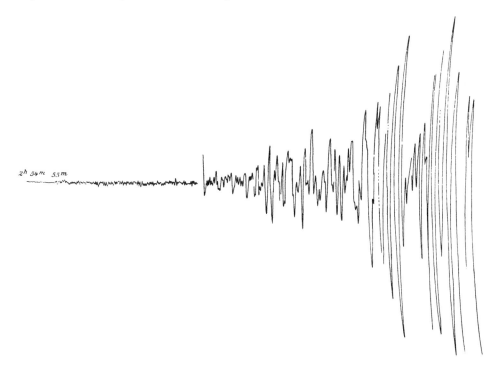

earthquake of 25 May 1923. In this case the actual location of the event was known from contemporary press reports, since the region of Kaj Darakht which was affected by the shock lies on one of the major routes in Khurasan. Yet, the ISS puts the epicentre more than 400 kilometres to the south of the area of maximum damage, near Kirman, and the BCIS puts it even further, near Farah in Afghanistan, 520 kilometres away. All that a relocation of the event could achieve was to shift the epicentre near Herat, 220 kilometres to the east of Kaj Darakht (Nowroozi 1976).

The case of the earthquake of 17 September 1923 was more complicated. Originally, the actual location of the meizoseismal region of the event was not known. The Russian press had reported that the shock was felt strongly in Ashkhabad and that it was perceptible in the region of Mashhad. This information suggested an epicentre somewhere in the upper Atrak valley, but field studies had failed to identify the epicentral region of this 6.3 magnitude earthquake. However, additional local information and press reports clarified the situation (see chapter 3: case histories). It emerged that the meizoseismal region of this event overlapped partly that of the much larger Kopet Dagh earthquake which occurred a few years later, this being a region that we had already studied in the field. A final site visit in 1975 proved that the epicentral region of the 1923 earthquake was to the northwest of Bujnurd. However, the ISS puts it near Turud, 310 kilometres

southwest of Bujnurd and so do Gutenberg and Richter. The epicentres re-calculated by IGS and Nowroozi (1976) do reduce the location error to 75 kilometres, but the two new locations remain 140 kilometres apart, the former to the southeast and the latter to the northwest of the meizoseismal region.

A quantitative assessment of the relative accuracy of instrumental locations can therefore be made by comparing original ISS epicentres of main shocks, excluding adopted estimates, with macroseismic epicentres of good and moderate quality. Figure 4.6 shows the average shift in distance of shallow depth earthquakes of $M \geqslant 5.5$ from their macroseismic locations (curve A). From this figure one can see that mislocation errors decrease from about 300 kilometres in the late

Figure 4.6. Location error distribution. (A): Average ISS–ISC location error for shallow depth earthquakes of $M \geqslant 5.5$, within the area extending from 24° to 40° N and 42° to 66° E, excluding *adopted* ISS epicentres. (B): Average ISS–ISC location error, as above, for $M < 5.5$. (C): Average relocation error of shallow depth events of $M \geqslant 5.5$, relocated by Nowroozi 1971, 1976, with standard error in epicentral position of less than 20 kilometres. (D): Average shift of relocated epicentre by Nowroozi from original ISS–ISC position. (*Note*: Average location error = average distance in kilometres per quinquennium between instrumentally determined epicentres and corresponding macroseismic epicentres of good or moderate quality. Average relocation error = average distance in kilometres per quinquennium between relocated epicentre and corresponding macroseismic location of good or moderate quality.)

Figure 4.5. Gutenberg's worksheet for the Ravar earthquake of 18 April 1911. (Millikan Memorial Library, California Institute of Technology.) See also table 4.6.

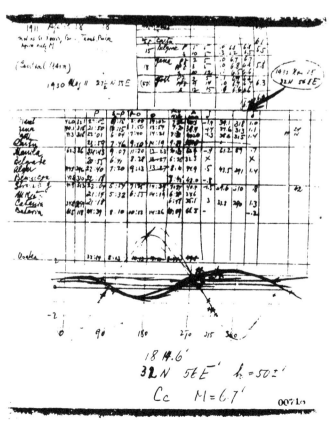

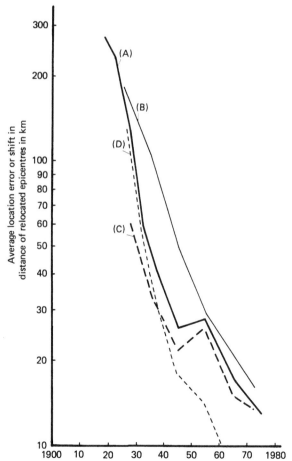

1910s to about 30 kilometres by the mid-1950s. After that period, shifts decrease further, on average to values of about 15 kilometres or less, and with instrumental locations now falling within the meizoseismal region, the use of macroseismic epicentres as a check on location accuracy ceases to be valid.

On the same figure we have plotted the average shift in distance for events of $M < 5.5$ (curve B). Their location error decreases less rapidly with time and although by the mid-1970s it reaches values of only 16 kilometres it remains significant at about two to four times the radius of the meizoseismal region of these small events, for which the macroseismic epicentre continues to be a valid check on location accuracy.

Thus, up to the mid-1950s ISS locations are on average in considerable error. This could have been much larger had we included adopted epicentres. This now raises the question of whether location errors can be reduced by relocating these events using modern computational techniques. An attempt was therefore made to relocate earthquakes in the Middle East prior to 1950 using the programme developed by Douglas et al. (1974). Out of 824 events that are known to have occurred in Persia during the first half of the century, 451 were tested singly. The input data of about one-third of the events tested were rejected as totally inconsistent without further investigation. The remaining 292, of all magnitudes, were relocated for a fixed focal depth of thirty-three kilometres. Inaccuracies of arrival times could not possibly permit valid depth determinations for the whole period, while for the period prior to 1924 relocation proved impracticable. Relocated epicentres of events after 1923 were found to have positions different from those originally computed by ISS, with an average shift of about seventy kilometres. At first sight this may be interpreted as an improvement of the locations, but on examination it was found that the average distance from their corresponding macroseismic locations remained significant at some 60 kilometres. In other words, relocation had improved the accuracy of the original ISS estimates far less than expected (Ambraseys 1978).

This observation led to a further effort to calculate again the location of a number of well-observed earthquakes for which there was sufficient macroseismic and instrumental data. By comparing readings from stations that had observed events in the same region over a decade or two, and by rejecting readings from stations with large travel time residuals, the results from the remaining stations confirmed the difficulty of improving on ISS locations before the mid-1930s. A few of the earlier re-calculated epicentres did show improvement but this was mainly due to the fact that macroseismic epicentres were used in the initial iteration, causing the rejection of about half of the station readings, or it was due to pure chance. Nowroozi (1971, 1976) who also re-calculated a considerable number of earthquakes in the Middle East, points out the difficulties in locating earthquakes of the early period, and grades his re-calculated epicentres according to their estimated precision in terms of the standard errors in latitude and longitude, focal depth and origin time, a precision of internal consistency rather than accuracy.

Figure 4.6 demonstrates the difficulty of significantly improving early ISS locations. Curve C shows the average shift in distance between shallow depth earthquakes of $M \geqslant 5.5$ re-calculated by Nowroozi with a standard error in epicentral position of less than 20 kilometres, and their macroseismic epicentres, which are of good and moderate quality. Curve D shows the shift in Nowroozi's relocated epicentres from their original ISS–ISC locations. We notice that although relocated epicentres (D) are considerably different from those calculated by ISS or ISC, their average shift with respect to their macroseismic epicentres (C) remains significant up to the late 1950s. For instance, for 1930 the average ISS location error is 85 kilometres (curve A) and the average shift of re-calculated epicentres from those located by ISS is 80 kilometres (curve D). This gives the impression that relocation significantly improved the accuracy of new epicentres. Yet the location error of relocated positions remains at 52 kilometres (curve C). For 1960, the ISS error is 22 kilometres, the relocation shift 10 kilometres and the relocation error 19 kilometres.

It is obvious, therefore, that the best method for improving or confirming instrumental epicentres of the earlier events is to attempt to find correlations with macroseismic data from first-hand field studies. The true epicentre of an earthquake is not, and should not, be in the centre of the meizoseismal region, and the meizoseismal region in turn may depend on population distribution (see figures accompanying the case histories in chapter 3). But at least macroseismic epicentres are not liable to the gross mislocation that is possible with the earlier instrumental locations. The most important point that emerges is that no-one's instrumental epicentre of a pre-1930 earthquake is likely to be so good as a macroseismic epicentre, while for smaller magnitude events this is likely to be true up to the late 1950s or 1960s particularly for regions covered by an ill-distributed network of seismic stations. After 1963, figure 4.6 shows that location errors decrease rapidly but remain on average in excess of about 10 kilometres which is just above the limit of macroseismic location.

4.4 The calculation of magnitude

The term 'magnitude' was first used to define the comparative size of an earthquake by Wadati (1931). He classified earthquakes in Japan by the logarithm of their maximum ground amplitude, as recorded instrumentally, thus defining the magnitude of an event in terms of the energy released. It was probably this work that prompted Richter in 1935 to devise a magnitude scale for earthquakes in California in order to obtain some idea of the comparative size of events in the region (Richter 1935).

Richter defined the local magnitude M_L as the logarithm of the maximum trace amplitude on a seismograph of certain type, corrected to a distance of 100 kilometres from the source. Between 1935 and 1956, in a series of papers, Gutenberg extended this definition to include distant observations at stations equipped with instruments other than the standard Wood–Anderson torsion seismograph originally used (Gutenberg & Richter 1942, 1956b). They defined M_S by the surface-wave magnitude calculated from surface-waves of periods of about twenty seconds, and by m_B the body-wave magnitude calculated from body-waves. In addition, they defined M_B and m_S as the surface and body-wave magnitudes that can be derived from empirical correlations between body- and surface-wave magnitudes respectively. In 1956 Gutenberg and Richter proposed the unified magnitude m which they calculated as the weighted mean of m_B and m_S (i.e. of calculated and derived body-wave magnitudes) which is often confused with m_B. The weighted mean of M_S and M_B they specified as M without subscript, the so-called Gutenberg–Richter magnitude M_{GR}, a designation which in fact they must already have used in their magnitude catalogues of 1941 and 1949.

It was after the publication in 1949 of the *Seismicity of the Earth and Associated Phenomena* (Gutenberg & Richter 1965) that the concept of magnitude began to receive general attention. During the 1950s seismic stations began to make their own magnitude determinations using procedures similar to that put forward by Gutenberg and Richter. For surface-wave magnitudes ground amplitudes were restricted to wave periods of about twenty seconds and for body-wave magnitudes the values of the distance–depth function Q were used.[9] In 1955 Soloviev relaxed the 20-second period constraint by replacing the ground amplitude in the magnitude formula with the ground velocity in terms of the amplitude–period radio (A/T) (Soloviev 1955), a procedure for calculating surface-wave magnitudes adopted by the USSR networks since about 1953.

Thus, by the end of the 1950s there were a multitude of formulae in use for surface- and body-wave magnitudes, in addition to the formulae already proposed by Gutenberg and Richter. The need for a magnitude standardisation on an international basis became obvious.

This need was answered during the following decade. At the Magnitude Symposium convened by Professor M. Båth during the General Assembly of the International Union of Geodesy & Geophysics (IUGG) in Zurich in 1967, it was agreed that magnitudes should be estimated on the basis of the maximum amplitude–period ratio for all waves for which calibrating functions are available, and that two magnitudes should be used, namely body-wave magnitude m, and surface-wave magnitude M (Båth 1969, Miyamura 1978). It was also agreed that for body waves the Q-values (Gutenberg & Richter 1956a) should be used, and that for surface-wave magnitudes the most appropriate formula was:

$$M = \log(A/T)_{max} + 1.66 \log(D) + 3.3 + C \qquad (4.1)$$

where $(A/T)_{max}$ is the maximum value of the ratio of the ground amplitude, A, of the surface-wave group in microns, and T, the corresponding period in seconds. D is the focal distance in degrees and C is the station correction. At the same time, a conversion formula, $m = 0.56(M) + 2.9$, was recommended for unifying the two magnitudes.

Equation (4.1) and the Zurich recommendations were to some extent already in use by the seismic networks in the USSR and in other Eastern European countries after 1962. At the General Assembly of the IUGG in Helsinki in 1960 similar proposals had been accepted by a number of national networks, notably that of the USSR, including the recommen-

dation to abandon Gutenberg and Richter's unified magnitude and to average magnitudes derived from different wave-types. Surface-wave magnitudes estimated by the USSR network (the nearest major network to Persia) have been consistent with (4.1) since 1962.

In 1963 the USCGS began a systematic magnitude determination programme in the Preliminary Determination of Epicentres. The recording of any earthquake, deep or shallow, for distances up to 105° (or the recordings of a nuclear explosion) has the advantage of always starting with a P-onset, whereas deep earthquakes do not usually show surface-waves of appreciable amplitude and therefore their maxima cannot always be identified or measured accurately enough. For this reason the USCGS chose to determine systematically body-wave magnitudes m_B in the manner defined by Gutenberg and Richter. The routine determination of m_B from the short-period vertical component initiated in 1963 still continues for distances down to about 5 degrees.[10] Estimates of surface-wave magnitude are not made by the USCGS on a routine basis except occasionally for the larger events. Each type of magnitude, m_B or m_S, reported by the USCGS is an average of individual station magnitudes which are determined from reported amplitudes and periods of representative waves. Before 1966 the USCGS estimated m_B as the logarithm of the average of $(A/T)10^Q$ instead of the average of the logarithm of this quantity. The effect of first taking the average is to increase the influence of the extreme values. The broader the range of the individual station estimates, the greater will be the difference in the results obtained by the two methods, so that in principle m_B estimates before November 1966 are somewhat over-estimated, particularly of the smaller events. Thus, in contrast with the USSR network, the bulk of the USCGS magnitude estimates are made for body-wave magnitudes available after 1963.

Where estimates of M_S by the two agencies are available, they are in good agreement. However, significant differences exist in the estimates of m_B, USCGS values being systematically under-estimated compared to the USSR network. The cause of this discrepancy is due to a number of factors, one of which may be the way in which the selection of the P-wave amplitude is made, as well as the use of short period, narrow-band seismographs which give a lower value for m as compared with the broad-band seismographs. If magnitude determinations are confined to the use of the first half-cycle of signal amplitude, then m_B estimates would be small and vary greatly. If, on the other hand, measurement for calculating m_B at a station is made within say ten to twenty seconds of the P-arrival (which is the procedure followed by Gutenberg and also the practice in the USSR, Bune & Vvedenskaya 1970), the radiation pattern is no longer accurately represented by the amplitude, but the results provide a better estimate of m_B when individual values are averaged.[11]

No magnitude estimates were made by the ISS up to 1963. Starting with 1964 its successor, the ISC, began to report body-wave magnitudes m_B like the USCGS. In calculating magnitudes, the ISC followed the procedure outlined by Gutenberg and Richter, similar to that followed by the USCGS, but based on more observations. The Centre has no control over the instruments used by the reporting agencies and stations, and it assumes that the majority of reported readings come from short-period instruments. Magnitudes reported by the ISC are derived from at least three individual estimates without station correction, and observations at distances less than 21 and more than 100 degrees are ignored. Where amplitude–period data for 20-second surface waves are available, M_S is calculated from equation (4.1) only in the distance range between 20° and 160°. But even though individual observations of M_S are calculated, no surface-wave magnitudes are as yet adopted by the ISC. With the exception of the period in early 1964 during which ISC was over-estimating m_B values by about 0.3 magnitude units due to a computer error, body-wave magnitudes determined by USCGS and by ISC with more observations differ very little.

To sum up the situation in the mid-1960s, therefore, we find by this time the establishment of three main groups publishing magnitudes calculated according to standard techniques.[12] The USSR network has determined surface-wave magnitudes consistent with the formula (4.1) since 1962. The USCGS has calculated body-wave magnitudes for events after 1963, and the ISC (since 1964) has also reported body-wave magnitudes.

4.5 Revision of magnitudes

For the period prior to 1963 the number of earthquakes in Persia for which magnitudes are reliably known is very small indeed. No magnitudes are available for events before 1909, and for the period up to 1930 Gutenberg and Richter have assigned 'magnitudes' to only half the events of magnitude greater than or equal to 6. For almost all smaller shocks magnitudes remain unknown. For the period 1931 to 1948 these authors assigned magnitudes to almost all the earthquakes greater than 6, but only to very few below this value. In the *Seismicity of the Earth* they assigned magnitudes to 63 earthquakes in Persia for the period 1909 to 1948. Of these, 29 are of class d ($5.3 \leqslant M \leqslant 5.9$), 24 are of class c ($6.0 \leqslant M \leqslant 6.9$), 9 of class b ($7.0 \leqslant M \leqslant 7.7$), and 1 of class a ($M \geqslant 7.8$). Four events are of intermediate depth.

Magnitudes estimated by individual stations in the period 1949–60 are quite numerous but very heterogeneous. The magnitudes of about 50 earthquakes in Persia calculated during that period are either non-homogeneous or it is not clear by which method they were derived. Reporting body-wave magnitudes as M, the use of peak-to-trough amplitudes and of inaccurate focal distances by near stations often lead to inconsistencies.[13] Also, the use of P-phase amplitudes given with only one decimal and a minimum value of 0.1 microns, was found to lead to an over-estimation of the smaller body-wave magnitudes. In fact for the whole period up to 1962, we have an agglomeration of magnitudes estimated at different times by different magnitude scales, often by a combination of scales,[14] or by scales that are not specified. Of every four earthquakes in Persia in that period, the magnitude of three is not known, while of every three known magnitudes, two at least are of questionable quality.

The question of correcting and unifying existing magnitudes was therefore considered. Some thought was also given to the possibility of assigning magnitudes to those remaining events which are without magnitudes on the basis of an empirical conversion from macroseismic data. However, it was finally decided that a substantial advantage could be gained by calculating the magnitude of all earthquakes before 1963 for which world-wide reported data was available, using an internationally-accepted technique in preference to correcting the conflicting values reported today and adding empirical formulae to the long list already existing. The intention was not so much to discover new large magnitude earthquakes after 1909, because these have already been sought out by Gutenberg and Richter[15] and by the USSR network. It was rather to produce a body of reliable data that could be used with confidence for the study of local and regional tectonics and seismicity.

There is a host of problems that can only be studied with reliable magnitude estimates. For instance, the use of non-uniform and incomplete magnitude estimates may cause a potentially serious distortion in magnitude–frequency recurrence relationships that are important in assessing seismic hazard. The contribution to this distortion from the low-magnitude range should not be under-estimated, particularly for regions of apparent low seismicity where recurrence formulae are usually computed on the basis of low to intermediate magnitude earthquakes. Also, the use of non-uniform magnitudes is likely to lead to erroneous conclusions about the coherence between M_S and m_B (Freedman 1967).

4.5.1 Early period

It is perhaps useful at this stage to review briefly the problems associated with the calculation of magnitudes for earthquakes before 1963.[16] The magnitude of an earthquake, whether estimated from surface- or body-waves at a station, is a parameter that is subject to considerable variations. Attenuation, crustal effects, and the pattern of radiated energy at the source account for much of this variation, which can be reduced by averaging a sufficiently large number of station estimates. But even this average may depend on the geometrical distribution of the seismographic stations. This is in addition to the basic problem of the accuracy of the reported amplitude–period data and of the calibration characteristics of the instruments, particularly with earlier recordings.

In the early days before 1906, most seismographs in operation were of low magnification and without sufficient damping. These were slightly damped horizontal penduli with optical registration. The Rebeur-Paschwitz type of instrument and its improved versions of Rebeur–Ehlert and Zöllner–Repsold, operating mainly in Germany, Austria and Russia, had a period of about 20 seconds and a damping constant of 0.05. The maximum gain of the two former types was on average about 50 for near shocks ($D \leqslant 10°$), and about 150 for distant shocks ($D \geqslant 20°$). The gain of the Zöllner–Repsold was 50 and 400 for near and distant events. A similar type of instrument, the Milne seismograph, was in operation in Europe, Asia, America and Africa. It had a shorter period of 17 seconds and a damping constant of about 0.08. Its maximum gain was between 10 and 15 for near shocks, and between 20 and 30 for distant events. A much heavier type of pendulum, the Bosch–Omori system, was in operation in Germany, Japan and in a number of other countries. Its mechanical registration system provided a heavier but more erratic damping of 0.20, and its maximum gain was almost the same as that of the Milne seismograph (Kirnos *et al.* 1961).

If we exclude a kaleidoscope of penduli that were operated by Italian stations (Agamennone, Brassart, Guzzanti, Stiattesi, Vincentini etc.), where hardly two stations were equipped with the same instrument, the largest network of standard instruments was that of Milne undamped seismographs. For instance, in 1903 out of 69 stations of the world's seismographic network, 28 were operating Milne, 9 Bosch and 8 Rebeur–Ehlert seismographs (Tams 1908*b*). The remaining 24 stations ran 18 different types of instruments. Excluding primitive Ewing duplex and Rocker penduli, these Milne stations constituted over one-third of the total number of seismological observatories in constant operation. In the following few years, the next largest group of seismographs was of the Bosch–Omori type. As can be seen from figure 4.7 and table 4.1, the number of Milne seismographs continued to increase, there being 38 in operation in 1912.

However, by the end of the first decade of this century more advanced, damped instruments began to gain ground, superseding the early instruments such as the Bosch–Omori, Vincentini and Ehlert, including the Milne. These were the Wiechert inverted pendulum and horizontal mechanical recorders of considerable moving mass of at least one ton, and operating in the period range of 5 to 15 seconds or more with a magnification from 150 to 300. Their damping constants

Figure 4.7. Variation with time of amplitude–period data supplied by seismic stations located about 90° from Iran. I: Number of stations reporting amplitude–period data regularly for events in Persia. II: Number of stations reporting such data less frequently or occasionally. III: Number of stations reporting regularly for events in the Eastern Mediterranean and Middle East.

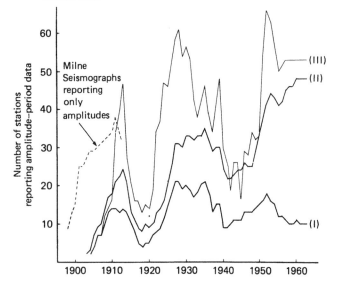

varied between 0.3 and 0.5. At an early stage Jena, Leipzig, Göttingen, Munich, Potsdam and Uppsala began to operate this type of instrument. The largest of them in Göttingen was a 17-ton, 1.5 seconds inverted pendulum, operating with a damping constant of more than 0.5 and a magnification of more than 2000. A decade later, the Milne seismographs (and to some extent the Wiecherts) were superseded by the more advanced Galitzin galvanometric recorders with magnetic damping, and the improved version of the Milne seismograph, the Milne–Shaw oil damped recorder (Berlage 1932).

The problem in the early period before 1913 with undamped or slightly damped instruments, is that it is extremely difficult to decouple the instrument response from the recording and recover the ground motion time-history. Also, the instruments' main characteristics of period and gain not only varied greatly with time and from station to station, but these constants were quite often not published, so that magnitude determination from recordings of such instruments is inevitably unreliable even from the best stations. From the early period, therefore, it is only the Milne network that presents the least problem with respect to the calibration of instruments and regular reporting of the results. Starting in the late 1890s, but systematically from 1899, times and amplitudes of maxima measured on records of Milne seismographs are given regularly in the *Circulars of the Seismological Committee*, published by BAAS, occupying a total of more than 1000 closely printed pages (BAAS 1900–13). The constants of the network's instruments were regularly reported and additional information, including prints of selected recordings made by Milne instruments as well as details on changes of the operation of the instruments, are contained in an equally voluminous set of reports and minutes of the annual meetings of the Seismological Investigations Committee since 1895 (*Reports on Meetings*, BAAS 1895–1914).

Figure 4.8, an azimuthal equi-distant plot centred on Yazd, shows the location of stations which operated Milne seismographs during the period 1899–1912. Yazd has been chosen as being near the centre of the region under study (32°N-54°E). The relatively large number of trace amplitudes recorded by Milne seismographs does allow the opportunity to test the reliability of the magnitudes calculated from them. In our case, trace amplitudes are available for 40 earthquakes in the period 1899 to 1913.[17] These were recorded by a fair number of instruments of the standard Milne network. For instance, the earthquakes of Qishm in 1902, Turshiz in 1903, Silakhur in 1909 and Ravar in 1911 were all recorded with 9, 7, 16 and 19 of the network's stations reporting trace amplitudes.

Table 4.2 lists all the events in Persia for which we have both trace amplitudes from Milne seismograms, as well as ground amplitude–period data from more advanced, damped instruments, mainly Wiecherts. From the latter data it has been possible to calculate surface-wave magnitudes using the standard formula (4.1), but the number of stations is too small to allow reliable averages of M_S. We can proceed, however, to fit to these average magnitudes their corresponding trace maxima recorded by the Milne seismographs, obtaining the following expression:

$$M_S{}^* = \log(2A_t) + 1.25 \log(D) + q \qquad (4.2)$$

where $(2A_t)$ is the double trace amplitude (peak-to-peak) in millimetres on Milne seismograms, D is the focal distance in degrees, and q is a constant which, for the events in table 4.2, is found to have a value of 4.06.[18] A comparison of the average magnitude determination using damped instruments (4.1) and undamped Milne seismographs (4.2) is shown in table 4.2. The correlation between M_S and M_S^* is reasonable with a standard deviation between the two magnitudes of 0.35.

For the majority of the early earthquakes of this century in Persia, data from the more advanced instruments are available so that the need to resort to the Milne trace amplitudes is minimal. The comparatively low accuracy of such magnitude estimates is therefore not very important. Equation (4.2) has helped us to assign magnitudes to only very few of the earthquakes of the early period.[19]

Gutenberg also found that trace amplitudes from the undamped Milne seismographs permit a reasonably acceptable assessment of magnitudes for shallow shocks. He made a somewhat similar attempt to assign magnitudes to the largest earthquakes for the world during the period 1896–1903 using Milne trace amplitudes, but with rather different results (Gutenberg 1956). To calibrate the Milne seismograph he used 16 large, shallow events of the period 1904–7 for which magnitudes had previously been determined by his own method and not by (4.1). The number of stations he used to calculate average M_S values is not known, but these being the largest shocks of the world of the period ($7.6 \leqslant M \leqslant 8.7$), he must

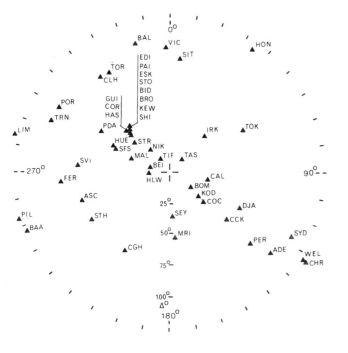

Figure 4.8. Network of seismographic stations operating undamped Milne recorders 1899–1912. For code identification see table 4.1.

have used between 5 and 10 stations per event (cf. Gutenberg 1956, table 1). He calculated the magnitude $M_S^* = M_S^{\overline{*}}$ by

$$M_S^{\overline{*}} = \log(A_{tm}/G) + 1.66 \log(D) + 1.82 + s \qquad (4.3)$$

where (A_{tm}) is the single trace amplitude in microns on Milne seismograms, G is the effective gain of Milne's seismograph, and s is a station correction. By adjusting G to make $M_S^{\overline{*}}$ equal to M, and by assuming that the maximum waves had a period of the order of 20 seconds, Gutenberg found that the undamped Milne instruments had an average gain of about 5. This is a puzzling result as this value is even smaller than the static magnification of Milne seismographs which is about 6.

The interpretation of this result is difficult. Kanamori & Abe (1979), who looked into this problem recently, conclude that the effective gain must depend on M. The results of their experiments with a newly-built Milne seismograph and the numerical test did show that the effective gain should be about 20 instead of 5, but that at least for the events of magnitude less than 7.8, the gain seems to be 5. On the other hand, a comparison of equations (4.1) and (4.2) shows quite independently that provided the shocks are small ($5.0 \leqslant M \leqslant 7.3$) with $q = 4.06$ in the period range $10 \leqslant T \leqslant 20$ seconds and $15° \leqslant D \leqslant 45°$, the most satisfactory value for G is 25. This is in agreement with the results reported by Kirnos, i.e. $10 \leqslant G \leqslant 15$ for $D \leqslant 10°$, and $20 \leqslant G \leqslant 30$ for $D \geqslant 20°$ (Kirnos et al. 1961). For larger events, or for $q = 4.52$, the most satisfactory value of G seems to be around 10, suggesting some dependency on M, but in a sense opposite to that suggested by Kanamori & Abe (1979).

There seems to be no way of comparing equation (4.2) with (4.3). The former has been derived from much smaller events than the latter, and for shorter distances. Moreover, the number of stations used in deriving (4.2) is too small to average out the various uncertainties in the characteristics of the Milne seismograph. Nevertheless, it does seem that Gutenberg's average gain of $G = 5$ really represents a very gross average that is influenced by effects other than those due to pure gain, and can only be applied to the largest events, when a substantial number of stations distributed all over the world are used. For smaller events occurring in the area limited to the region of Persia, the azimuth and distance of the stations used in deriving (4.2) are inevitably limited to a relatively small range, so that Gutenberg's world-wide average of $G = 5$ probably does not apply.

4.5.2 Advanced period (1903–62)

Effective magnitude assessment of earthquakes in Persia begins in 1903 with the recordings by Wiecherts at Potsdam and Göttingen of the earthquakes at Anguran (9 February, $M = 5.6$), east-central Khurasan (22 March, $M = 6.2$), central Gilan (24 June, $M = 5.9$), Igdir (6 August, $M = 6.0$) and Kashmar (25 September, $M = 5.9$). In fact these two stations supply the only amplitude–period data for magnitudes in the Middle East in 1903. By 1907, the number of stations reporting data for amplitudes and periods increases to six, Jena, Leipzig, Strasbourg and Uppsala added to the list of reliable stations which by 1910 increase to eighteen.

Table 4.3 lists the stations to the bulletins of which we have had access, and which were found to contain useful data for magnitudes of earthquakes in Persia during the period 1903 to 1963. The list includes (i) stations that reported amplitude—period data regularly, and (ii) stations that reported such data less frequently or occasionally. Table 4.3 also lists 25 stations that furnished either reliable ground amplitudes or trace amplitude—period data that are often helpful in confirming magnitude assessments. The table does not include 43 stations that operated rather insensitive instruments at distances less than 55° from Persia or stations at greater distances. Data from these stations have been used, however, to supplement the compiled material. From this table one can see the completeness of, or gaps in, the information furnished by the station bulletins consulted, as well as the frequency with which amplitude—period data are reported, and the span of time covered in the retrieval of magnitude data. The table also shows the number of amplitude—period readings finally extracted from each station for the calculation of M_S, as well as the mean value of the reported periods.

Figure 4.7 shows the variation with time of the number of stations that reported amplitude—period data regularly (I), and of those that reported less frequently or occasionally (II). For the sake of comparison, the same figure also shows the variation of the number of stations that reported amplitude—period data for the Eastern Mediterranean, i.e. the Balkans, Turkey and the Near East (III).

From this figure we notice that from 1903 to 1913 the number of stations reporting useful amplitude—period data for magnitudes increased very rapidly, not only for Persia, but also for the Eastern Mediterranean: from two stations in 1903, to twenty-three and forty-seven stations in 1913 for Persia and the Eastern Mediterranean respectively. However, this rapid improvement was arrested by the First World War during which some stations suspended operation and others ceased to report information other than onset times, amplitude—period data becoming very scanty during the period 1914—19. A new rapid increase in magnitude data reporting begins in 1920, which lasted for almost a decade. During that period not only does more detailed information become available, but there was also a drastic improvement in instrumentation, particularly in timekeeping. Unfortunately, this improvement was not accompanied by an equal amelioration in the reporting of other features of the seismic record, particularly of amplitude—period data. On the contrary, in about 1930 there begins a sharp relative decrease in the amount of amplitude—period information and in the description of various phases which in earlier station bulletins were invaluable for the determination of magnitude and the identification of intermediate depth shocks. As a matter of fact, the advent of the electromagnetic recorder marks the beginning of a new decline in amplitude—period information. Thus, between 1930 and the beginning of the Second World War, in spite of the rapid increase in the number of stations which during that period more than doubled, and notwithstanding the improvement of equipment and increase in their sensitivity by a factor of many hundreds,

the number of stations reporting amplitude—period data continued to decline. This is reflected by all three curves in figure 4.7, and it may be explained by the fact that with more sensitive instruments the volume of observational data had increased to the point that stations had to impose some limitation on the quantity of data to be published,[20] with a view to reducing both the expense and time expended on reading all seismograms. It is not unlikely, however, that for some of the secondary and a few primary stations the major factor contributing to this decline was not the increase in volume of observational data, but rather the inadequate calibration of their more sophisticated seismographs.

The Second World War marks another low in amplitude—period data. In contrast with other parts of the world where stations continued to operate, many stations in Europe and in the East ceased to operate altogether. Consequently, for the period 1939 to 1947 and particularly from the near stations, important information is lacking, the bulk of the data available being for the larger shocks.

Beginning with 1948 there is another increase in the amplitude—period data, but this time of a different pattern. With the exception of a few stations, about a dozen in all, which continued to report amplitude—period information of all events recorded on a regular basis, the majority of stations (the number of which increased rapidly) became selective and reported this type of information occasionally, either for local shocks or for large events. This is shown in figure 4.7 where up to the late 1940s curves (I), (II) as well as (III) followed each other, whilst after about that year they begin to diverge. For the smaller events surface-wave magnitudes in the 1950s and early 1960s become scarce, whilst for body-waves and for the larger surface-wave magnitudes data is more ample than before.[21]

For all earthquakes for which we have been able to retrieve amplitude—period data, magnitudes were calculated using the procedure put forward by the Committee on Practice (Willmore & Kárník 1971), an annotated extension of the procedure recommended in 1967 at the Magnitude Symposium in Zurich.

Individual surface-wave magnitudes, therefore, were calculated using equation (4.1) which is valid for distances between 2 and 150 degrees,[22] provided T is less than the values shown in table 4.4. With the almost total lack of local stations, there were very few cases that involved shorter distances, and these were treated as Lg(Sg)-phases. Most estimates of M were made from LH amplitudes which constitute the bulk of the available information. However, where LV amplitudes were available a separate estimate of M_{LV} was made. In estimating M_{LH}, the maximum value of the amplitude—period ratio $(A/T)_{max}$ was calculated irrespective of the value of T, but for periods of less than 10 seconds and more than the values shown in table 4.4, the calculated magnitudes were treated separately. In the ratio $(A/T)_{max}$, the amplitude used for the calculation of M_{LH} is the resultant of the two horizontal components of motion, provided of course that their times do not differ by more than T. Otherwise, M_{LH} is estimated from one component and its value is increased by 0.1. In (4.1) the distance D is measured from the macroseismic epicentre in most cases.

Body-wave magnitudes were estimated using the Gutenberg distance—depth factor $Q(D,h)$ for distances in excess of $20°$, i.e. from $m = \log(A/T)_{max} + Q(D,h)$, and for whichever phase, PZ, PH, PPZ, PPH or SH amplitude data was available. With body-waves we had little choice but to combine magnitude estimates from all phases and take an average in which PH predominate before the mid-1950s and PZ after that period.

Errors in the estimate of body-wave magnitudes are likely to be greater than those of surface-waves. This is mainly due to the lack of information on focal depths, which for most events are known only approximately. Thus, the ambiguity of the effect of focal depth on Q cannot be resolved. However, almost all stations reporting body-wave data are at distances between $25°$ and $50°$, while the best stations that furnish the bulk of body-wave data are at distances of $38° \pm 5°$. In this range the Q factor does not change very much with depth (h), so that the error in m due to uncertainties in the value of h should not exceed ± 0.2 on average.

For the period between 1903 and 1963, the total number of individual surface- and body-wave magnitudes that can be calculated in this manner is 2700 and 400 respectively. These belong to about 500 earthquakes of magnitude equal to or greater than 4.5, a sufficiently large sample to allow the assessment of station corrections for about 50 stations. The station correction C_i is defined by

$$C_i = \sum_i (M_n - M_i)/n$$

where M_i is the magnitude of a particular earthquake at a particular station (i) and M_n is the average value of the magnitude, of the same event, estimated from a number of stations;[23] n is the number of earthquakes for which M_n and M_i estimates are available.

For most stations C_i was found to vary erratically with time, an observation already made by Kárník from a much larger body of data for the European area (Kárník 1968). These variations are particularly noticeable around the years 1914, 1928, 1940, the largest of them occurring systematically in the early 1950s. However, the data being insufficient, no attempt was made to find a physical explanation for these variations. The fact that some of them can be proved to be associated with either the improvement or replacement of seismographs and also with changes in the staff of the station, makes it unlikely that the study of these variations would be of scientific interest in the present case.[24]

Table 4.3 shows the station corrections calculated for the periods 1903 to 1950 and 1951 to 1962. Their values are relatively small, comparable to those calculated by Kárník (1968) for the European area. However, their standard deviations are on average about 200% of their actual values. Bombay, De Bilt, Hamburg, Irkutsk, Strasbourg, Tiflis and Uppsala provide the most stable station corrections with deviations of less than 100%.

The data available is insufficient to provide station corrections for body-wave magnitudes.

Magnitudes were obtained by averaging values calculated from individual station readings, the number of which varied with magnitude and the year of occurrence (see figure 4.7). For surface-wave magnitudes M_{LH}, the average number of stations was 5 and the maximum 30. Body-wave magnitudes were derived from a smaller number of individual estimates. Station corrections for surface-waves had little effect on the final average except when the number of stations was relatively small.

The comparison of these re-calculated magnitudes with those estimated by individual stations or agencies for the period between 1903 and 1963 is rather difficult. This is mainly due to the lack of continuous observations. For the earlier period the only body of data with which we can compare our estimates is that of 31 shallow earthquakes in Gutenberg & Richter (1965) covering the period 1909 to 1948.[25] For surface-wave magnitudes there is hardly any difference between these two sets of estimates, M_{GR} values being slightly but systematically higher by just over 0.1 units of magnitude. For body-waves no comparison is possible.

Figures 4.4 and 4.5 show facsimiles of Gutenberg's unpublished worksheets for the 1909 Silakhur and 1911 Ravar earthquakes. His calculation notes and worksheets[26] suggest that in assessing magnitudes Gutenberg exercised a considerable degree of personal judgement, not only in the selection of the data, but also in the way in which he combined the results from different phases to arrive at a final estimate. Tables 4.5 and 4.6 show legibly Gutenberg's data of figures 4.4 and 4.5, side by side with the data we used to estimate the magnitudes of these two earthquakes. These tables, as is true for most of his calculations, show that Gutenberg was systematically using fewer stations than were available at the time. As tables 4.5 and 4.6 show, this is not because he restricted himself to amplitudes of 20-second periods for surface-wave magnitudes. Had he in fact chosen to do so, he would have used at least Hamburg and Leipzig in his calculations for the Silakhur earthquake, and Potsdam in the case of Ravar, which he did not.

To understand Gutenberg's system we have to make our own interpretation of his worksheets. It seems that he used a phase-dependent station correction which he added to the $\log(A/T)$, as for example the values in column (6) of tables 4.5 and 4.6. Also, he must have taken the Q-values from an early version of the distance—depth function (column 8). As for the two horizontal components of the ground displacements, he did not always combine these vectorially.

These are not very serious inconsistencies, and for events with $6.5 \leqslant M \leqslant 7.5$ the differences in magnitude are small. However, for smaller magnitudes our estimates are systematically lower than Gutenberg and Richter's by 0.2 units of magnitude.[27] It seems that at least in part this difference is the result of Gutenberg often determining M from all body-waves and separately from the surface-waves, and then taking the mean, a general habit for shallow focus earthquakes before 1967. For the smaller events the contribution of m in this average is of course to increase the M value. This difference may also be due to the fact that for deeper shocks Gutenberg always used the average of body-wave magnitudes only. For example, because he considers it to be of intermediate depth, he assigns $M = 6.7$ to the Ravar earthquake, which is essentially a broad-band body-wave magnitude (cf. figure 4.5). The surface-wave magnitude which he calculated but did not use is $M_S = 6.0$.

The comparison between magnitudes that we calculated and those estimated by individual stations up to the present is also difficult because several stations are still changing their method of calculation. For instance, starting in 1955 Uppsala and Kiruna began to report magnitude estimates using Gutenberg and Richter's method. From 1955 onwards a phase-dependent station correction, the value of which was between 0 and 0.5, was added to these estimates. Up to 1968 magnitude estimates reported by Kiruna and Uppsala were calculated from the average of all available phases, PH, PZ, PZ', PPH, PPZ, PPZ', SH and LH. It is only after 1967 that these stations complied with the Zurich recommendation.[28]

But although individual station magnitudes before 1963 differ, their agreement with the estimates that we calculated using amplitude–period data and equation (4.1) from different stations is very good. For instance, for the period 1951 to 1977 and using amplitude–period data from Uppsala and Kiruna, we find that M_S estimates from these two stations are mutually consistent, figure 4.9. The former station systematically under-estimates M_S with respect to the latter, by about 0.2 units of magnitude. In a similar way we find that for the smaller magnitudes, 4 to 5, Moscow's M_S estimates are higher than Uppsala's by about 0.2. For larger magnitudes,

however, this difference decreases, becoming zero at about 6.5, figure 4.10. No significant differences were found between M_S estimates, derived from amplitude–period data from Strasbourg, Pasadena and Uppsala. This observation, however, is based on a rather limited number of observations and it refers to the larger events.

However, the same is not true for body-wave magnitudes. As has already been pointed out, there are serious discrepancies between m_B-values estimated by different stations and agencies (Bune & Vvedenskaya 1970). Figure 4.11 shows that body-wave magnitudes (m_{pz}) calculated from Uppsala and Kiruna data are systematically larger by almost 0.5 units than those estimated by the USCGS (cf. Båth 1979a). Moreover, there is a cut-off at about $m = 5.3$, below which the correlation becomes diffused. This can be explained by the fact that in the distance range of Uppsala and Kiruna from Persia, which is from $25°$ to $50°$, the Q-value for normal shocks is about 6.6. Because the minimum amplitude of the P-phases reported in the bulletins of these stations is never less than 0.1 microns, and also because the periods associated with these phases do not exceed about two seconds, Uppsala and Kiruna cannot discriminate Persian events of m_B of less than about 5.3.

4.6 Semi-empirical assessment of magnitude

The data available on surface- and body-wave magnitudes (about 600 pairs in all) is sufficient to allow the derivation of an empirical relationship between m and M for shallow depth earthquakes. For the whole region the regression gives:

$$m = 0.62(M) + 2.30 \tag{4.4}$$

Figure 4.9. Comparison between Kiruna and Uppsala surface-wave magnitudes for earthquakes in Iran. Period of observation July 1951 to April 1977.
$M_{KIR} = 0.35(0.14) + 0.98(0.02)M_{UPP}; \quad n = 150; s = 0.12,$
$M_{UPP} = -0.26(0.15) + 1.01(0.02)M_{KIR}; \quad n = 150; s = 0.14.$

Figure 4.10. Comparison between Moscow and Uppsala surface-wave magnitudes for earthquakes in Iran. Period of observation, 1958 to 1976.
$M_{MOS} = 0.59(0.18) + 0.91(0.03)M_{UPP}; \quad n = 106; s = 0.15,$
$M_{UPP} = -0.47(0.21) + 1.07(0.04)M_{MOS}; \quad n = 106; s = 0.17.$

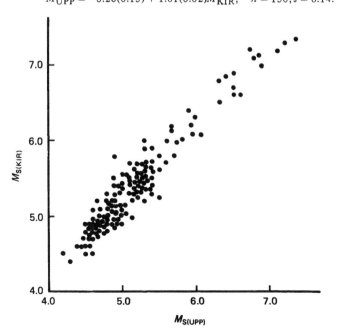

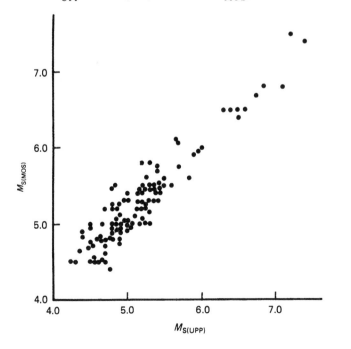

which is in fairly good agreement with the relationship recommended at the IUGG Assembly in Zurich in 1967 and with the relationships derived for the Eastern Mediterranean (Båth 1979b: 30). This formula under-estimates m by more than 0.4 for earthquakes in eastern Iran, perhaps a genuine effect of multiple mechanism, and over-estimates m for events in the Kopet Dagh and northeastern Iran. However, the relatively poor quality of m (see p. 125) precludes the use of equation (4.4) to identify regional mechanisms and radiation features, but allows the assessment of one of these magnitudes when the other is known.

Another empirical way in which the magnitude of an earthquake can be assessed is by making use of the number of stations that recorded the event and the maximum distance at which the earthquake was registered.

If we assume a world-wide network of uniformly distributed stations equipped with the same type of, say, pre-1958 seismographs, then the average maximum distance D_{max} at which a shallow earthquake will be recorded should be a measure of the surface-wave magnitude of the event. This is very roughly equivalent to saying that stations at D_{max} will write barely legible amplitudes so that the term $\log(A/T)$ in the magnitude formulae will attain its minimum value, on average, the same for all stations at D_{max}. For these stations, therefore, M_S will not be a function of D_{max} alone, or, since the distribution of stations is uniform, of the number of stations N_s at which the event was recorded.

Such a perfect distribution and uniformity of seismographs is hardly realistic. Nevertheless, there is evidence that as a first approximation M_S may be obtained from D_{max}, N_s or from a combination of both.

For instance, Gutenberg observed that earthquakes of

Figure 4.11. Comparison between Kiruna/Uppsala and USGS body-wave magnitudes for earthquakes of the period 1960–80 in Iran.

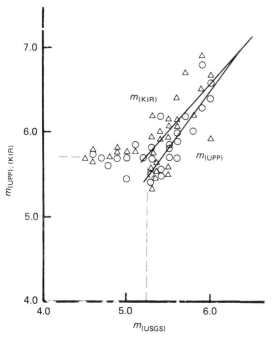

class a and b ($M \geqslant 7.0$) were recorded at all stations, presumably for the period prior to 1948. Class c ($6.0 \leqslant M \leqslant 6.9$) was recorded up to distances of 90°, class d ($5.3 \leqslant M \leqslant 5.9$) up to about 45°, and class e ($M \leqslant 5.3$) not beyond 10° (Gutenberg & Richter 1965). Also, Kirnos et al. (1961) assessed the maximum distance at which different types of pre-1958 seismographs can write a useful record with legible P- and S-phases (cf. Båth & Duda 1979), figure 4.12. Kárník, Rothé and Miya-

Figure 4.12. (See text for explanation.) Key:
ZL = Zöllner, horizontal pendulum, mechanical registration, magnetic damping.
CA = Cancani, vertical pendulum, mechanical registration.
ML = Milne, horizontal pendulum, optical registration, without damping.
GT = Galitzin, heavy horizontal pendulum, mechanical registration, magnetic damping.
RP = Rebeur-Paschwitz, horizontal pendulum, optical registration, without damping.
RE = Rebeur–Ehlert, critical horizontal pendulum, optical registration, without damping.
ZR = Zöllner–Repsold, horizontal pendulum, optical registration, without damping.
OB = Omori-Bosch, horizontal pendulum, mechanical registration, without damping.
GS = Galitzin seismograph, galvanometric registration, magnetic damping, horizontal & vertical.
GK = Standard seismograph, c. 1950.
GR = Magnitude limits from Gutenberg & Richter 1965.

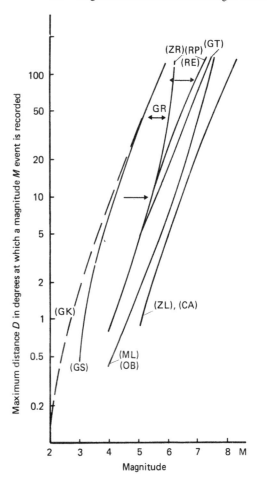

mura in fact used N_s to estimate magnitudes (Kárník 1968, Rothé 1969, Miyamura 1976a, 1976b, Alsan et al. 1975).

In order to assess the magnitude of earthquakes for which there is insufficient instrumental data, we attempted to find a relation between the magnitude of an event and the number of stations N_s and distance D_{max} at which it was recorded. The relationship sought is of the form

$$M_S = a + b \log(N_s) + c \log(D_{max}) \qquad (4.5)$$

where M_S has been estimated from amplitude–period data and N_s and D_{max} can be taken from the ISS/ISC bulletins.

Because the number of stations in the ISS/ISC has been changing continually since 1913 (see figure 4.7), and also because the overall sensitivity of the world-wide network of seismographs has changed at least four times in this century, it was found necessary to determine the constants in equation (4.5) at intervals of a few years and also to increase the number of events by enlarging the area of study to comprise the region between 24° to 40°N and 20° to 90°E.

The results of the analysis show that M_S may be assessed from D_{max} and N_s, but with a precision of not less than 0.4 units of magnitude, much smaller than that which Rothé claims to have achieved for the period 1953–65. Taking more than one time interval for the period 1917–30, the constants in equation (4.5) were found to be:

$a = 2.70, \quad b = 0.40, \quad c = 1.4 \quad$ for $4\frac{3}{4} \leqslant M \leqslant 6.0+$;
for the period 1931–38,

$a = 3,38, \quad b = 0.93, \quad c = 0.33, \quad$ for $4.0 \leqslant M \leqslant 6.0$;
for the period 1939–48,

$a = 3.54, \quad b = 0.72, \quad c = 0.39, \quad$ for $4\frac{3}{4} \leqslant M \leqslant 6.0$.

For the period 1949–62 these constants vary rapidly with time and they have to be taken at intervals of not more than two years, the value of c decreasing to almost zero. For the remaining period, 1963–1976,

$a = 1.73, \quad b = 1.57, \quad c = 0, \quad$ for $3.5 \leqslant M \leqslant 5.5$.

Figures 4.13 to 4.16 show the error between calculated from

(4.5) and estimated surface-wave magnitudes. It is of interest to note that with time, as the sensitivity of the world-wide network increases, the contribution of D_{max} to the assessment of M_S, and also the maximum magnitude that can be assessed by (4.5), decreases, the former to zero and the latter to about 3.5. A similar trend has been observed for other parts of the Middle East, but the constants a and b differ so that the numerical values given above apply only to the region of Persia.

4.7 Assessment of magnitudes of historical events

The magnitude of early earthquakes may be assessed from the size of the area over which the shock was felt, the degree of damage wrought in the epicentral area of the event, and very roughly from the duration of aftershock sequence which can then be calibrated against macroseismic information

Figure 4.14. Comparison between estimated magnitudes M_S and those calculated from $M = 3.38 + 0.93\log(N_s) + 0.33\log(D_{max})$, for the period 1931–8.

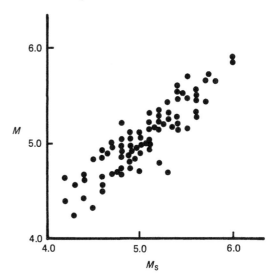

Figure 4.13. Comparison between estimated magnitudes M_S and those calculated from $M = 2.70 + 0.40\log(N_s) + 1.14\log(D_{max})$, for the period 1917–30.

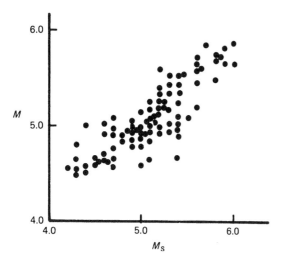

Figure 4.15. Comparison between estimated magnitudes M_S and those calculated from $M = 3.54 + 0.72\log(N_s) + 0.39\log(D_{max})$, for the period 1939–48.

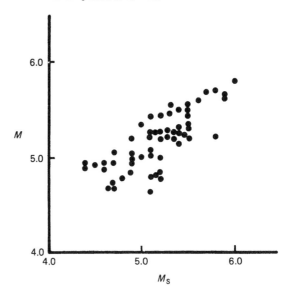

about similar twentieth-century earthquakes for which instrumental data is available.

The size of the area over which an earthquake was felt can be expressed in terms of the radius of perceptibility of the shock, (r_3). This is the mean epicentral distance of the limits of the area within which the shock was felt with an Intensity equal to, or greater than about III (MM). It can be shown that provided the prevailing differences in the depth of focus, regional crustal structure, and low-velocity superficial geology in the region are not markedly large, (r_3) may be taken as a rough measure of the radiated seismic energy, and consequently as a measure of the magnitude of the event, M (Sponheuer 1960). As a matter of fact, (r_n) retains this property for any of the lower levels of an Intensity assessed by a reasonable criterion or law, and it is often used as a means for estimating magnitude.

In our case, because of the necessity to maintain the same criteria in the definition of the macroseismic parameters for events that occurred before as well as after 1900, their homogeneous assessment requires that the radius of perceptibility should be taken at a mean epicentral distance corresponding to an Intensity IV± rather than III (MM). The relatively low density of habitation in certain parts of the region under study, and the overall coarse and uneven system of communications dictate a higher than normal threshold of perceptibility level, which at IV± (MM) was found to be adequate.

With the radius of perceptibility (r') re-defined as the mean epicentral distance at which the shock was felt with an Intensity IV± (MM), it can be shown that there is a distinct

Figure 4.16. Comparison between estimated magnitudes M_S and those calculated from $M = 1.73 + 1.57(N_s)$ for the period 1963–76.

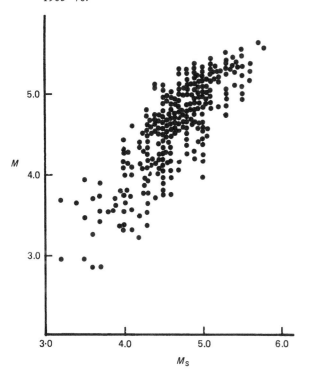

correlation between magnitude $M = M_S$ and (r') in kilometres expressed by:

$$M = -0.67 + 2.87 \log(r') \qquad (4.6)$$

which is shown in figure 4.17. This figure is based on just over 300 shallow depth twentieth-century earthquakes for which both M and an indication of (r') are known. It shows that the spread of the magnitude, as one would expect, is rather large, with a standard deviation of 0.3, but it is by no means unacceptable. Moreover, this spread does not increase with increasing data-points, suggesting that equation (4.6) shows a definite trend and that the real spread is probably not larger than indicated by the available data.

An improvement in the estimate of M can be achieved by including in its correlation with (r') the degree of damage wrought to man-made structures in the epicentral area which can be expressed in terms of the epicentral Intensity I_0 (MM). Indirectly this takes care of scaling effects on (r') due to differences in focal depth $(0 \leqslant h \leqslant 60 \text{ km})$ and variations of the attenuation.

Magnitude scales based solely on macroseismic parameters such as I_0 and (r_3) have been derived for some parts of the world. In Japan, the scale derived by Kawasumi (1951) and by the Central Meteorological Observatory (1952) for events of the period 599 to 1949, i.e.:

$$M_K = I_J = 2.33 + 2 \log(r_3) - 0.0167(r_3)$$

was in use up to the late 1940s, subsequently calibrated against magnitudes estimated by Gutenberg and Richter, M_{GR}, giving $M_{GR} = 4.85 + 0.50(M_K)$, (Kawasumi 1954).[29]

For Persia, the 262 earthquakes of this century for which values of M, (r') and I_0 are known, are well fitted by:[30]

$$M = -0.74 + 1.98 \log(r') + 0.28(I_0) \qquad (4.7)$$

Figure 4.18 shows the error between magnitudes calculated from (4.7) and estimated from amplitude–period data. The standard error of this correlation is 0.23, which is better than the data, particularly the low accuracy with which I_0 has been estimated, would entitle us to expect.

Equations (4.6) and (4.7) may therefore be used to assess the magnitude of historical events for which (r') alone, or both (r') and I_0 are known. These two calibration formulae give estimates of M with an overall accuracy of one-quarter of a magnitude unit or better. The apparent advantage of using (4.7) for historical events is that accurate (r') values are not needed. An error of 30% in the radius of perceptibility would produce less error in the estimated magnitude than an error of one degree in Intensity. On the other hand, this puts great importance on an accurate assessment of the Intensity I_0, which can be a matter of uncertainty for some historical events.

The possibility of making use of the epicentral Intensity alone to estimate magnitudes has been considered by many authors (Gutenberg & Richter 1956b, Lee 1958, Kárník 1968, Toppozada 1975). Attempts have been made to use I_0 to determine magnitudes of historical events in Greece, Italy, China and elsewhere. However, we should regard these with suspicion. Estimates of magnitude based solely on the conversion of Intensities into magnitudes are notoriously unreliable, and the very last resort for assessing early magnitudes. In our case, the best fit of the 511 data-points of M versus I_0 for shallow earthquakes of the last 79 years gives:[31]

$$M = 0.77(I_0) - 0.07 \qquad (4.8)$$

However, the scatter is so large that one magnitude estimate could be associated with several different Intensity levels. This is shown in figure 4.19 where mean values of magnitude are represented by full circles, and the spread of data above and below the mean is indicated by vertical error bars. The fact that the spread becomes larger for those Intensities for which more data is available, suggests that the real spread is probably even larger than shown for the last 79 years. For the first 30 years of this century the standard deviation of M is 0.2. Ten years later it reaches 0.3, and by the end of 1978 it exceeds 0.6.

Figure 4.20 shows a plot of magnitude versus Intensity I_0 for Persia and for other regions. From this figure we notice that for all practical purposes these formulae give similar results, except for $I_0 \leqslant V+$ where equation (4.8) predicts somewhat lower magnitudes. This is mainly due to the differ-

ent magnitude scales (M_S, M_L, m, M_{GR}) used in the derivation of the other magnitude–Intensity formulae.

The next problem we have to resolve is that of estimating the radii of perceptibility of the historical events. The use of equations (4.6) and (4.7) for the assessment of magnitudes of earthquakes before 1900 requires a knowledge of the radius of perceptibility of these events. For many of the historical earthquakes, this value (r') is not known. Using an Intensity scale of five grades that quantify reported effects, we have, however, generally been able to assess at least one mean epicentral distance for an Intensity greater than $i = 5$ (corresponding to IV± MM). These distances then have to be converted into a common radius of perceptibility (r'). This can only be achieved by estimating the average rate at which higher Intensities attenuate with distance.

If we define by (i_n) and (i_j) two Intensities associated with a particular earthquake for which we know the mean epicentral distances (r_n) and (r_j),[32] from the data available we find that for all practical purposes:

$$(r_n/r_j) = \exp 1.08(i_n - i_j)$$

Figure 4.17. The relation between surface-wave magnitude and radius of perceptibility for shallow earthquakes during the period 1903–77, i.e. $M = -0.668(0.14) + 2.874(0.066)\log(r')$. For magnitudes of less than about 4, the energy class TKSE-scale has been used ($K = 1.8(M) + 4.3$).

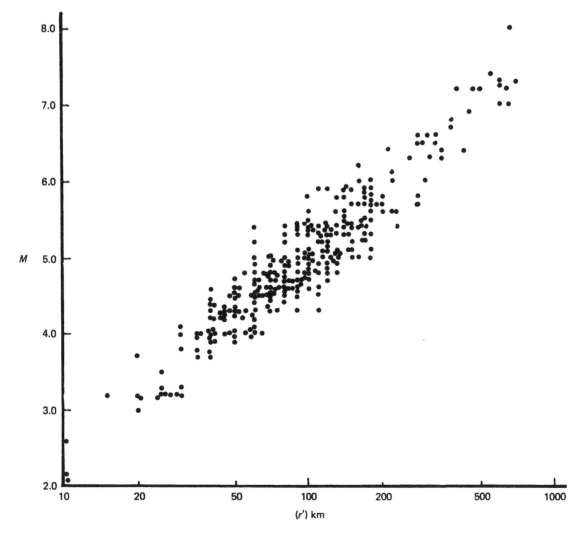

Figure 4.18. Comparison between estimated magnitudes M_S and those calculated from $M = -0.74 + 1.98\log(r') + 0.28(I_0)$ for earthquakes in the period 1903–78.

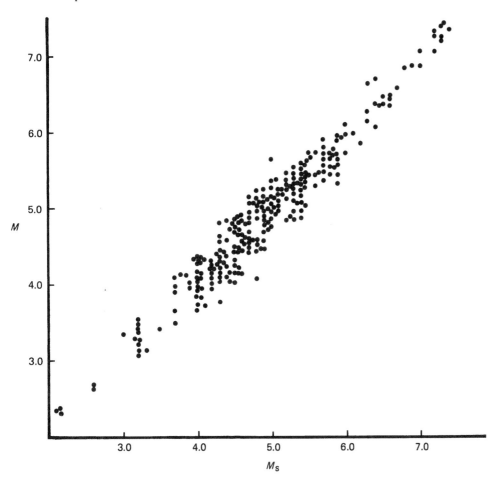

Figure 4.19. The relation between magnitude of shallow earthquakes and epicentral intensity: $M = 0.77(I_0) - 0.07$ for the period 1903–78. Horizontal bars and figures show the spread of the data and number of observations per Intensity used.

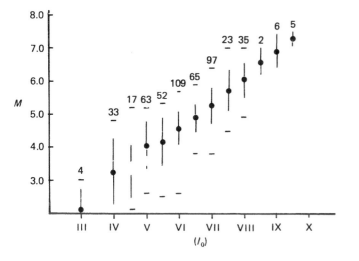

Figure 4.20. Comparison between different $(M - I_0)$ relationships.

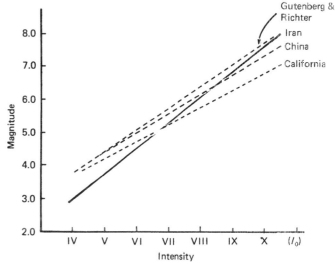

or more specifically, that if (r_n) is the radius of Intensity (i_n), then the radius of the next lower Intensity (i_{n+1}) will be given by:

$$r_{(n+1)} = 3.0 \, r_{(n)} \tag{4.9a}$$

The next lower Intensity radii are given by:

$$r_{(n+2)} = 8.6 \, r_{(n)} \tag{4.9b}$$

and

$$r_{(n+3)} = 20.2 \, r_{(n)} \tag{4.9c}$$

Equations (4.9a) and (4.9b) are based on 170 pairs of radii.

Thus, equations (4.9) may be used to obtain estimates of the radius of perceptibility, $(r') = r_{(5)} = $ IV$\pm$ (MM), of earthquakes for which we have at least one known radius of Intensity higher than $i = 5$.

If we now turn to the estimation of (r'), and at the same time of its corresponding epicentral Intensity (i_0), we have to consider the rate at which Intensity dies out with distance. Assuming that Weber–Fechner's law is applicable (Sponheuer 1960), i.e. that the Intensity decay between any two locations is proportional to the logarithm of the energy density change, and also that the decay of the energy density is due to both geometric spreading and absorption, a simple attenuation relationship may be expressed by:

$$(i - i_0) = -a + b(r) + c \log(r) \tag{4.10}$$

By fitting this equation to the radii of 82 isoseismals of 26 pre-1900 shocks for which we have three or more radii per event, we find that $a = 2.32$, $b = 0.0012$, and $c = 2.10$. The standard deviation of (r_i) for this set of data is, for all values of $(i - i_0)$, almost constant and equal to 34% of the mean. The use of a much larger body of data (of 239 radii from all pre-1900 earthquakes) gave almost identical values for the constants in (4.10), i.e. $a = 2.29$, $b = 0.0013$, and $c = 2.07$, with a standard deviation of (r_i) of 38% of the mean.

Attenuation relationships similar to (4.10) relating Modified Mercalli Intensities (I) and epicentral distance (r) have been recently published for some parts of the world. However, because of the difference in the structure of the Intensity scales used, these relationships cannot be compared with (4.10) unless, of course, a scaling factor is sought to convert our Intensity (i) into (I).

But both the (MM) and our simplified (i) Intensity scales, as well as any other such scale, have been conceived in terms of the same observable qualitative effects of earthquakes. Their main difference lies in the number of Intensity grades chosen to describe the whole spectrum of these effects. Consequently, there must be a scaling factor which will stretch out or compress uniformly the gradings of one scale to make it fit the range of the other. And if the grades, say of the MM scale, fortuitously approximate an even progression of some kind, so will the reduced grades of the other scale.

As we have seen, our simplified Intensity scale consists of five grades that quantify effects that have been observed or reported in documentary sources. In the (MM) scale these range between Intensities IV$\pm$ and XI. This then implies that as a first approximation we may take the value of the scaling factor at 1.5, or more explicitly that $i_5 = $ IV, $i_4 = $ V+, $i_3 = $ VIII, $i_2 = $ VIII+, and $i_1 = $ X, so that equation (4.10) may now be written down in terms of the (MM) scale as follows:

$$(I_0 - I) = -3.44 + 0.002(r) + 3.10 \log(r) \tag{4.11}$$

Figure 4.21 shows a plot of equation (4.11). Three other curves based on twentieth-century data from the San Andreas, Cordilleran and Eastern USA provinces, are also shown for the sake of comparison (Howell & Schultz 1975). From this figure we notice that up to an epicentral distance of approximately 160 kilometres, the attenuation law derived for historical events in Persia (4.11) fits well that deduced for the San Andreas region. Beyond that distance, Intensities in Persia attenuate less rapidly than in California in spite of the fact that the former region is more sparsely populated. This slower attenuation in Persia, with coefficients of anelastic attenuation between 0.001 and 0.004 km^{-1}, implies somewhat larger focal depths, azimuthal variations in the Q-values at distances beyond 2°, and most probably a systematic under-estimation of epicentral and over-estimation of lower Intensities. These causes, however, do not alter the observation that in Persia attenuation due to geometric spreading rather than due to absorption accounts for much of the rapid Intensity decay. However, the spread of data is so broad that this is not a sensitive method of discriminating between individual decay processes.

Figure 4.21. Average attenuation characteristics of Intensity of shallow earthquakes in Persia; horizontal dashed lines indicate spread of data-points.

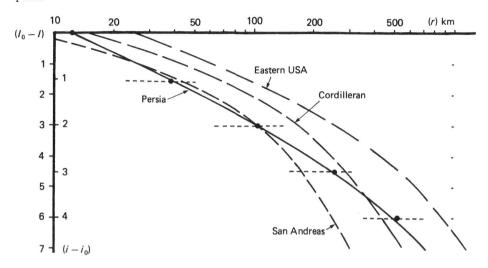

Table 4.1. *Network of undamped Milne seismographs in operation: 1899–1912*[1]

Station	Code	Period of operation	Distance from Iran (D) and azimuth of station (θ) in degrees		Number of events in the region recorded
			D	θ	
Adelaide	ADE	1909–1912	104	123	1
Ascension Isl.	asc	1910–1912	76	251	1
Azores (Ponta Delgada)	PDA	1903–1912	64	300	1
Baltimore	bal	1901–1911	107	345	1
Batavia	DJA	1899–1909	63	117	1
Beirut	bei	1904–1912	16	282	10
Bidston	bid	1901–1912	46	316	15
Bombay (Colaba)	BOM	1899–1912	21	123	8
Bromwich, West	bro	1909–1912	45	314	1
Buenos Aires	BAA	1908	125	248	0
Calcutta (Alipore)	CAL	1899–1912	32	99	7
Cape of Good Hope	CGH	1899–1912	74	210	10
Cheltenham USA	CLH	1904–1905	96	324	0
Christchurch	CHR	1901–1911	131	123	0
Cocos (Keeling Isl.)	CCK	1911	60	130	0
Colombo	COC	1906–1912*	35	131	2
Cordova (Pilar)	PIL	1899–1909	128	253	0
Cork	cor	1912	47	315	0
Edinburgh	EDI	1901–1912	46	319	8
Eskdalemuir	ESK	1909–1912	46	318	2
Fernando Noronha	fer	1911–1912	89	265	0
Guildford	gui	1910–1912	44	313	1
Haslemere	has	1906–1912	44	312	5
Helwan	HLW	1899–1912	19	270	10
Honolulu	HON	1901–1912	118	34	0
Huelva (Rio Tinto)	hue	1911–1912	49	295	1
Irkutsk[†]	IRK	1901–1909	41	45	16
Kew	KEW	1899–1912	44	313	9
Kodaikanal (Madras)	KOD	1899–1912	31	130	4
Lima	LIM	1907–1911*	131	283	0
Malta (Valetta)	mal	1906–1912	33	288	3
Mauritius	MRI	1899–1912	52	175	5
Nikolaev[†]	nik	1902–1908	23	318	4
Paisley	pai	1902–1912	47	319	2
Perth	PER	1901–1911	87	131	0
Porto Rico	por	1904–1905*	104	302	0
San Fernando	SFS	1899–1912	49	293	12
Seychelles (Mahe)	sey	1911–1912	36	178	0
Shide	shi	1899–1912	44	312	16
Sitka	SIT	1904	91	5	0
St Helena	sth	1911*	75	239	0
St Vincent Cp. Verde	SVI	1910–1912	75	274	0
Stonyhurst	STO	1909–1912	45	316	3
Strasbourg[†]	STR	1899–1904	38	309	0
Sydney	SYD	1906–1912	112	117	0
Tashkent[†]	TAS	1902–1906	15	48	2
Tiflis[†]	TIF	1903–1911	12	325	30
Tokyo	TOK	1899–1910	69	60	0
Toronto	TNT	1899–1912	93	328	1?
Trinidad (St Clair)	TRN	1901–1912*	105	293	0
Victoria	VIC	1899–1912	99	358	0
Wellington	WEL	1900–1912*	132	123	0

[1] See figures 4.7 and 4.8.
[†] Stations included in the BAAS Circulars but reporting additional data in their own station bulletins. IRK and TIF trace amplitudes are occasionally in error by a factor of two.
*Discontinuous data reporting.

Table 4.2. *Comparison of magnitudes determined from damped instruments (M_S) and undamped Milne seismographs (M_S^*)*

Date	Time (GMT)	M_S	$n^\dagger$	M_S^*	n	
1903 Feb. 9	0518	5.6	(1)	5.6	(2)	
Mar. 22	1435	6.2	(1)	6.1	(11)	
Jun. 24	1656	5.9	(1)	5.8	(6)	
Aug. 6	0350	6.0	(2)	5.8	(8)	
Sep. 25	0120	5.9	(1)	5.8	(7)	
1904 Nov. 9	0328	6.4	(2)	5.6	(1)	
1905 Jan. 9	0617	6.2	(2)	6.0	(7)	
Jun. 19	0127	6.0	(2)	5.5	(4)	
1906 Jul. 6	0045	5.3	(3)	5.8	(1)	i
1907 Mar. 29	0738	4.8	(2)	5.1	(1)	
Mar. 31	1414	5.1	(2)	5.4	(2)	
Apr. 10	0938	5.0	(1)	5.2	(2)	
Jul. 4	0921	5.7	(5)	6.1	(9)	i
Nov. 4	1344	5.0	(1)	4.9	(2)	
1908 Jul. 26	0332	5.3	(2)	5.1	(1)	
1909 Jan. 23	0248	7.4	(13)	7.2	(16)	
Apr. 11	0402	5.8	(6)	6.1	(7)	
Nov. 1	0916	5.3	(5)	5.5	(5)	
1910 Dec. 4	1402	5.4	(4)	5.8	(1)	i
1911 Apr. 18	1814	6.4	(8)	6.7	(19)	
1912 Feb. 24	1436	5.3	(2)	5.6	(2)	
1913 Mar. 24	1034	5.8	(1)	5.5	(6)	

$^\dagger n$ = number of stations used.
i indicates events probably of intermediate depth.

Table 4.3. *List of stations used reporting amplitude–period data: 1897–1962*

Station	Code	D	θ	Period investigated[1]	Number of reported (A/T) values[2]	Station corrections ≤1950	Station corrections >1950
Agra	AGR	21	97	(1938–9)agT(1940–)F	12(12)	−0.1	+0.1
Akhalkalaki	AKH	13	321	(1903–11)arutd			
Algiers	ALG	42	291	(1921–31)agT(1932–)F	18(16)	+0.5	
Alma Ata (Vernyi)	AAA	21	51	(1907–12)Ar(1927–36)AgT(1944–8)Ag(1951–)F	10(4)		
Andijan	ANR	17	54	(1929–36)AgT(1944–8)ag	16(3)		
Ashkhabad	ASH	7	30	(1947–8)Ag(1950–)dF	14		
Athens	ATH	25	292	(1900–11)Ar(1909–27; 1950–)agTd	28(14)b	−0.1	
Baku	BAK	9	340	(1906–11)arut(1927–39)AgT(1944–8)Ag(1951–)AgT	115(13)	−0.1	
Balakhani	bal	9	340	(1903–11)aru			
Barcelona	FBR	42	298	(1919–25)agtdF(1926–35)agTF(1949–)F	4(21)	−0.1	
Batum	btm	14	318	(1903–10)Aru			
Beograd	BEO	29	306	(1911–14)agTd(1923–54)AgT(1955–)dF	87(15)b	+0.0	
Bokaro	BOK	29	113	(1957–)agtF	2(18)		−0.1
Bombay	BOM	21	123	(1923–56)agt(1957–)F	103(14)	+0.2	+0.3
Borjom	BOR	13	322	(1903–11)aru			
Budapest	BUD	31	311	(1902–6)arud(1907–14)Ar(1925–54)agtd	2(18)		
Calcutta	CAL	32	99	(1938–54)agTd(1955–)F	14(12)	−0.5	−0.3
Catania	CAT	32	291	(1897–1907)arut(1908–11)agt(1912–14)F	4(10)		
Catanzaro	ctz	31	293	(1905)ArT			
Chatra	CHA	29	92	(1953–7)agt(1958–)F	2(14)		−0.2
Colombo	COC	35	131	(1938–50)ar			
Copenhagen	COP	37	322	(1927–8)AgT(1929–)T	10(15)	+0.1	
De Bilt	DBN	40	314	(1904–7)Aru(1908–44)AgT(1945–)agt	96(19)	−0.1	+0.1
Dehra Dun	DDI	21	88	(1938–50)art(1956–)agtF	16(12)		
Delhi	NDI	20	94	(1943–50)agt(1950–)F	10(16)	−0.3	−0.3
Derbent	dbt	11	337	(1906–11)arud			
Eger (Cheb)	CHE	36	313	(1942–54)agT	25(13)	+0.1	+0.1
Erivan	ERE	11	320	(1944–8)Ag			
Feodosia	FEO	19	317	(1947–8)Ag			
Florence (Ximen.)	FIR	35	302	(1902–3)Ar(1904–14)Artd(1924–6)agt	6(9)		
Frunse	FRU	20	50	(1927–36)agT(1944–8)Ag	9(4)	−0.1	
Göttingen	GTT	37	315	(1903–14)AgT(1922–40)agTd	67(18)B	+0.1	
Granada (Cartuja)	CRT	47	293	(1907)ar(1908–13)agtF(1914–16)AgT(1917–53)agtF	1		
Grozny	GRO	13	332	(1944–8)Ag	4(3)		
Hamburg	HAM	38	319	(1902–3)Aru(1907)Ar(1908–15)AgT(1919–31)agT(1935–)agtdF	69(14)b	−0.3	
Hohenheim	HOH	37	310	(1905–7)ArT(1905–13)agT(1914–18)AgT(1925–33)agtdF	7(21)b	+0.1	
Hurbanovo (Ogyalla)	HRB	31	311	(1949–)agtF	9(11)	−0.1	−0.0
Hyderabad	HYD	26	118	(1938–60)agt(1961–)F	71(13)	−0.1	−0.1
Irkutsk	IRK	41	45	(1912–14)AgT(1925–39)agT(1944–8)Ag(1950–)	58(13)	−0.3	−0.2
Ischia	PDI	31	297	(1897–1903)Arud			
Jena	JEN	36	314	(1905–13)AgT(1926–52)agT	104(16)B	+0.1	+0.2
Kabansk	KAB	43	46	(1906–7)aruT			
Karlsruhe	KRL	38	310	(1926–40)agtF(1950–)F	3(13)	−0.1	
Kew	KEW	44	313	(1951–)agTd	17(12)		−0.0
Kiruna	KIR	41	341	(1951–)AgT	224(16)B		−0.1
Kodaikanal	KOD	31	130	(1938–52)agt(1953–)F	24(15)	−0.0	+0.5
Krakow	KRA	31	316	(1905–8)aruT(1909–15)agt(1960–)	7(12)	+0.1	−0.0
Krasnoyarsk	krs	36	37	(1903–8)aruTd			
Kremsmünster	KMR	34	310	(1900–3)Aru(1907–9)Ar			
Kuchino	KUC	26	339	(1927–35)AgT	39(17)	+0.2	
Leipzig	LEI	36	314	(1902–5)aruT(1906–10)agT(1925–30)AgT(1931–)F	48(17)b	+0.1	
Lemberg (Lvov)	LVV	28	317	(1899–1902)Aru(1951–)agtF	5(9)		
Leninakan	LEN	12	320	(1946–8)Ag(1949–)agtdF	7(12)		
Leningrad	LNN	32	338	(1926–8)AgT	1		
Ljubljana (Laibach)	LJU	33	307	(1903–7)arut(1913)agt(1925–39)agt	18(16)b	−0.0	
Lome	LOM	55	254	(1957–)F	4(16)b		
Makeyvka	MKY	20	327	(1926–8)AgT	6(15)	+0.2	
M'Bour	MBO	66	273	(1957–)agTF	8(18)b		−0.3
Mileto	MLE	31	293	(1909)ArT			
Mineo	MNE	32	290	(1909–)ArT			
Moncalieri	MNC	38	304	(1906–8)AruT(1927–9)AruT			
Moscow	MOS	26	339	(1935–9)AgT(1944–51)Agd(1952–)agT	130(11)	+0.1	+0.2
München	MNH	36	309	(1907)Ar(1908–14)agTd(1922–4)agtd	2		
Padova	PAD	35	305	(1903–9)Arut(1911–13)F	6(8)		
Paris	PAR	42	309	(1910–16)agT(1921–36)AgT(1937–)agTF	60(18)b	+0.0	+0.2

Table 4.3 (*cont.*)

Station	Code	D	θ	Period investigated[1]	Number of reported (A/T) values[2]	Station corrections ≤1950	>1950
Piyatigorsk	PYA	15	328	(1944−8)Ag			
Port Blair	PBA	41	111	(1958−)agtF	1		
Potsdam	POT	36	317	(1897−1901)Ar(1902−9)AgT(1910)T(1911−53)agT(1954−)F	43(15)b	−0.1	
Praha	PRA	34	313	(1940−54)agTd(1955−)	57(16)b	+0.1	+0.1
Pruhonice	PRU	34	313	(1960−)agt	18(15)		+0.1
Pulkovo	PUL	32	337	(1910−14)agT(1927−39)AgT(1950−)agtd	97(14)	+0.1	+0.0
Quatro Castello	QCI	35	302	(1903−7)AruTd			
Raciborz	RAC	32	315	(1953−)agrd	2(22)		
Ravensburg	RAV	37	308	(1926−33)agtd	2		
Reykjavik	REY	55	329	(1927−50)arTdF(1952−)agtF	2		
Rome (R.P.)	RDP	37	299	(1903−9)ArT(1912−14)agrd	10(9)		
Samarkand	SAM	13	50	(1929−36)agtF(1944−8)Ag(1952−)dF	10(10)		
San Fernando	SFS	49	293	(1915−19)agTd(1931−4)arut(1934−5)agt	18(12)	−0.2	
Sarajevo	SAR	30	303	(1905−10)ar(1911−12)ArT(1921−38)AgTd	14(12)	−0.1	
Semipalatinsk	SEM	27	39	(1944−8)Ag			
Shillong	SHL	34	91	(1956−)agtF	5(17)		−0.2
Simferopol	SIM	20	316	(1936−9)agrd(1947−8)AgF	2		
Simla	SMI	20	86	(1905−6)ar(1907−)arT			
Skalnate Pleso	SPC	30	314	(1949−54)agt	10(14)	+0.0	−0.0
Sochi	SOC	16	320	(1946−8)Ag(1949−)F			
Sofia	SOF	26	303	(1905−11)AruT(1935−46)agtd	7(10)		
Stalinabad (Dushambe)	DSH	14	57	(1944−8)Ag(1954−)F	5(9)		
Strasbourg	STR	38	309	(1897−1904)aru(1905)ArT(1906−16)Agt(1919−35)agT(1936−)agt	89(17)b	−0.1	−0.0
Stuttgart	STU	37	310	(1930−)agtdF	1		
Sverdlovsk (Ekaterin)	SVE	25	8	(1906−11)Aru(1927−39)AgTd(1944−8)Ag(1950−)agt	112(15)	−0.0	+0.1
Tananarive	TAN	51	188	(1944−)agtF	6(18)	+0.3	+0.0
Taranto	TAR	31	297	(1909)ArT			
Tashkent	TAS	15	48	(1902−10)Ard(1927−39)AgTd(1944−8)Ag(1946−)agt	206(11)b	−0.1	−0.2
Tchimkent	TCH	16	46	(1934−6)agT(1944−8)Ag	2(4)		
Tchita	tst	47	46	(1904−7)aru			
Tiflis	TIF	12	325	(1900−11)Aru(1905−16)artF(1933−7)AgT(1952−)agT	67(10)	+0.2	+0.1
Trieste	TRI	34	306	(1899−1907)Aru(1911−39)agtdF	10(9)		
Uppsala	UPP	37	330	(1906−)AgT	272(16)B	+0.2	+0.1
Vladivostok	VLA	61	55	(1930−9)agT(1944−8)Ag(1950−)agtF	15(13)	−0.2	
Warszawa	WAR	31	320	(1946−52)agT(1953−)agtF	39(13)b	−0.2	−0.2
Wien	VIE	33	311	(1909−12)AgT(1913−20)agtd(1927−41)agtdF	8(18)	−0.1	
Yalta	yal	20	315	(1944−8)Ag			
Yurev (Dorpat)	TTU	32	333	(1897−1907)Aru			
Zagreb (Agram)	ZAG	32	306	(1906−8)aruT(1913, 1922−55)AgTd	12(10)	+0.1	

Note: In addition, amplitude−period data from Pasadena (PAS), Palisades (PAL), Berkeley (BER), College (COL), Osaka (OSK), Matsushiro (MAT) and Roxburgh (ROX) were used for the larger events.

(1) (Ag), (Ar) and (T) imply that during the period of observation shown in brackets, ground amplitudes, trace amplitudes and periods of maxima respectively are reported regularly. (ag), (ar) and (t) indicate that these quantities are frequently not reported. Discontinuous reporting is shown by (d), and (F) marks periods of observation of little (A/T) information. (u) indicates data from undamped or slightly damped recorders.

(2) Figures in brackets are the mean values of the periods of the maxima reported (in seconds). (B) or (b) indicates that body-wave data are frequently or infrequently reported respectively.

(3) Distance in geocentric degrees and azimuth in degrees East of North, of station with respect to centre of the region (32°N − 54°E).

Table 4.4. *Distance—period relationship*

D (degrees)	2	4	6	8	10	15	20	30	40	60	80	100	140
T_{min} (seconds)	4	5	5	6	7	8	9	10	12	14	16	16	18

Table 4.5. *Modified transcript of Gutenberg's worksheet for the Silakhur earthquake of 23 January 1909 (see figure 4.4).*
Numbers in parentheses are author's additions to Gutenberg's data. Asterisks indicate amplitudes from one component only.

	Surface-wave Amplitude (1)	Period (2)	Station phase (3)		Body-wave Amplitude (4)	Period (5)	log (A/T) (6)	(7)	Q-value (8)	(9)	Body-wave magnitude (10)	Surface-wave magnitude (11)	(12)
(1)	750		Cartuja	P	12	4	0.6		7.0		7.6	7.5	
				S	56	7	1.0		6.6		7.6		
(2)	500 (420)*	(26)	Osaka									7.7	(7.7)
(3)	1000 (2500)	(30)	Wien									7.3	(7.5)
(4)	600 (515)	(17)	Götting.	PH	6 (6)	4 (4)	0.5	(0.2)	6.8	(7.0)	7.3 (7.2)	7.2	(7.4)
				S	50 (45)	11 (11)	0.8	(0.6)	6.2	(6.6)	7.0 (7.2)		
				Pz	(5)	(4)		(0.1)		(6.7)	(6.8)		
(5)	5000		Jena	Pz	4	3	0.4		6.6		7.0	8.0	
				S	35	10	0.8		6.3		7.1		
(6)	1000 (1000)*	(30)	Strasb.	P	10 (8)*	5 (5)	0.3	(0.2)	6.8	(7.0)	7.1 (7.2)	7.4	(7.3)
				PP	10	5	0.3		7.0		7.3		
(7)			Batavia	S	20	6	0.6		7.0		7.6		
(8)	350 (326)	(13)	Uppsala	S	36 (30)*	11 (11)	0.5	(0.4)	6.3	(6.6)	6.8 (7.2)	6.9	(7.4)
(9)	2000 (1740)	(15)	De Bilt	P	8 (8)	4 (4)	0.2	(0.3)	6.9	(7.0)	7.1 (7.2)	7.5	(7.8)
				S	200 (150)*	12 (12)	1.0	(1.1)	6.5	(6.7)	7.5 (7.9)		
(10)	500		Pulkovo									7.0	
(11)	(7780)	(15)	Athens										(8.1)
(12)	(880)*	(16)	Hohenheim	P	(2)	(3)		(−0.2)		(7.0)	(6.8)		(7.7)
(13)	(880)*	(21)	Hamburg	S	(40)	(15)		(0.4)		(6.6)	(7.1)		(7.3)
(14)	(1000)*	(22)	Leipzig	Pz	(2)	(4)		(−0.3)		(6.7)	(6.4)		(7.6)
(15)	(250)*	(16)	Potsdam	PH	(12)	(3)		(0.6)		(6.9)	(7.5)		(7.0)
				S	(46)	(5)		(1.0)		(6.5)	(7.5)		
(16)	(150)*	(14)	Krakow										(6.9)

Table 4.6. *Modified transcript of Gutenberg's worksheet for the Ravar earthquake of 18 April 1911 (see figure 4.5.).*
Numbers in parentheses are author's additions to Gutenberg's data. Asterisks indicate amplitudes are from one component only.

	(1)	(2)	(3)		(4)	(5)	(6)	(7)	(8)	(9)	(10)	(11)
(1)	30		Paris									6.1
(2)	100 (73)	(10)	Hamburg	S	(29)	(8)		(0.6)		(6.7)	(7.3)	6.4 (6.5)
(3)	25		Cartuja									6.1
(4)	15		Belgrade	P	5	5	0.0		6.8		6.8	5.5
				S	10	5	0.3		6.2		6.5	
(5)	18		Jena	Pz	2	5	0.0		6.7		6.7	5.8
				PPz	3	5	0.0		6.8		6.8	
				S	10	12	0.1		6.5		6.6	
(6)	c. 50 (70)	(24)	Göttingen	Pz	2 (2)	5 (5)	−0.1	(−0.4)	6.7	(6.7)	6.6 (6.3)	6.3 (6.5)
				PPz	9 (9)	14 (14)	0.0	(−0.2)	6.7	(6.6)	6.7 (6.4)	
				S	12 (12)	15 (15)	0.1	(−0.1)	6.6		6.6? (6.6)	
(7)	(45)*	(12)	De Bilt									(6.5)
(8)	(61)*	(25)	Uppsala									(6.6)
(9)	(35)*	(20)	Potsdam									(6.2)
(10)	(59)*	(28)	Osaka									(6.7)
(11)	(60)*	(44)	Trieste									(6.1)

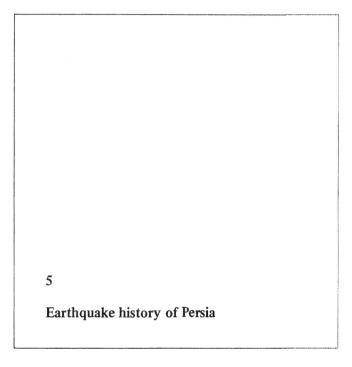

5

Earthquake history of Persia

5.1 General review

The preceding chapters describe the manner in which information about earthquakes in Persia has been sought out and classified, using a variety of sources and techniques. The total number of earthquakes thus identified for the whole period of observation amounts to just over 6000 and figure 5.1 shows the cumulative time distribution of these events.

Before turning to the interpretation of this data, it may be convenient to pause and draw together some comments on the reported seismicity of the periods discussed in chapter 1.

Figure 5.1. Cumulative time distribution of number of earth-quakes in Persia.

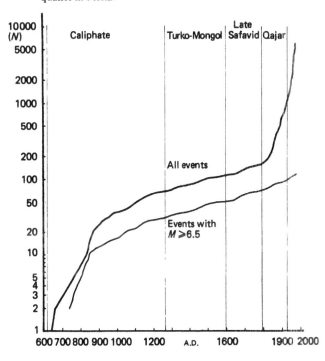

The main factors seen to influence the survival of data are the quality of the contemporary literary record, the prevailing historical circumstances, the geographical location of events and their magnitude. The temporal and spatial distribution of reported earthquakes reflects the combined effects of these conditions.

Figure 5.1 reveals a broad uniformity in the cumulative time distribution of earthquakes recorded up to around 1800. Seen by historical divisions, we identify a comparatively high level of reported activity from the ninth to the twelfth centuries, the core of the Caliphate period, giving way to a generally lower level in the Mongol and Turkoman period. This decline has to be seen in terms of inferior reporting of events in Persian as compared with Arabic sources. Their coverage of earthquakes is erratic and their silence difficult to interpret consistently. In the period to 1258, the small number of places mentioned as affected is a function of the lack of source material of local origin. Our dependence on information reported in Baghdad gives the distribution of earthquakes a bias in favour of Iraq and the western Zagros as opposed to eastern regions, with a concentration of shocks in the area between Baghdad, Hamadan and Mosul. Information is also available for areas along routes most directly connected with Baghdad. Flourishing regional routes are not sufficient to ensure the survival of data on a wider level. This is particularly apparent after the fragmentation of the empire in the eleventh century. From 1065 to 1258, roughly the Saljuq period onwards, macroseismic data are lacking for places not on the Khurasan highway across northern Iran.

A small scatter of places named is also characteristic of succeeding periods, although they are in higher proportion to the number of earthquakes recorded. This demonstrates the value of indigenous local histories for events in more remote rural areas (such as Kuhistan and the Caspian provinces), but the improvement is illusory for it only benefits our knowledge of few earthquakes and the number of specific localities otherwise mentioned remains small. There is a continuing dependence on the main cities for macroseismic data, the distribution of earthquakes following the lines of the more important routes and clustering around the larger towns along them. In the Mongol and Turkoman period, this distribution is fairly even throughout the region. In the seventeenth and eighteenth centuries, there is a strong bias of information for the northwest of Persia. This is the result of a concentration of source material on the region between Van, Erivan and Tabriz. During this period the greater variety of available sources leads to a fuller record of events, which in the eighteenth century are all confined to the western half of our area of study.

This picture given by the documentary sources, of the occurrence of an earthquake worthy of notice somewhere in the region every four to nine years up to 1800, would seem to be reasonably comprehensive if we assume that most of the events mentioned were shocks of moderate or larger magnitude, unless specifically stated to the contrary, and generally affected a considerable number of people.

The tenor of our review of the documentary source material has been that the main factor determining the survival of macroseismic data is not so much the magnitude or frequency of occurrence of past earthquakes, as their historical context or particular geographical location. It is clear, however, that in given historical circumstances the magnitude or size of an earthquake considerably influences its chances of attracting attention. To ensure their being reported, smaller events require a more perceptive recording system, such as the monastery documents, colophons and correspondence materials that are available for other regions like Turkey and the Balkans, or such as is later provided by consular diaries, newspapers and instrumental records. But even larger events, forming our main interest, have different thresholds of perception in different areas and at different times. These thresholds are determined by the factors we have already discussed.

Since we are obliged to study seismic activity in Persia for the whole period before 1800 almost exclusively in terms of the effects of earthquakes in the major urban and provincial centres, an idea of the varying thresholds of perception of events in these places is important. It not only helps to identify the likely size of reported shocks and illuminate the situation in the wider district around the towns, but also gives a useful measure of the extent of apparent seismic quiescence. In some areas such as southeast Iran, a large magnitude earthquake would only have been mentioned if it had specifically affected Kirman, a town whose perception of less destructive or more distant shocks was low. The same applies to other towns, such as Tabas. In other more remote areas, such as south Sistan and the Makran, indications of seismic activity, however great, might only be found in oral legends. Districts such as those of Ray or Nishapur were sufficiently important over a long span of time for details of the more destructive earthquakes to be generally available with some consistency, while in others, such as Shiraz, a comparable level of perception of events was only achieved occasionally and gaps in the record must be interpreted differently. Where historical circumstances were particularly conducive to the survival of data, the available information can be assessed with greater confidence. The record of shocks in Baghdad and Hamadan under the Caliphate is almost independent of their intensity, which was generally low. The sensitive perception of events in Tabriz in subsequent periods also permits a simple distinction between destructive and relatively minor shocks. On the other hand, the absence of such details from Isfahan can be taken as evidence of a genuine quiescence over the period studied. Other areas come into focus at different periods, for example the Herat region and northern Kuhistan from the fourteenth to the end of the sixteenth century, or the Caspian region during the seventeenth century. The absence of any evidence of historical seismicity in western Afghanistan would appear to be significant, especially as it is supported by a similar lack of evidence in the more recent period.

After 1800 we have much fuller information for most parts of the region. Distinction between large and small shocks is generally straightforward, allowing more precision and less chance of distortion in assessing the size and location of events. The bias of information in documentary sources remains on large towns, but this is counteracted by information that can be derived from local sources and field studies. It would be justifiable to regard our data for the last century and a half as an extremely comprehensive sample of the distribution and intensity of seismic activity throughout the area under study.

In chapter 1 we have examined the extent of the survival of information on earthquakes in Persia as part of the literary output of the oriental and, to a lesser extent, occidental historians and travellers. This has involved a discussion of some of the ways in which information has been preserved and of the main characteristics of these recorded data (in historical sources, official archives, press and other reports). We have also seen the implications of some of the deficiencies in the sources and in the system of acquiring and propagating this information, which changed with time due to other factors, such as political and economic developments which affected population densities and communications. These aspects of the source material have a significant bearing on any classification that can be made of the recorded events, and knowledge of these factors gives the investigator an appropriate stand from which to view a changing situation. For instance, the survival of any information over a long period in a remote part of Persia, although not in itself necessarily conclusive, suggests a comparatively significant event. Conversely, lack of information about earthquakes in a region for which there is no lack of source material, can be seen to reflect a period of relative seismic quiescence. Knowledge of the historical background of an event allows the investigator to take into account other more subtle factors such as natural exaggeration of the observer, local traditions, population changes, the attitude of the author towards the importance of what he is recording and cultural changes brought about by other factors that cannot be quantified.[1]

While most of the historical information is by itself sufficiently full to be analysed systematically, the investigator has to calibrate these data against quantifiable information on similar, twentieth-century events. To do this he has to resort to the field study of recent earthquakes and through this process acquire an intimate knowledge of local conditions, which in Persia have not changed much over the centuries until very recently. This is discussed in chapter 2, while chapter 3 and its appended commentaries present a substantial sample of the material thus acquired from both documentary sources and field studies.

In chapter 4 we have combined this macroseismic information with seismological data to arrive at a 'yardstick' with which to classify uniformly all earthquakes in terms of their size (magnitude) and distribution in space and time. Tables 5.1 and 5.2 summarise quantitatively this information which, together with the material contained in this chapter, can now be used to assess the long-term seismic activity in Persia.

5.2 Distribution of seismic activity: qualitative evaluation

The question now arises of how complete is the sum total of this information. Is it likely that areas of relatively high seismicity have passed unnoticed because of their low population density or literary output, thus distorting significantly the delineation of the long-term seismicity? This is a question often raised with regard to historical studies by the seismologist, to which there is no numerically precise answer. The use of statistical methods for testing the fullness of the data which assume stationarity (such as a sample completeness test) will of course defy the purpose of the whole exercise, which among other things aims at disclosing whether earthquake processes are stationary or not. The assumption of Poisson arrivals and that magnitudes and locations of earthquakes are independent of previous occurrences, naturally defies any test designed to show whether seismic activity alternates quasi-periodically between contiguous zones or occurs in clusters. For short-term samples of observations and areas of limited extent where there are some grounds for assuming stationarity, completeness tests may prove useful for what they are worth, but for long-term observations or for large areas the problem is how to establish stationarity *a priori* and discriminate between random events and non-random processes.

The view is often expressed, perhaps too lightly, that population density is the controlling factor in the survival of historical data. Although rather simplistic, this view is generally valid. Figure 2.1 shows the population density distribution in Iran in 1869 and less than a century later. Comparing these figures with those showing the communication lines of earlier periods (figures 1.1, 1.2, 1.4 and 1.5), it is obvious that given an adequate contemporary literary record, it is unlikely that a great, major or even large earthquake with a radius of perceptibility of a few hundreds of kilometres could easily have passed unnoticed. This is more specifically true of the more densely populated regions. Some remote areas in Persia were comparatively less so a few centuries ago, but even semi-desert regions can yield information. In Persia such regions are relatively small and bordered by towns or cities more likely to provide data. Large and major events in these regions, important for the assessment of seismicity and tectonics, can in most cases be identified from the size of the area over which they were felt, even though the exact location of their meizoseismal region may be in doubt.

A general answer to the question set above, therefore, is that it emerges from study of the data that many small and moderate magnitude earthquakes have been missed out, particularly in desert regions. For the whole area and period studied, it is indeed unlikely that all moderate and many large magnitude shocks would have been recorded. It remains very probable that any major or great earthquake has been noted, although not necessarily fully identified. It is reasonable to suppose, therefore, that the available data for the whole region is incomplete for all magnitudes.

However, to get some measure of the extent of this incompleteness, and to see to what extent it distorts the pattern of seismic activity, it is necessary to break down this general picture to look at specific areas and time intervals. Figures 5.2, 5.3 and 5.4 show the areas affected by destructive shocks in Persia during the last thirteen centuries in three consecutive periods, viz., eighth to eighteenth, nineteenth and twentieth centuries respectively. Each figure locates the areas

Figure 5.2. Areas affected by destructive earthquakes during the period AD 700–1799: a = major events; b = large events; c = moderate events. Darker shading shows areas of Intensity $i \leqslant 2$, lighter shading shows $i \leqslant 3$. The following sites marked + are associated with earthquakes before the Islamic period, or with later traditions of seismicity:

At	Ak Tepe	Ks	Kuh-i Sultan
Ch	Churik	Na	Nasratabad
Ga	Ganaveh	Ni	Nisa
Gt	Godin Tepe	Sa	Sagzabad
Gz	Gaud-i Zireh	Sa	Saimareh
Kh	Khunik	Tk	Tang-i Khas
Ki	Kishmar	Zk	Zalzaleh Kuh
Ku	Kuhbanan	Za	Zaribar

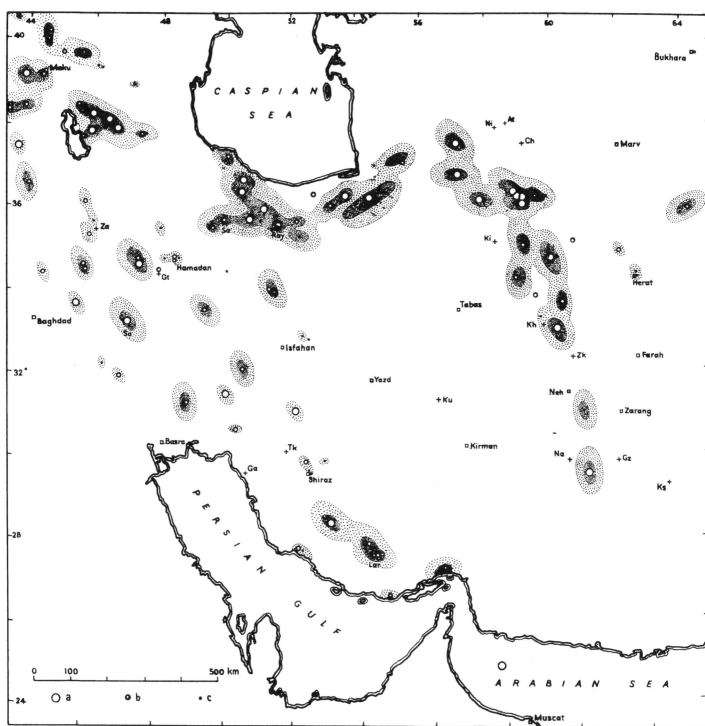

Figure 5.3. Areas affected by destructive earthquakes during the nineteenth century, superimposed on areas affected before 1800.

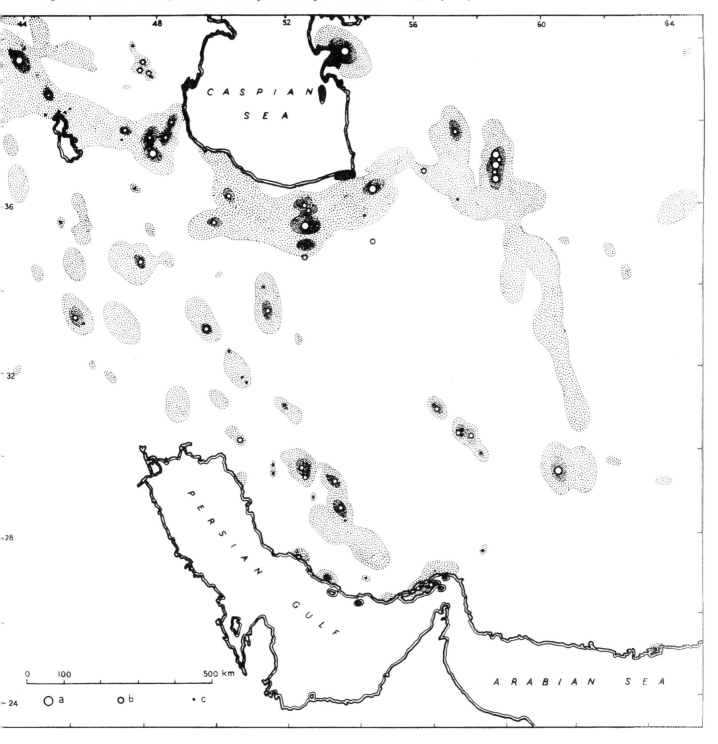

Figure 5.4. Areas affected by destructive earthquakes during the twentieth century, superimposed on areas affected before 1900. Note that this is not a seismic zoning map and it does not suggest that blank areas are unaffected by earthquakes or free from heavy damage caused by smaller shocks.

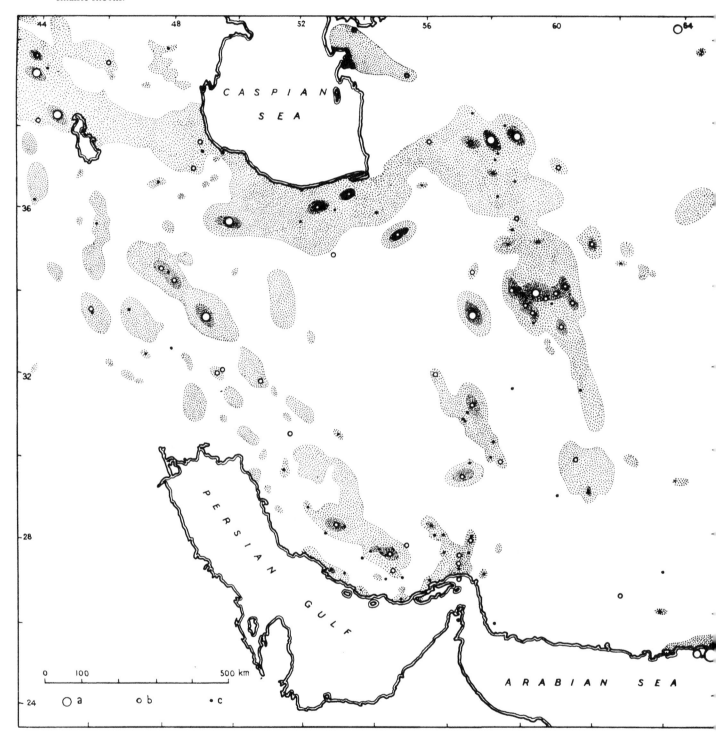

of the strongest effects ($i \leqslant 2$) which in most cases can be taken as identical with the epicentral areas of the events. Some of these areas are necessarily approximate, corresponding to the locations mentioned as more seriously affected, some of which are also shown in the figures that accompany chapter 3. Nevertheless, it is unlikely that the macroseismic epicentres that can be derived from descriptive accounts and from field evidence will be in gross error with respect to the true location of the event, particularly for the larger magnitude shocks where the focal area is correspondingly much larger than in small events. Each figure also shows the areas damaged ($i \leqslant 3$) by all events of the period considered and of the preceding period, thus showing the extent to which seismic gaps in the historical periods are filled in with the passage of time. It is important to emphasise that these three figures show only areas of the highest intensity shaking, associated with the larger earthquakes. It does not therefore follow that the blank areas are free from earthquakes, or from high intensity shaking from smaller magnitude, local shocks.

If we exclude the regions of the Makran and the Kopet Dagh, which have a poor and rather incomplete long-term history, we see a steady improvement of information with time. Data remain incomplete for small and moderate magnitude events up to the early 1900s, but it becomes increasingly unlikely that any major or great earthquake has escaped notice. For some individual zones, completeness up to this level is achieved much earlier, which is even more the case in regions to the west of Persia, i.e. the Eastern Mediterranean and Anatolia.

From the juxtaposition of the historical and twentieth-century earthquakes on figures 5.2 to 5.4, we notice that quite a few of the earthquakes of the last eighty years, after the advent of seismology, have happened in the same places as much earlier events. For instance, with the exception of the Tabas earthquake of 1978, all important earthquakes of this century occurred in regions devastated by historical events.[2] This close parallel of past and recent seismicity both confirms that modern earthquake occurrence is largely consistent with a long-term seismic pattern and equally, indicates that historical events also provide a relevant and representative sample of seismic activity. The two sets of data complement each other.

On the other hand, it also emerges from these figures that the distribution of the larger twentieth-century earthquakes in many areas forms a negative or mirror image of the distribution of the larger historical events. Thus, apparent seismic gaps are revealed both for the historical period and for the twentieth century, particularly for the first half of the 1900s. The incompleteness of any seismic map deduced solely from recent events is evident by comparing figure 5.4 with figures 5.5 and 5.6 prepared by de Ballore (1906: 210) and Sieberg (1932: 812) respectively, and with maps based on secondary sources.[3] As the most interesting events are those which have happened where their occurrence could not be predicted, or, alternatively, the most interesting areas are those where earthquakes could be expected but have not happened, these seismic gaps must be examined.

An overall illustration of the time–space distribution of

earthquakes in the Northern, Eastern and Zagros zones shown in figure 5.9 is depicted in a simplified diagram in figures 5.7 and 5.8. Earthquake epicentres are projected on the parallel running along the Northern zone and on the meridians passing through the two other zones, and plotted as a function of time. The magnitude of each event is shown by the length of an equivalent fault break, derived from equation (5.6) (see below). 'Holes' in the pattern of such plots show seismic gaps in space and time. Since the source dimensions and slip associated with small to moderate magnitude earthquakes are unimportant as compared with those of large or major shocks, the identification of a gap should be based on the long-term occurrence of the larger earthquakes. As a matter of fact, small events do not seem to be pertinent to the prediction of gaps and the occurrence of large earthquakes, particularly in the Zagros and in other zones (p. 152).

A number of seismic gaps revealed in the historical period occur in the more remote areas of Persia, such as the western Lut and further southeast (figures 5.2, 5.3, 5.4, 5.7 and 5.8). Lack of evidence of earthquake occurrence does not, of course, necessarily imply that no earthquake occurred. It simply means that there was no major or great event and the lack of information for moderate or large shocks can be attributed to a number of different factors, as we have seen. It is quite natural that little macroseismic information should be preserved for desert or semi-desert areas with very low population density and poor literary output and it is thus not to be expected that a moderate magnitude historical event in Turud or Khuvar, for example, would have been documented in the oriental sources, nor that large earthquakes in southern Sistan or Baluchistan and the Makran would be widely reported.

In the case of the Kirman area, the silence is rather more surprising for the town was a comparatively important political and commercial centre in the southeast. The lack of data for Kirman, which is clearly illustrated by comparing figures 5.2 with 5.3 and 5.4, should certainly be seen partly in terms of an inadequate coverage of the area in historical sources before the nineteenth century.

Figure 5.5. Seismic map of Iran after de Ballore 1906.

Figure 5.6. Seismic map of Iran after Sieberg 1932.

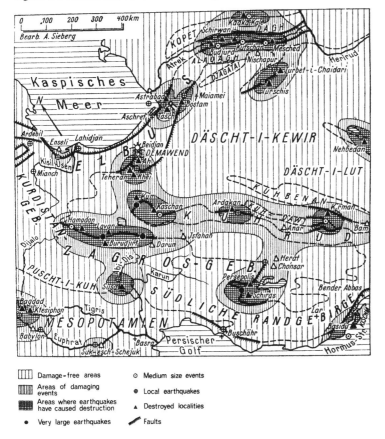

Figure 5.7. Space–time distribution of earthquakes in the Northern and Eastern zones of Iran. Length of bars correspond to the magnitude of the events (from equation (5.6)).

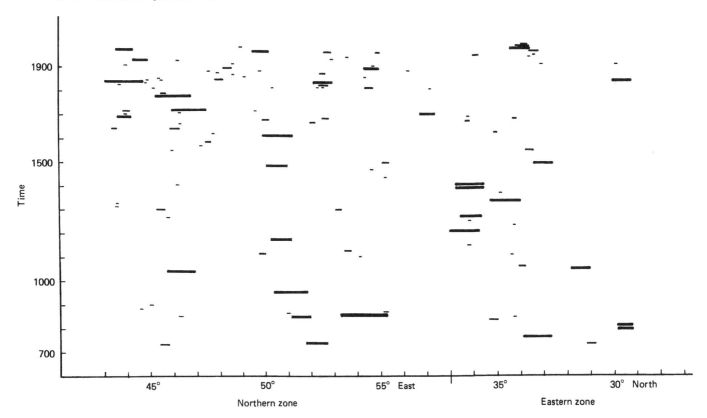

Nevertheless, our lack of information for these remote desert areas may reveal a genuine comparative seismic quiescence for the historical period. While it is unlikely that any small or moderate events would have been recorded, it remains very probable that any large earthquake would have been noted. Even if its epicentral area was in a desert region, a large magnitude earthquake should have been perceived quite strongly in some of the main centres of population, for towns all stand on the fringes of the desert areas, as is clear from Iran's geography. Even if this does not apply quite so much to the remote and ill-documented southeast of the country, it is certainly applicable to the more central areas of Iran.

As for the apparent gaps in the distribution of the nineteenth- and twentieth-century events, these fall into four main areas: central Azarbaijan and Tabriz, Shahr Ray (Tehran) and much of the Alburz, western Khurasan and Nishapur, and eastern Khurasan and Kuhistan. These gaps seem to be genuine and in the well-documented areas of Tabriz, Shahr Ray and Nishapur, long periods of seismic quiescence can also be observed in the past. These gaps between large earthquakes are not filled in by smaller shocks, with the exception of the Tabriz region, where, as in the Zagros zone, locally destructive earthquakes are recorded at comparatively long intervals, the interim periods being filled by the occurrence of several smaller events with long aftershock sequences.[4] In this respect, the Tabriz region also has some of the characteristics of earth-quake occurrence in the Zagros zone, where the recurrence of relatively large events seems to be delayed or their magnitude is reduced by a continuous process of small shocks, with the

difference that large events in the region of Tabriz are of greater magnitude than in the Zagros.

If the pattern of seismicity in the Alburz is discontinuous, but with gaps filled in gradually by relatively large events, the same cannot be said for the Zagros, where as can be seen from figures 5.2, 5.3, 5.4 and 5.8, the pattern is diffused. In terms of moderate to large earthquakes, there is no apparent trend or lineament to the events, except that much of the activity falls on or to the southwest of the main Recent fault. Nor is there any pattern of recurrence, except in the southern Zagros around Qir, Lar and Bandar 'Abbas. These regions may form a separate entity. While it is true that a great deal of the information in the historical sources is not adequate for a confident classification of all the events retrieved, it seems on the whole that major earthquakes in the Zagros are not common.

Although more details are available than for the eastern parts of Persia, historical evidence for the Zagros paradoxically gives a more certain impression of relatively low seismicity than does the total silence about Kirman. As we have already suggested, it is true that the remote areas of the central Zagros, or the southwestern Dasht-i Kavir, are unlikely to provide literary records or large events, but nevertheless these should have been experienced, at lower levels of Intensity, by one or more of the several major towns in this relatively densely populated region. The histories of Shiraz, Isfahan, Hamadan and Qazvin, all in their time important capital cities, certainly give few grounds for supposing that large destructive events occurred in their close vicinities and have escaped notice,

Figure 5.8. Space–time distribution of earthquakes in the Zagros zone. Length of bars correspond to the magnitude of the events (from equation (5.6)).

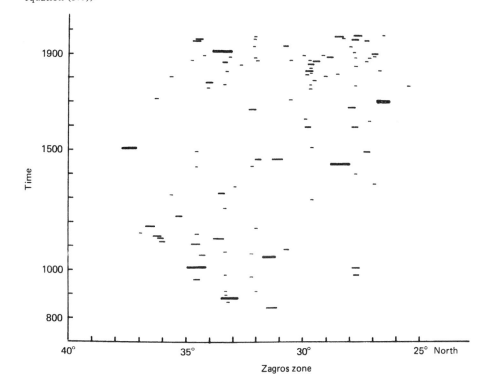

although it remains true that their perception and recording of more distant events is irregular over the historical period.

The general pattern and relative completeness of historical seismicity, depicted in figures 5.2 to 5.4 and 5.7 and 5.8 is supported by a different type of information, namely, general references to earthquakes which are usually not to be associated with a particular event, but rather with local traditions of high or low seismicity, reflecting past earthquake activity in a particular region.

For instance, this information suggests that throughout the period surveyed, the seismicity of the Zagros has been continuous with occasional local paroxysms. As we have seen, literary evidence of earthquakes in Kurdistan goes back to the second millennium B.C. and in the Zagros to the ninth century A.D., and even earlier if we consider archaeological evidence. The Jibal, Saimareh, Izeh, Siraf and Dinavar were particularly known for the occurrence of earthquakes, which often triggered large landslides.[5] The formation of the Zaribar Lake in Kurdistan and of Irene, south of Durud, are attributed to such phenomena. The formation of Tang-i Khas is also considered to be the result of an earthquake. Even in relatively recent times, many parts of the Zagros are known to the tribes to suffer one or two shocks every year, which cause rockfalls but rarely any serious damage. Early and recent sources attest to frequent earthquakes in Qal'eh-yi Tul, Qal'eh Huma, Falat and Yazdikhwast, where shocks are frequent but damaging only once every thirty years.[6] Further to the south, tradition has it that the source of the stream that irrigated the Jannabeh (modern Ganaveh) plain was caused to change by an earthquake which resulted in the desolation of the region just north of Bushire.[7] The action of earthquakes as well as a general subsidence of the land around the port was also

noticed at Bushire.[8] In Hurmuz (Bandar 'Abbas) a tremor every year was the norm (Valle 1677).

In contrast with the Zagros, indications are that damaging earthquakes seldom occurred in the adjacent regions of central Iran such as Isfahan and Yazd, a fact that has contributed conspicuously to the preservation there of the largest number of minarets in Iran (figure 5.10). Early and more recent sources confirm that whenever a shock was felt in these regions, it has been when a serious earthquake had occurred at considerable distance.[9]

The situation regarding the seismicity of the Alburz is quite different. There, we have evidence of large and major earthquakes but also of long periods of quiescence. Strabo, writing late in the first century B.C. places Rhagae (Ray) among the regions of the then known world that were notorious for their frequent occurrence of great earthquakes. Ten centuries later Arab sources single out Tabaristan, Qumis and Amul, while six centuries later this very same region is considered to be subject to shocks that merely cause alarm and seldom have fatal results.[10] These general statements fit the pattern shown in figure 5.7 well. Indications are that the Alburz zone has again become active, but today it is in a period of low seismicity that has lasted about one century.

For the region of Azarbaijan and for Tabriz we have a similar situation. Mustaufi, referring to the period between the mid-eleventh and mid-fourteenth century, says that although earthquakes are frequent in Tabriz, they are not serious. Relatively long periods of such local quiescence were, however, followed by periods of high activity, reflected in figure 5.7.[11]

To the east, in Khurasan and Kuhistan, traditions refer to large earthquakes affecting extensive regions with great loss of life. According to Khalifeh (p. 146) Nishapur had been destroyed by earthquakes eighteen times from when it was first founded until the eleventh century, a statement that indicates a tradition of high seismicity in early times. This is all the more interesting in that there is little evidence of earthquakes in that region since the end of the seventeenth century, cf. figure 5.7. Further to the south, Mustaufi (*Nuzhat*: 143) relates that there was once a cypress tree at Kishmar, in the province of Turshiz (Kashmar), which protected the district from the earthquakes that frequently occurred all around. The story has a legendary character, but is nonetheless interesting. Archaeological evidence of seismic activity east of Birjand together with local traditions imply large earthquakes in Kuhistan. The mountain range that separates the Duruh plain from the Kand Ghinau valley is called *Zalzaleh Kuh*, and just to the north at Sahlabad there is still evidence of large earthquakes,[12] while in Sistan both old and new traditions are extant. Mustaufi (*Nuzhat*: 201) relates that in the time of the Ghaznavid sultans, gold was found in Sistan but that the mine was ruined by a great earthquake some time in the eleventh century. In northern Baluchistan, local tradition alludes to large-scale faulting at Gaud-i Zireh and Kuh-i Sultan, comparable with that associated with the Chaman earthquake of 1892.[13] The Baluchi tribesmen, late in the last century, related that during their lifetime, on some occasions after severe earthquakes, deep fissures appeared in the ground over

Figure 5.9. Zones of earthquake activity in Iran, with location of major and great events.

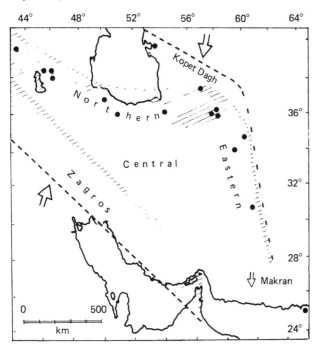

great distances and that they had similar accounts handed down to them by their fathers.[14] After one of these occurrences, presumably on the Chaman fault, the water supply of the springs along the fault break had largely increased, and elsewhere, most probably along the Kuh-i Sultan range, earthquakes were responsible for extensive shattering of the mountains. Popular tradition of earthquake occurrences in Sistan and northern Baluchistan is substantial.

For the Kopet Dagh area little has been found. Muraviev (1871: 16) attributes the ruin of Kizil Alan, near Gumush Tepe, to an earthquake. He also relates a Khivan tradition, that the Amu Darya altered its course to the northward as a result of an earthquake late in the thirteenth century (Muraviev 1871: 102, 104, cf. Tolstov 1960). Blaramberg (1850) suggests that changes in the eastern coast of the Caspian, particularly at Cheleken, are the result of early earthquakes, an observation

based on local information but also echoing Hanway's description of the Caspian coasts and of their changes that may be attributed to a fluctuation of the sea level rather than to seismic activity (Hanway 1753: i, 155; Ouseley 1819: iii, 316). Further inland, archaeological evidence but very little oral tradition exists for destructive earthquakes at Nisa, Ak Tepe and Churik. Local tradition in the Quchan region, however, does suggest that Ghulaman and Quchan were often destroyed by earthquakes, but it is possible that this is inherited from early nineteenth-century sources that in fact refer to the events of *c*. 1833 and 1851. Our information on this area appears to be incomplete anyway.[15]

There is little to illuminate early seismicity in the province of Kirman, apart from legends about historical earthquakes in the region of Kuhbanan. One of these alludes to a destructive shock that ruined Kuhbanan some time in the

Figure 5.10. Distribution of extant, free-standing structures in Iran, built before the middle of the sixteenth century (full circles). Open circles show sites of ruined free-standing structures. The natural periods of oscillations are indicated where known.

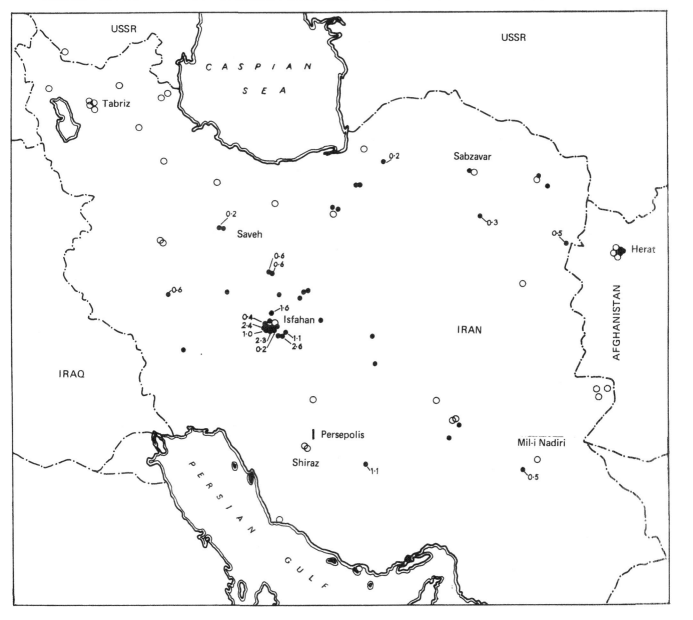

twelfth century, causing ground deformations to the east and west of the town, particularly at a locality about twelve kilometres away and still called *gaud-i zalzaleh-zadeh*. Another local legend refers to a fourteenth-century event seven kilometres south of Kuhbanan, near Givar at a place called today *zalzaleh sang*. On examination, there is little to substantiate these legends (see pp. 24, 169 (note 16)).

We have found no general references to earthquakes or any relevant legends for Mesopotamia, the north coast of Arabia and Oman, or for the western parts of Afghanistan. For central Iran and for the Dasht-i Lut there is also little evidence, except that some of the early general references to catastrophic earthquakes in Khurasan might refer in fact to events in Kuhistan and the neighbouring Great Desert, on the borders of which the recent earthquake of Tabas occurred (16 September 1978).

The locations referred to in these general statements are shown in figure 5.2, from which we notice that quite a few of them coincide with regions affected by later events, while others are still quiescent.

These general observations, together with the macroseismic evidence shown in figures 5.2 to 5.4, suggest that large earthquakes seem to follow a wide zone which runs from Azarbaijan, along the Alburz to northern Khurasan, where it turns south, following the eastern part of the Lut all the way to northern Sistan. This is a well defined, broad zone, within which most major historical and recent earthquakes have occurred, and which for the sake of brevity we have previously called the Iranian Crescent (Ambraseys & Melville 1977), figure 5.9. In contrast, the Zagros and its extension into Laristan, which has normally been considered a highly seismic zone, does not seem to have been responsible for major earthquakes.

As a matter of fact, the distribution of early structures, such as minarets, that survive today seems to attest to the occurrence of large earthquakes in this wide zone. Figure 5.10, which locates the sites of free-standing minarets built up to the Turkoman period (sixteenth century), shows how few of these early structures have survived today in Azarbaijan, in the Alburz and in eastern Iran, where only their stumps can be found. In contrast, the oldest and the tallest minarets are standing in central Iran.[16] It is recognised of course that causes other than earthquakes (such as deliberate damage,[17] neglect and high winds[18]) have contributed more to the decimation of these early structures. However, large earthquakes recurring in this zone should have deterred people from rebuilding destroyed minarets or attempting to construct new, tall ones, as observed by some early travellers.[19] Other types of early structures, such as bridges and free-standing columns, are more difficult to use as a criterion. The survival of the former depends more on floods and war action, and of the latter on wanton destruction rather than on earthquakes. The decimation of the columns at Persepolis can hardly be attributed to earthquakes.[20] Taking these structures as a whole, the observed systematic trend for their survival suggests that statistically, seismicity should have been higher within the Iranian Crescent than elsewhere in the region.

5.3 Distribution of seismic activity: quantitative evaluation

The shading in figures 5.2 to 5.4 shows the zones affected by destructive earthquakes during the last thirteen centuries, and figures 5.7 and 5.8 depict the space—time distribution of seismicity in these zones. These figures, which are based solely on macroseismic data and constructed quite independently of any knowledge of regional tectonics, depict quite clearly an overall pattern of seismic distribution and demonstrate the extent to which seismic gaps in historical and modern periods are filled in with the passage of time. On a large scale these zones may be defined as (I) the Eastern zone that comprises eastern Khurasan, Kuhistan and northern Sistan, (II) the Northern zone that runs along the Alburz and includes western Azarbaijan, (III) the Zagros zone that extends from west of Lake Rizaiyeh all the way to Bandar 'Abbas, and (IV) the Central region of Iran. Two other zones, (V) the Kopet Dagh and (VI) the Makran—Baluchistan, extend well beyond the region under study and as such they are incomplete (figure 5.9).

Of these zones, that of the Zagros aligns with the synonymous suture zone, a well known interplate feature (figure 5.11). It is interesting, however, that the other two zones, the Eastern and Northern, do not follow a pattern that is apparent from the surface tectonics, nor do they depend on the major known Quaternary faults shown in this figure. The most important events in these two zones, some of them associated with recent faulting, are connected only with relatively minor faults of recent age that cut either across throughgoing, regional Quaternary lineaments or into individual blocks. As a matter of fact, in these intraplate zones, no large or major earthquake has utilised a known major Quaternary fault, while in the interplate Zagros zone there have been no major or great earthquakes (figure 5.9). Also in the Central zone, there is no evidence of large events, but the seismicity of the Kopet Dagh, as depicted by the activity of the last 150 years, seems to be comparable with that of the Eastern zone. The long-term seismicity beyond these zones, in western Afghanistan and Turkmenistan to the northeast of the Kopet Dagh, is negligibly small for all practical purposes. Also to the west of the Zagros, in the Arabian plate, there is no evidence of significant intraplate activity. To the south, the seismicity of the segment of the Makran included in this study is controlled by the earthquakes of 1483 and 1945, and its long-term activity, reminiscent of that of the Aegean Arc, cannot be examined before the completion of the historical study of the region east of 64°.

One way of displaying graphically the overall, long-term seismic activity of the whole region and the interaction of the different zones with time, is to plot the cumulative strain $\Sigma(E)^{1/2}$ or seismic moment ΣM_0, against time (figure 5.12). The cumulative strain may be calculated from Båth's (1958) formula for the seismic wave energy E,

$$\log(E) = 12.24 + 1.44(M) \tag{5.1}$$

and the surface-wave magnitude (M) may be computed or assessed as in chapter 4. The seismic moment (M_0) was calculated from:

$$\log(M_0) = 16.7(1.7) + 1.4(0.3)(M) \tag{5.2}$$

which is the best fit in the range $6.0 \leqslant M \leqslant 7.5$ to the moments of Eastern Mediterranean and Near Eastern earthquakes computed by North (1973) and others more recently.[21]

From figure 5.12 we notice that the long-term rate of straining in Persia as a whole is remarkably constant, varying between 1.0 and 3.0×10^{10} (erg)$^{1/2}$/year, with no evidence of excessively long periods of overall quiescence, curve A. Moment rates shown by curve B are also relatively constant, varying between 40 and 90×10^{24} dyn cm/year. The gradual increase of strain rate shown by curve A in the last three centuries is mainly due to the larger number of smaller shocks that become identified during that period, particularly from the more remote parts of Persia. However, these smaller events contribute very little to the total moment, the rate of which remains on average constant at about 60×10^{24} dyn cm/year.

At first sight figure 5.12 suggests that in terms of moment rates the overall seismicity over the past millennium does not differ much from that over the last eight decades. However, this is fortuitous. Had we chosen a larger area of interplate activity that included, say, Anatolia or the Makran and the Hindu Kush, or had we taken a much smaller area covering only one of the individual zones or activity, moment rates would vary quasi-periodically with time with recurrence intervals of quiescence of several hundred years. This is demonstrated in figure 5.13, which shows that in both the Eastern and Northern zones, moment rates remain approximately constant only for a few centuries, changing conspicuously by a factor of ten as they pass from periods of high to low and back to high activity. For the Kopet Dagh the data cover a relatively short period and they show only one high period. In contrast, in the Zagros zone moment rates remain roughly constant throughout the last eleven centuries and they are very low. In the Eastern and Northern zones, periods of activity of two to three centuries alternate with longer

Figure 5.11. Association of earthquake activity during the last thirteen centuries with active tectonics. 1: solid lines show major faults of Quaternary age, dashed lines show faults of late Tertiary age; 2: location, extent and sense of displacements of faulting associated with events since A.D. 700, superimposed on 3: areas affected by destructive earthquakes ($i \leqslant 3$) since 700, cf. figure 5.4.

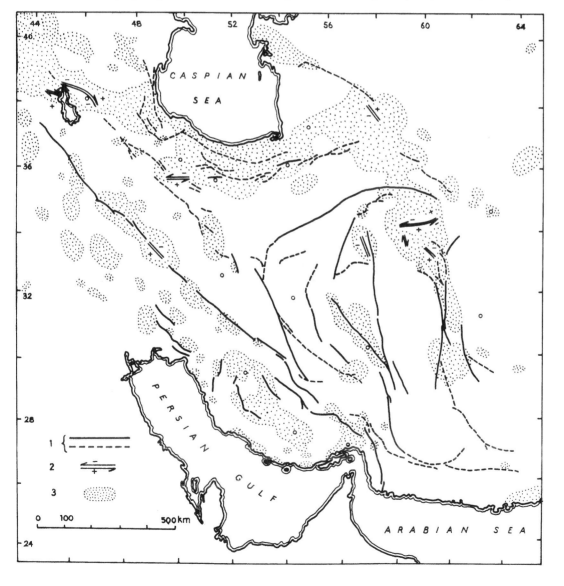

periods of relative quiescence of three to five centuries, the recurrence period of clustered activity of the two zones being out of phase by about two centuries.

These rather long periods of low activity should be genuine. They are shown also in figure 5.7 and it is rather unlikely that so many of the large earthquakes that would be needed to fill in these gaps of low seismicity during well-documented periods have been missed out, or that magnitude estimates could have been systematically low only for the low periods shown in figure 5.13. In order for this to be so one would need an under-estimation of at least one magnitude unit to remove the quiescent periods observed, which, after all, are reflected in our historical sources and have been observed elsewhere.

Alternating periods of activity and quiescence of similar or longer duration have been observed in Anatolia, Portugal and China,[22] but some of them must always be viewed with an eye to the levels of magnitude at which regional earthquakes were reported in the available documentary sources. It is beyond our competence to comment on similarities or differences between long-term seismicity in Persia and China. The area covered by the Chinese earthquakes catalogues of historical events is about five times larger than that of Persia and the time span it covers is almost two and a half times longer (Academia Sinica 1956, 1977). However, from what we have

been able to glean through secondary sources of information, the published historical record of Chinese earthquakes seems to suffer from lack of homogeneity for the period it covers. This may well be due to the important changes with time of the regions covered by the historical sources and also due to the method that the authors of the catalogue have used to assess the relative size of historical events, which most probably was based solely on epicentral Intensities (Lee 1958, see figure 4.2; York *et al.* 1976, Lee *et al.* 1976).

Returning to figure 5.13, we may now assess the average slip rates for the different zones. Assuming throughout predominantly high angle thrusting ($a = 45°$) in a 20-kilometre thick crust with $\mu = 3 \times 10^{11}$ dyn/cm^2, the slip rate may be obtained from:

$$\dot{u} = \dot{M}_0 \sin(2a)(2\mu L h)^{-1} \tag{5.3}$$

where $\dot{M}_0$ is the long-term moment rate (figure 5.13), L is the length and h (20 km) is the thickness of the zone.

For the Zagros the moment rate calculated from moderate magnitude or larger events is 7.3×10^{24} dyn cm/year which corresponds to a slip rate of only 0.4 mm/yr, or, for the more active periods, not more than 1.2 mm/year. This rate of shortening is comparable to that calculated on different premises by North (1973)[23] and Niazi & Taheri (1981), but it is only a few per cent of the value calculated from plate motions by McKenzie (1972) and Minster & Jordan (1978). The slip rate of 0.4 mm/year may be somewhat underestimated here since we have no moment—magnitude relationship to account for the bulk of the earthquakes with $M < 6.0$, and we have assumed that during each earthquake slip occurs throughout the Zagros. It is probable that slip takes place only during large earthquakes along certain parts of the zone, while elsewhere it occurs aseismically. It is also probable that in much earlier periods than those examined, movements in different parts of the Zagros and at different times during the past, say ten to fourteen million years ago, occurred in stages, occasionally at much greater rates. The present-day slip rates are almost identical to that deduced by Falcon (1974) for the uniform uplift of the Zagros mountains since the early Pliocene, i.e. 1.0 mm/year, which implies that the actual slip rates across the Zagros zone are greater than these deduced from purely seismological considerations.[24] It appears, therefore, that the Zagros being an interplate zone is subjected to a continuous deformation, much of which is taken up aseismically. The seismic activity within this 1600-kilometre long and 250-kilometre wide zone is high but diffused, as we have seen, and consists of a large number of relatively small magnitude, shallow depth earthquakes that occur with little apparent association with tectonic lineaments, no obvious pattern of recurrence, and with little evidence of large-scale faulting. As a matter of fact, with the exception of the 1909 Silakhur event, no earthquake in this zone can be said with confidence to have been associated with primary tectonic deformations. A few other cases of alleged faulting, already discussed in chapter 3, seem to be connected with secondary effects of local readjustment caused by movements in deeper structures. This situation implies an incompetent basement and lends support to Jack-

Figure 5.12. Cumulative strain and seismic moment for Persia (excluding the Makran) for events of magnitude equal to or greater than 6.5.

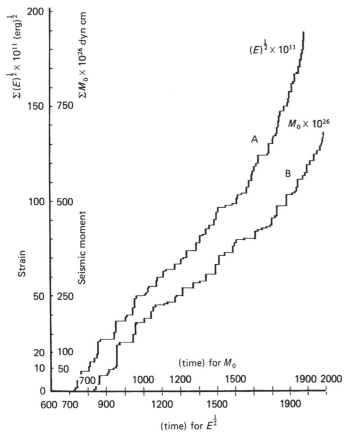

son's model for the existence of reactivated basement faults in the Zagros zone.

With no evidence for the existence of an oceanic lithosphere beneath the Zagros (Niazi *et al.* 1978), Jackson (1980) postulates the reactivation as thrusts of listric normal faults in a stretched basement on which thick sediments were deposited. This reversal of motion does not necessitate either abnormal thickening or subduction of continental crust in the early stages of suture, and the process involves an already fractured basement, less capable of building up large amounts of unreleased energy. This contraction mechanism will presumably work until the reverse motion on normal faults restores the basement back to its original thickness, in the process the pre-existing basement faults inducing intense folding and ultimately faulting of the sediment cover.

There is no reliable data for the Central zone, which seems to behave as a relatively rigid[25] mosaic structure, occasionally broken down further but with a lack of major or great earthquakes.

Average slip rates of the Northern and Eastern zones are somewhat larger, but comparable with that of the Zagros. These rates, from figure 5.13 and equation (5.3), are 1.7 and 1.3 mm/year respectively.

However, in contrast with the Zagros, movements of these zones are spasmodic with rates of 3.2 and 4.6 mm/year during periods of broadly clustered activity and of only 0.5

and 0.2 mm/year during periods of relative quiescence. The Northern zone seems to have been in a low period for more than one century, while the activity of the Eastern zone seems to have been building up again to a high. In the Kopet Dagh, the average slip rate corresponding to the last 150 years of high activity is about 4.0 mm/year, comparable to that of the two neighbouring zones. Slip rates beyond these zones in western Afghanistan, Turkmensitan and in the Arabian plate are to all practical purposes zero. The historical data for the Makran zone are incomplete, as we have seen. However, if we consider only the period after 1483 and a 500-kilometre length of this subduction zone, the slip rate that obtains from equation (5.3) is of the order of 4.0 mm/year with a recurrence period of a great earthquake of about 500 years. The figures are comparable with those deduced by Page *et al.* (1979), but they depend heavily on the extrapolation of equation (5.2) to large magnitudes.

Deformations of the Earth's crust in Persia and much of the associated seismicity appear to be in a major part the by-product of the behaviour of the Zagros zone, at present shortening as a result of the collision between Arabia and Iran. We have seen that on a large scale the Zagros transmits the pressure to the northeast, acting as a pressure gauge. It does this mainly through the Central zone, a relatively rigid block it thrusts into the Iranian mass. The Central zone has a triangular shape, with the Northern zone and the Alburz ranges forming

Figure 5.13. Cumulative seismic moment variation with time for different zones in Persia. Figures in brackets show average movement in 10^{24} dyn cm/year. L and d are the length and width of the zones respectively.

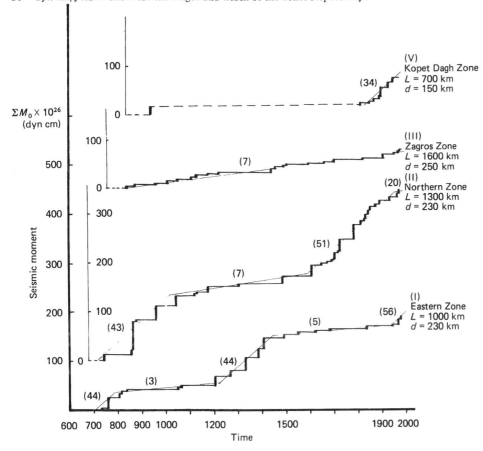

its northern boundary, and the Eastern zone with the Tabas ranges west of the Lut forming its eastern boundary. Also, we have seen that these two zones are taking up much of the motion of Arabia towards Eurasia with major earthquakes, occasionally associated with oblique-thrust faulting, while the Central zone remains relatively inactive, a situation reminiscent of an indentation process. As a matter of fact, figures 5.9 and 5.11 do suggest a gross analogy between observed active seismotectonics and a simplified plastic indentation configuration, similar to that proposed by Molnar & Tapponnier (1977) for Asia. Undoubtedly, the comparison is crude, lacking in detail, but the mechanism it suggests, namely of the Central zone penetrating into the Iranian mass, seems to provide a plausible explanation for the observed alternation of seismic activity between the Northern and Eastern zones.

Whereas plate boundaries with oceanic crust are characterised by well defined narrow zones of seismic activity and rigid tectonics, seismicity is usually diffused in continental regions, and on a large scale 'plastic' deformations play an important role. One of the reasons for this is that continental regions have been affected by pre-existing geological structures which left residual features that pre-dispose the crustal material to react in a seemingly irregular and complex fashion to an otherwise simple regional pattern of stresses. Thus, with time each increment in internal deformation in the Northern, Eastern and the Kopet Dagh zones alters the shape of the slip boundaries so that present stress fields in these zones require knowledge of the previous tectonic history, a time–space sequence that cannot be sought out solely from seismological observations covering a few decades. From such a process, all that we may deduce for the characteristics of these zones of intense deformation is a considerable degree of work softening, sufficient to explain the alternating activity between these two zones as the result of the Central zone jostling into the Iranian mass. Whereas during penetration into a strain hardening material stress and strain fields develop and expand uniformly with a perfectly symmetric distribution, penetration into a work softening material is associated with the development of inhomogeneities in addition to pre-existing ones, which in turn induce kinematic instabilities. These are due to the progressive development of a series of velocity discontinuities in the material penetrated, that give rise to an alternating variation of the stresses that act on the sides of a penetrating wedge. This type of unstable kinematics, exhibiting a quasi-cyclic fluctuation of the forces acting on one side of the wedge, have been observed in strain softening materials (see for instance Butterfield & Andrawes 1972 and Blair-Fish & Bransby 1972). They may explain why the Northern and Eastern zones alternately assume dominant roles in resisting the motion of the Central zone, and caution against the use of a random process to describe the generic cause of the observed fluctuations in seismicity in these two zones.[26]

Returning to figure 5.13, the activity it shows implies that in a particular zone of the dimensions examined here, what controls the relative frequency of occurrence of earthquakes with different magnitudes may or may not be a random process, and that large earthquakes may occur in broad clusters instead of being more uniformly spread in time. In turn, quasi-cyclic activity suggests that for the very short geological time we studied (thirteen centuries) there may be a maximum magnitude or seismic moment for each zone, limiting the upper bound of the size of its earthquakes. The chances that the tectonics that dictate the observed behaviour of a particular zone are changing during the period of our observations are of course very small.

Conspicuous alternation of activity between contiguous regions over periods of a few centuries has also been noticed in the Eastern Mediterranean and Anatolia, where the historical record is far more complete than in Persia and covers a longer period of twenty centuries. And here again, from other considerations, the evidence is that the generic process is also non-random. The temptation to devise a space-time model of a branching process to simulate the observed activity is considerable, following for instance Kagan and Knopoff (1979) or Vere-Jones (1978). At this stage it is doubtful that anything very useful would be gained by fitting models to sequences of the type shown in figure 5.13 or that modelling, without understanding the factors which might govern the time scale of stress transfer from one zone to another, would add much that is not already apparent from this figure. One could envisage some kind of alternating process between two or more zones, a slow flip-flop mechanism, whereby each large earthquake in one zone increases the probability of a reversal of roles, with the main resistance transferring from one zone to another. Also, one could attribute such an alternating mechanism to inherent or continuously produced nonhomogeneity, but it is difficult to postulate its time scale without studying the long-term behaviour of a much larger region that contains more interacting zones. There is some evidence that the time scale between reversal of roles from one zone to another becomes longer with increasing dimensions of the areas between zones. However, verification of this hypothesis for continental regions will have to await the completion of the assessment of the historical data for the whole Eastern Mediterranean region.

The preceding discussion shows the possibilities for a high degree of spatial and time inhomogeneity of seismicity in Persia, and the consequences of this in the assessment of a magnitude–frequency distribution for a particular region. If a stationary generic process is assumed (which is always the case with short-term seismological observations), while in fact the seismicity is quasi-periodic or broadly clustered, the deduced frequency distribution function $f(x)$ will either over-estimate or under-estimate future activity, depending on whether the sample of observations was taken from a high or low period of current seismicity.

If we consider the so-called Gutenberg–Richter relation:

$$\log N_M(x) = a - bx \qquad (5.4)$$

in which $N_M(x) = P(x \geqslant M)$ is the cumulative frequency of all events exceeding or being equal to a given magnitude M, with $N(x) = 1 - f(x)$, then figure 5.14 shows the short-term, twentieth-century frequency–magnitude plot for the whole region under study. It is apparent that a linear expression such as equation (5.3) cannot possibly fit the whole range of observed data, particularly in terms of single frequency.[27] Not only does such fitting over-estimate significantly the large magnitudes, which are not supported even by long-term observations, but also converted into a cumulative strain or seismic relation, it diverges with decreasing recurrence frequencies.

The data used in figure 5.13 suggest that magnitude has a maximum or upper bound value M_{mx} which is different for different zones, at least for the active periods identified. Figure 5.15 shows equation (5.4) plotted for each of the constituent zones separately, i.e. for the Eastern, Northern, Zagros, and Kopet Dagh zones, for the periods of high activity. Equation (5.4) fits well the data for all zones except the Zagros, where the data points depart from linearity for the higher magnitudes, probably due to inclusion of the Laristan and Bandar 'Abbas areas. The same figure also shows the maximum magnitudes observed during these periods. Thus, the Zagros shows a very high rate of activity which consists of relatively moderate magnitude earthquakes, with a maximum value of about 7.3. In contrast, the Northern and Eastern

zones show a much lower rate during their periods of activity, but this is associated with relatively larger magnitudes with maximum values of 7.6 and 7.9. The same figure shows the magnitude–frequency relation for the region that obtains from the four zones taken together. With decreasing magnitudes this curve would tend to coalesce asymptotically with equation (5.4) of the zone with the highest rate of activity (largest b-value, in this case the Zagros zone); with increasing magnitudes and with individual zones ceasing to contribute for $M > M_{mx}$, the curve would tend to turn sharply to lower frequencies. The shape of the composite magnitude–frequency curve derived in this manner from historical periods of high activity, as can be seen from figure 5.15, is very similar to the plot obtained from twentieth-century data alone.

The main difference between these two graphs is in the range of higher magnitudes, where the curve derived from historical data is constrained by the maximum values of the magnitudes observed in each zone. These are not the so-called regional maximum possible magnitudes, the validity of which

Figure 5.14. Frequency–magnitude relation for the period 1900–79 (• single frequency with $dM = 0.5$: ○ cumulative frequency with $dM = 0.1$). Best fit for single frequency, $\log(N) = 6.88 - 0.86(M)$.

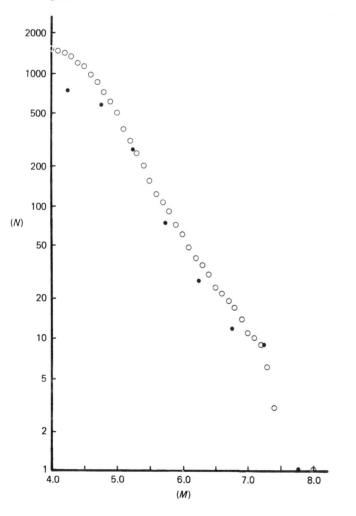

Figure 5.15. Recurrence relations. For the data on which this figure is based see tables 5.1 and 5.2. The dotted line is based on unpublished data for magnitudes <5.

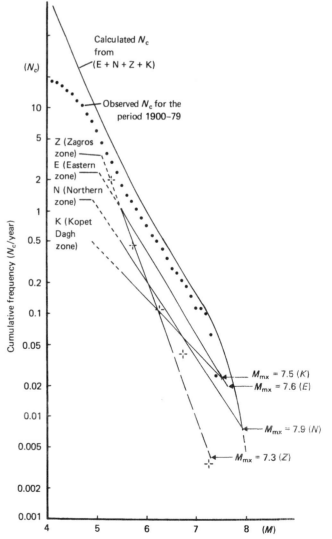

is questionable and which cannot be calculated from theoretical considerations, but rather the intrinsic maximum values observed within a particular region and period of time. Extreme value methods are often used too lightly to truncate equation (5.4) with an upper bound derived, say, from a finite interval distribution (Gumbel III). The results from such a method are unsuitable (Knopoff & Kagan 1977), and they can be grossly misleading for short-term or incomplete observations pertaining to tectonically inhomogeneous regions such as Iran or the Eastern Mediterranean. In our case an independent estimate of the truncating magnitudes for the different zones for periods of high activity may be assessed from faults of known or inferred length and mobility. An estimate of magnitudes from the dimensions of rupture and dislocation for Eastern Mediterranean and Middle Eastern events may be made from:

$$M = 1.1 + 0.4 \log(L^{1.58}R^2) \tag{5.5}$$

where L and R are the length of the fault break and maximum relative displacement respectively, in centimetres (Ambraseys 1976: 73). Maximum displacements R were found to be about three times larger than average displacements u immediately after an earthquake, so that from equations (5.2), (5.5) and the moment equation, we have as a first approximation that average dislocations are of the order of 10^{-5} of the length of the rupture, or

$$\log(L) = 0.7(M) - 3.24 \tag{5.6}$$

where L is the length of faulting in a twenty-kilometre thick crust, associated with a shallow earthquake of magnitude M. In equation (5.6) the fault length is in kilometres, and the equation is based on data for $M \geqslant 6.0$.

Thus, from the maximum linear dimensions of historical meizoseismal regions in the Northern, Eastern and Zagros zones, we may infer fault breaks 200, 150 and 50 kilometres long respectively. These, from equation (5.6), correspond to events of magnitude 7.9, 7.7 and 7.1, almost identical to the maximum values observed (figure 5.15). Longer fault breaks cannot be ruled out, but neither the known tectonics nor the historical evidence of the last thirteen centuries suggest this.

It would appear, therefore, that a truncated, single frequency magnitude distribution of the type (5.4) fits long-term observations for intraplate zones of the limited dimensions met with in Iran reasonably well. However, the same does not seem to be true for larger regions of plate dimensions and consisting of a number of such individual zones of alternating activity. Distribution of seismicity in such regions would not only appear to be non-linear with respect to magnitudes (particularly in their higher range), but also becomes a function of time. Similarly, even the zones identified here are relatively large, covering on average areas of about a quarter of a million square kilometres. Probabilities are relatively small, therefore, that during a given period of high activity a few centuries long, a specific area of a few thousands of square kilometres will be affected by a large event in the zone concerned. Considering intervening inactive intervals, return periods of a millennium such as those suggested in the Ray, Qumis and Nishapur areas and in Sistan, are not surprising. Apart from a longer period of historical observation, for which information is not foreseeable, this general picture can only be checked by a more precise identification of individual subzones and the further breaking down of the broad zones examined here into smaller units. Conversely, the study of a much larger area, such as the whole of the Middle East and Eastern Mediterranean, may help to solve the problem of time scale between reversals of activity from one zone to another.

Note on tables 5.1 and 5.2

Table 5.1 lists events that occurred before 1900. The list is comprehensive up to around 1840, giving all the earthquakes for which information has been retrieved. Those described in §§ 3.2 and 3.3 are listed without references; selected sources of information for the other events are given in the footnotes to the table. After 1840, the list is selective, with earthquakes of magnitude less than 5.0 and aftershocks being omitted unless they are of particular interest. The magnitude M of events less than around 5.5 is generally not shown. The time of the shock is given in local time unless marked with an asterisk, in which case it is in GMT. When no precision is possible, the time of day is indicated by 06 (between midnight and 6.00 a.m.), 12 (6.00 to midday), 18 (midday to 6.00 p.m.) or 24 (6.00 to midnight). Where data are insufficient for even a tentative epicentral location, the area affected is named.

Table 5.2 is a selective list of earthquakes, excluding aftershocks, in the twentieth century, all shocks of magnitude less than 5.0 being excluded; references are only given for events not described in § 3.3.

The key to both tables is as follows:
M = surface-wave magnitude
m = body-wave magnitude
(M) = surface-wave magnitude derived from equation (4.7)
M^* = surface-wave magnitude calculated from equation (4.3)
i_0 = epicentral Intensity (see § 2.5 and note 4.32)

r_0 = radius of meizoseismal region in kilometres
r' = radius of perceptibility, in kilometres, as defined in § 4.7
i = focal depth probably greater than normal
q = quality of macroseismic epicentre location (see §§ 2.5–2.6)
Q = quality of instrumental determination (see §§ 4.1–4.7).
From (1) Nowroozi 1971 & 1976; (2) Earthquake Files, Institute of Geological Sciences, Edinburgh; (3) Quittmeyer & Jacob 1979; (4) Regional Catalogue of Earthquakes, International Seismological Centre; (5) Gutenberg & Richter 1965; (6) International Seismological Summary; (7) US Geological Survey/US National Oceanic & Atmospheric Administration; (8) Strasbourg; (9) *Atlas*; (10) British Association for the Advancement of Science: i.e. $A = 5$, $B = 10$, $C = 20$, $D = 25–90$, $E \geqslant 100$ km.

No references are given in the footnotes to the tables for information derived from local sources during field trips, such as oral information and family documents; neither to unpublished reports and statistics of casualties compiled by regional offices of the Red Lion & Sun Organisation, to technical reports prepared by Iranian and foreign engineering firms, to data in *Lloyd's List* nor to private correspondence.

Newspapers and journals are referred to either by their date (*Iran*: 1300.7.15) or by their issue number (*Khwandaniha* 1336: 26).

Table 5.1. *Historical period prior to 1900*

Date	Time	Epicentre N°	E°	i_0	r_0	r'	M	q	
3rd Millenn.		?5.6 – 49.9							
2nd Millenn.		38.1 – 58.5							
17th c. B.C.		34.5 – 48.0							
4th c. B.C.		35.5 – 51.8		1			7.6+		
1st c. B.C.		38.0 – 58.2							
2nd c. A.D.		39.8 – 44.6							
A.D. 628		35.5 – 44.4							
658		30.5 – 47.8		5				d	
734		31.0 – 60.5		2			6.5	c	
735		39.7 – 45.6		2			6.5	c	
743		35.3 – 52.2		2	30		7.2	c	
763		33.3 – 59.3		1			7.6	e	
805 Dec. 2		29.5 – 60.5		1+			7.0	c	(1)
815		29.5 – 60.5		1+			7.0	c	
819 Jun.		36.4 – 65.4		1		510	7.4	b	
840 Jul.		35.2 – 60.4		2			6.5	e	(2)
840		31.3 – 48.8		2			6.5	c	
849		34.3 – 62.2		3			5.3	d	
855		35.6 – 51.5		2		540	7.1	b	
856 Dec. 22		36.2 – 54.3		1	45	900	7.9	b	
857									(3) Fars
858		38.1 – 46.3		2+			6.0	c	
859		36.2 – 54.3						e	
859		33.2 – 44.6		4		100	5.0	c	
863 Feb. 13		40.0 – 44.6		3			5.2	c	
864 Jan.		35.7 – 51.0		3			5.3	c	
872 Jun. 22		33.2 – 47.2		1+		330	6.8	c	
874		37.2 – 55.2		2			6.0	c	
881 Oct.		33.3 – 44.4		5				d	(4)
893 Dec. 24	24	40.0 – 44.6		2	8	160	6.0	b	
902 Jun.		33.3 – 44.4		5				d	
906 Apr.		39.7 – 45.0		2		170	6.1	b	
912 May		32.0 – 44.4		4					
914 Apr.		39.7 – 64.4		3				d	
943 Aug.		37.6 – 57.0		1			7.6	c	
956		34.8 – 48.1		3		120	5.3	c	
958 Feb. 23		36.0 – 51.1		1	50	700	7.7	b	
958 Apr.		34.5 – 45.9		2		250	6.4	b	
973 Sep.		32.2 – 46.3		4				d	(5)
977 Nov.		33.3 – 44.4		5				d	
978 Jun. 17		27.7 – 52.3		3			5.3	c	
1008 Apr. 27	18	34.6 – 47.4		1+			7.0	c	
1008		27.7 – 52.3		2			6.5	c	
1042 Nov. 4	18	38.1 – 46.3		1			7.6	c	
1052 Jun. 2		36.2 – 57.7		1+			7.0	c	
1052		31.5 – 50.0				400	6.8	b	
1058 Dec. 8	18	34.3 – 44.7				300	6.4	b	
1063 Dec.		32.2 – 46.3		5				d	
1066 May		33.9 – 59.2		2			6.5	e	
1072 Jan. 20	06	33.3 – 44.4		5				d	(6)
1085 May		30.7 – 50.3		3			5.8	c	
1087 Nov.		34.8 – 48.5		2+			5.9	c	
1094 Feb.	06	33.3 – 44.4		5				d	
1102		36.3 – 54.1		4				d	
1102 Feb. 28	18	34.4 – 62.2		3			5.3	d	
1107 Sep.		34.6 – 47.4		2			6.5	c	(7)
1118 Apr. 3	06					190	5.9		W. Zagros
1119 Dec. 10	18	35.7 – 49.9		2			6.5	c	
1127		36.3 – 53.6		2	19	400	6.8	b	
1130 Feb. 27	18	33.6 – 45.7				410	6.8	c	
1135 Mar.	06	33.3 – 44.4		5				d	(8)

Table 5.1 (*cont.*)

Date	Time	Epicentre N°	E°	i_0	r_0	r'	M	q	
1135 Jul. 25		36.1 − 45.9				220+	6.1	c	
1135 Aug. 13	24	36.1 − 45.9				300	6.4	c	
1144 May 29	24	33.3 − 44.4		5				d	
1145		36.2 − 58.8		3			5.3	c	
1150 Apr. 1	12	34.5 − 45.9		2+		190	5.9	b	
1156 Feb.									N. Iraq
1177 May		35.7 − 50.7		2		650	7.2	b	
1179 Apr. 29	18	36.5 − 44.1		2		330	6.6	b	
1191		34.8 − 48.5		4				d	
1194 Mar.		32.0 − 44.3		3+		100	5.0	c	
1209	12	36.4 − 58.7		1	30		7.6	b	
1226 Nov. 18	06	35.3 − 46.0		2		280	6.5	b	
1238		34.3 − 58.7		3			5.3+	c	
1251		36.2 − 58.8		3			5.3+	c	
1252		33.3 − 44.4		5				d	(9)
1270 Oct. 7	12	36.2 − 58.8		1+			7.1	c	
1273 Jan. 18	18	38.4 − 45.8						c	
1301		36.1 − 53.2		2	18		6.7	c	
1304 Nov. 7	24	38.5 − 45.5		2		200+	6.7	c	
1305 Apr. 16									Azarbaijan
1310		35.6 − 46.1		3			5.3+	c	
1316 Jan. 5		33.5 − 49.4		2		200	6.2	c	
1319		39.1 − 44.5		3			5.3	c	
1320		40.5 − 43.6		2		100+	5.9	b	
1336 Oct. 21	06	34.7 − 59.7		1	33		7.6	b	
1344		32.9 − 52.3		3		180	5.7	c	
1345				5					Tabriz
1361		26.9 − 56.2		3			5.3	c	
1364 Feb. 10		34.9 − 61.7				230	5.8	c	
1389 Feb.	06	36.2 − 58.8		1			7.6	b	
1400		27.7 − 54.3		3			5.3+	c	
1405 Nov. 23		36.2 − 58.8		1			7.6	b	
1406 Nov. 29		39.4 − 46.2		4				d	
1410									Balkh
1428		35.9 − 63.8		2			6.5	c	
1430		34.5 − 48.0		2+			5.9	c	
1430		32.2 − 46.4		3			5.3	d	
1436		37.2 − 55.2		3			5.3	c	
1436									Azarbaijan
1440		28.4 − 53.1		1+			7.1	b	
1457		31.9 − 46.9				230	6.1	c	
1459		31.1 − 52.1				340	6.6	c	
1459									Azarbaijan
1470		37.1 − 54.6		3		150	5.5	c	
1483 Feb. 18		(24.9 − 57.9)				800	7.7	e	(off-shore)
1485 Aug. 15	18	36.7 − 50.5		2	28	650	7.2	a	
1493 Jan. 10	06	33.0 − 59.8		2	20	400+	7.0	b	
1495		34.5 − 50.0				200	5.9	c	
1497		27.2 − 56.3		2			6.5	c	
1498		37.2 − 55.2		2			6.5	c	
1503		37.4 − 43.8				450	6.9	b	
1506		29.6 − 52.5		4				c	
1549 Feb. 15	24	33.7 − 60.0		2	15	350	6.7	c	
1550		37.8 − 46.0		3				d	
1567		39.0 − 47.2		3				d	
1591		29.8 − 52.4				200	5.9	c	
1593		37.8 − 47.5		2		180	6.1	c	
1593 Sep.		27.7 − 54.3		2			6.5	b	
1608 Apr. 20	12	36.4 − 50.5		1	26	600	7.6	b	
1619 May	12	35.1 − 58.9		2			6.5	c	
1621 May 21	24	37.4 − 47.7		5				d	(10)

Table 5.1 (*cont.*)

Date	Time	Epicentre N° — E°	i_0	r_0	r'	M	q	
1622 Oct. 4	12	27.1 – 56.4	3		100		b	
1622		39.6 – 45.4	5				e	(11)
1623		29.9 – 52.9	3				c	
1641 Feb. 5	18	37.9 – 46.1	2	19	400	6.8	b	
1648 Mar. 31	24	38.3 – 43.5	2	18	300	6.5	b	
1659		39.4 – 46.3	4				c	
1664		38.1 – 46.3						(12)
1665		35.7 – 52.1	2			6.5	c	
1666		32.1 – 50.5	2		300	6.5	c	
1673 Jul. 30		36.3 – 59.3	2		310	6.6	b	
1677		27.9 – 54.2	2		260	6.4	b	
1678 Feb. 3	06	37.2 – 50.0	2			6.5	c	
1678	24	34.3 – 58.7	2			6.5	c	
1679 Jun. 4	06	40.2 – 44.6	2		200	6.2	a	
1687		36.3 – 52.6	2			6.5	e	
1687 Apr.		36.3 – 59.6	3				d	
1695 May 11	24	36.8 – 57.0	2	25	500	7.0	b	
1696 Apr. 14		39.1 – 43.9	1+	20	400	7.0	b	
1703		26.6 – 54.9			300+	6.8	c	
1705		30.5 – 47.8	5				d	
1709		37.3 – 49.6	5				d	(13)
1713		37.3 – 49.6	5				d	(13)
1714		36.2 – 44.0	5				d	(14)
1715 Mar. 8	06	38.4 – 43.9	2	14	320	6.6	b	
1716 Jan. 23		40.2 – 44.5	5				d	(15)
1717 Mar. 12	06	38.1 – 46.3	2+			5.9	c	
1720		38.1 – 46.3	5				d	(16)
1721 Apr. 26	07	37.9 – 46.7	1	24	700	7.7	b	
1752		29.6 – 52.5	5				d	(17)
1755 Jun. 7	12	34.0 – 51.4	2+			5.9	b	
1759 Nov. 13		39.2 – 45.4	5				d	(18)
1765 Apr. 23	08	29.6 – 52.5	4				d	
1765		25.4 – 65.8					c	
1766		27.7 – 54.3	3				e	
1769 May 1	12	33.3 – 44.4	5				e	
1778 Dec. 15	24	34.0 – 51.3	2	14	210	6.2	b	
1780 Jan. 8	24	38.2 – 46.0	1	31	700	7.7	a	
1780			2			6.5		Khurasan
1784 Mar. 1	24	29.5 – 52.4	4				c	
1786 Oct.	18	38.3 – 45.6	2		230	6.3	b	
1802		35.6 – 45.4	3				d	
1803		36.4 – 48.8					d	
1804		36.3 – 57.2	3		150		c	
1805		36.2 – 52.4	2+		100		c	
1806 May		38.1 – 46.3	4				d	(19)
1806		29.0 – 50.8	4				d	(20)
1807 Jul. 11	12	38.3 – 45.2	3	15	130		b	
1808 Feb. 23		28.5 – 53.6	3		100		c	(21)
1808 Jun.		38.5 – 44.9	3		20		d	(22)
1808 Jun. 26	18	35.3 – 54.5	2		320	6.6	b	
1808 Oct. 9	24	36.2 – 52.4	5				d	(23)
1808 Dec. 16	18	36.4 – 50.3			200	5.9	b	
1809	12	36.3 – 52.5	2	15	290	6.5	b	
1810		38.0 – 57.2	2			6.5	c	
1812 May 14	24	38.1 – 46.3	5				d	(24)
1812 Jun. 23	14	38.1 – 46.3	5				d	(25)
1812		29.8 – 52.4	3				c	
1812		38.9 – 45.6	4				d	
1815 Jun.		35.9 – 52.2	4	30			d	
1819 Jan.		38.0 – 46.5	4				c	
1820 Jun.		38.1 – 46.3	5				d	(26)

Table 5.1 (*cont.*)

Date	Time	Epicentre N°	E°	i_0	r_0	r'	M	q	
1823 Dec.		38.1 −	46.3	5				d	(27)
1824 Jun. 2		29.7 −	51.6	3	15	140		b	
1824 Jun. 25	05	29.8 −	52.4	2	14	250	6.4	b	
1824 Aug. 28		29.8 −	52.4	4				b	(28)
1824 Dec. 30		29.8 −	52.4	5				b	(29)
1825		36.1 −	52.6	2	24	360	6.7	b	
1825 Oct.		29.6 −	52.5	4				c	
1826 Jun. 14		29.6 −	51.6	5				d	(30)
1827		33.2 −	46.1	3		150		b	
1828 Aug. 6	24	39.8 −	46.7	5				d	(31)
1828 Aug. 14	12	39.8 −	46.7	5				d	(31)
1829 Mar.	18	26.7 −	55.3	4				d	(32)
1829 Apr. 24	12	38.5 −	44.9	5				d	(33)
1830 Mar. 27	12	35.7 −	52.5	2	37	570	7.1	b	
1830 Apr. 6	24	35.9 −	52.6	3				b	(34)
1831		38.1 −	46.3	5				d	(35)
1832 Dec. 8		38.7 −	48.8	5				d	(36)
1833		37.3 −	58.1	3	16	320	6.2	b	
1834		39.7 −	43.7	3		180+	6.0	b	
1837 Jun.		38.2 −	44.8	3		150		c	
1838		29.6 −	59.9	1		300+	7.0	b	
1840 Jan. 22	18	38.7 −	48.8	5		20		d	(37)
1840 Jul. 2	19	39.5 −	43.9	1	27	520	7.4	a	
1843 Apr. 18	08	38.7 −	44.9	2+	19	190	5.9	b	
1844 May 12	18	33.6 −	51.4	2	14	250	6.4	a	
1844 May 13	19	37.4 −	48.0	2	21	430	6.9	b	
1845 Jul. 9	13	38.4 −	45.5	3		100		b	(38)
1849		27.7 −	57.6	3				c	(39)
1851 Apr. 9	16	40.0 −	47.3			240	6.2	c	(40)
1851 Apr. 19	17	25.1 −	62.3	3				d	(41)
1851 Jun.	12	36.8 −	58.4	2	24	450	6.9	b	
1852 Apr. 25	21	35.9 −	54.3	3		100		c	(42)
1853 May 5	12	29.6 −	52.5	2	12	200	6.2	b	
1853 Jun. 5		31.3 −	51.9	2+		110	5.5	b	(43)
1853 Jun. 11		32.6 −	50.3	3		140	5.5	b	(44)
1854 Oct. 1	15					200	5.9		(45) Caspian
1854 Nov.		30.5 −	57.3	2		130	5.8	c	(46)
1856 Oct. 4	20	38.2 −	46.5	4		50		a	
1857 Sep. 6		38.3 −	45.4	3	5	90		b	(47)
1859 Jun. 4	06	37.7 −	47.1	3		100		b	(48)
1861 Feb. 28	17	39.2 −	47.9	3+		150		c	(49)
1861 May 24	16	39.4 −	47.5i	3		260	6.0	c	(49)
1862 Dec. 19	05	39.3 −	47.8i	3		280	6.1	c	(49)
1862 Dec. 21	10	29.5 −	52.5	2+	12	250	6.2	b	
1863 Dec. 30	22	38.2 −	48.6	2+	13	240	6.1	b	
1864 Jan. 17		30.6 −	57.0	2		150	6.0	b	
1864 Dec. 7	20	33.3 −	45.9	2		260	6.4	b	(50)
1865 Jun.		29.6 −	53.1	2		150	6.0	b	(51)
1865		27.2 −	53.1	2	6	100	5.6	b	(52)
1868 Mar. 18	17	39.6 −	47.6	3		260	6.0	c	(53)
1868 Aug. 1	20	34.9 −	52.5			300+	6.4	c	(54)
1871 Aug. 4	14	30.6 −	57.0	3		120		c	(55)
1871 Dec. 23	18	37.4 −	58.4	2	19	650	7.2	b	
1872 Jun.		34.7 −	47.7	2	5	170	6.1	b	(56)
1874 Feb.		31.9 −	50.8	3				a	(57)
1875 Mar. 21	15	30.5 −	50.5	3	23	190	5.7	b	(58)
1875 May		31.2 −	56.3	2		130+	6.0	b	
1876 Sep. 28	03	33.1 −	49.7	2+	16	160	5.8	b	(59)
1876 Oct. 20	15	35.8 −	49.8	2+	19	150	5.7	b	(60)
1877		30.1 −	57.6	2	5	100	5.6	b	(61)
1879 Mar. 22	04	37.8 −	47.9	2	20	380	6.7	a	

Table 5.1 (*cont.*)

Date	Time	Epicentre N° E°	i_0	r_0	r'	M	q	
1880 Jul. 4		36.5 – 47.5	3	16	150	5.6	a	
1880 Aug.	12	27.1 – 54.1	3	15	130		a	(62)
1880		32.0 – 50.7	3		120		a	(63)
1883 Apr. 28	17	37.0 – 56.2	3	15	190	5.8	b	(64)
1883 May 3	12	37.9 – 47.2	2	10	210	6.2	b	(65)
1883 Oct. 16	24	27.7 – 52.3	3	20	210	5.8	a	(66)
1884 May 19	24	26.9 – 56.0	3	12	120		a	(67)
1890 Feb. 7	06	34.2 – 51.3	3	5	70		a	(68)
1890 Mar. 25		28.8 – 53.5	2	14	250	6.4	a	(69)
1890 Jul. 11	24	36.6 – 54.6	1	19	380	7.2	a	
1891 Dec. 14	24	29.9 – 51.6	3	6	100		b	(70)
1892 Aug. 15	12	29.1 – 52.7	3	7			c	(71)
1893 Nov. 17	15*	37.0 – 58.4	2	21	570	7.1	a	
1894 Feb. 26	24	29.5 – 53.3	2+	12	180	5.9	b	
1895 Jan. 17	11	37.1 – 58.4	2	11	380	6.8	a	
1895 Jul. 8	22*	39.9 – 53.7i	2	45	900	7.5	b	(72)
1895		34.2 – 51.3	3	5	50		c	(73)
1896 Jan. 4	16*	37.8 – 48.4	2	18	340	6.7	a	
1897 Jan. 10	21*	26.9 – 56.0	2		250	6.4	b	(74)
1897 May 22		31.3 – 56.3	3		150	5.5	c	(75)
1897 May 27	21*	30.6 – 57.0	3	6	120		b	
1898 Jan. 15	04*	36.6 – 54.7	3	12	130		b	(76)

References to table 5.1
  (1) Tarikh-i Sistan:160.
  (2) Michel:iii,105; Bar Hebraeus:139/152.
  (3) Al-Tabari:iii/3,1443; Ibn al-Athir:vii,53; al-Suyuti:26.
  (4) Ibn al-Athir:vii,269; al-Suyuti:28; Sani' al-Dauleh, Muntazam.
  (5) Ibn al-Athir:viii,476.
  (6) Ibn al-Jauzi:viii,272; al-Suyuti:35.
  (7) Al-'Umari:fol.65r.
  (8) Ibn al-Jauzi:x,44.
  (9) Ibn al-Fauti:262; Taher 1979:199/128.
 (10) Poser 1675 sub anno.
 (11) Mallet 1852:70; Stepanian 1942.
 (12) Hajji Khalifeh:137; see note 3.183.
 (13) Rabino 1917:291; Sutudeh 1970.
 (14) Al-'Umari:fol.225v.
 (15) Hakobyan 1956:522.
 (16) Hakobyan 1951:294.
 (17) Al-'Umari:fol.238v.
 (18) Hakobyan 1956:526.
 (19) Jaubert 1821:137.
 (20) Morier 1812:57; Ouseley 1819:i,194; Ritter 1840:viii, 781.
 (21) Dupre 1819:i,452; Ritter 1840:vii,756.
 (22) Tancoigne 1820:319.
 (23) Trezel 1821:418.
 (24) Freygang 1816:278.
 (25) Ouseley 1819:iii,407.
 (26) Lumsden 1822.
 (27) Arago 1859.
 (28) IO LPS 9 83 no.23.
 (29) Mallet 1854:164.
 (30) Alexander 1827:112.
 (31) Mallet 1854:194.
 (32) Whitelock 1838:177.
 (33) Armstrong 1831:167.
 (34) Conolly 1838:i,15; Bell 1840:579.
 (35) Smith & Dwight 1834:321.
 (36) Byus 1948:31.
 (37) Mallet 1854:292.
 (38) Burgess 1942:82; Abich 1857:51.
 (39) Abbot 1855:51.
 (40) Vaqa'i'-yi ittifaqiyyeh 1267:22; San' al-Dauleh, Mir'at:ii, 65; Musketoff & Orloff 1893.
 (41) Merewether 1852:284.
 (42) Vaqa'i'-yi ittifaqiyyeh 1268:67; Iran 1336:145; Sani' al-Dauleh, Mir'at:ii,110.
 (43) Brugsch 1862:ii,125; Windt 1891:161; Ambraseys 1979.

 (44) Ambraseys 1979.
 (45) Brugsch 1862:ii,477; Perrey 1864:23.
 (46) Vaqa'i'-yi ittifaqiyyeh 1271:201; Bunge 1860:annex; Ambraseys et al. 1979.
 (47) Vaqa'i'-yi ittifaqiyyeh 1274:347,351; Gobineau 1859: 509; Perrey 1860:99,105; 1862:25.
 (48) Eastwick 1864:197; Dalyell 1862:62; Vaqa'i'-yi ittifaqiyyeh 1275:436.
 (49) Brugsch 1862:ii,381; Perrey 1864:83,94,101; 1865:106; 1866:32.
 (50) Perrey 1866:91; 1867:49; Arch.minist.Aff.Etrang. (Paris): Dossier Coupur. Perse 1850–70.
 (51) Cf. Perrey 1867:74; Ballore 1906:209.
 (52) Lorimer 1915.
 (53) Perrey 1872a,b.
 (54) Ruznameh Daulati:1285.6.7,11; San' al-Dauleh, Muntazam:iii,308; Ballantine 1879:174.
 (55) Iran 1288:41.
 (56) Fuchs 1886:484; Khwandaniha 1336:26; Schmidt 1879: 31.
 (57) Garthwaite (in preparation); Houtum Schindler 1879:53.
 (58) Rivandeneyra 1880; Schmidt 1879:335.
 (59) Iran 1293:301; Houtum Schindler 1879:60; Ambraseys 1979:58.
 (60) Iran 1293:299,300,307; Rodler 1890.
 (61) Houtum Schindler 1881b; Ambraseys et al. 1979.
 (62) Muvahid 1970; Lorimer 1915.
 (63) Stack 1882:ii,67; Bell 1889a,b; Rodler 1888:201; Bishop-Bird 1891:375; Sawyer 1891:48; Ambraseys 1979.
 (64) Sani' al-Dauleh, Matla':98; Gazet.Persia 1910:i; Iran: 1300.7.15.
 (65) Wilson 1896:168,224; Kavkaz 1883:123–4; Iran: 1300.7.15; al-Qusi:195; Musketoff & Orloff 1893:484.
 (66) IO R 15 1 193; Iran:1301.1.25; 1300.12.18.
 (67) Iran 1301:547; Kababi 1963; Lorimer 1915.
 (68) Iran:1307.7.5; Windt 1891:121.
 (69) Iran 1307:714–15.
 (70) Iran 1309:762.
 (71) Iran:1310.2.19.
 (72) References in Atlas and instrumental data.
 (73) Stürken 1906:121.
 (74) Anonymous 1899; Lorimer 1915; Sykes 1902b:127; The Times:1897.3.13; Kababi 1963:129; Agamennone 1897; Kortazzi 1900; Rudolph 1903.
 (75) Iran 1315:914; the date of the event is given as the last ten days of Dhu 'l-Hijja 1314, probably 20/22 May 1897.
 (76) Habl al-Matin:1315.10.20.

Table 5.2. *Recent period: 1900–79*

Date	Time (GMT) hrs : mins	Macroseismic epicentre N° E°	q	Instrumental epicentre N° E°	Q	M	m	i_0	r_0	r'	
1900		31.85 − 47.16	d					3			(1)
Feb. 24	0030	38.45 − 44.87	a			5.4*		3	9	120	
1901 May 20	1229	36.39 − 50.48	c			5.4*		3	3	110	(2)
1902 Jul. 9	0338	27.08 − 56.34	b	27.00 − 56.00	10E	6.4*		3	20		(3)
1903 Feb. 9	0518	36.58 − 47.65	b			5.6		3	10	200	(4)
Mar. 22	1435	33.16 − 59.71	c	35.00 − 60.00	10E	6.2		2			
Jun. 24	1656	37.48 − 48.96	b	39.00 − 49.00	10E	5.9		3	16	140	(5)
Sep. 25	0120	35.18 − 58.23	a	34.00 − 58.00	10E	5.9	5.3	2	13	170	
1904 Nov. 9	0328	36.94 − 59.77	b			6.4		3		430	(6)
1905 Jan. 9	0617	37.00 − 48.68	c	38.00 − 46.00	10E	6.2		3		300	(7)
Apr. 25	1401	27.67 − 56.03	b			5.8		3	30	180	(8)
Jun. 19	0127	29.89 − 59.98	b			6.0	6.8	3	10		(9)
1907 Apr. 17	0836	37.74 − 57.85	b			5.8*		3	8		(10)
Jul. 4	0912	27.18 − 56.28	d	27.00 − 55.00i	10E	5.7	6.0	5			(11)
1908 Jul. 26	0332	34.59 − 61.49	c			5.3		3		130	(12)
Sep. 28	0628			38.00 − 44.00	10E	6.0					
1909 Jan. 23	0248	33.41 − 49.13	a	33.00 − 53.00	5E	7.4	7.2	1+	20	600	
Oct. 27	1845	30.09 − 57.58	b			5.5		3	5	75+	(13)
1911 Apr. 18	1814	31.23 − 57.03	a	32.00 − 56.00	5E	6.2	6.7	2	17	330	
Apr. 29	0533	30.36 − 57.58	b			5.6	6.4	3	7		(14)
Sep. 13	0322	27.67 − 54.44	b			5.5		3	10		(15)
1913 Mar. 24	1034	26.80 − 53.70	b	27.60 − 53.91	2C	5.8		2+	13	120	(16)
1914 Feb. 6	1142	28.67 − 64.75	c	29.50 − 65.00i	5C		5.9	3+		130+	(17)
1917 Jul. 15	1758	33.48 − 45.82	b	33.50 − 46.50	10D	5.6	6.3	3	17	380	(18)
Aug. 29	13--	37.37 − 58.05	b			(5.7)		3	12	180	(19)
Oct. 24	11--	36.94 − 54.31	b			(5.3)		2+	10	100	(20)
Nov. 28	1442	37.18 − 57.88	b			5.9		3	20	170	(21)
1918 Mar. 24	2314	35.08 − 60.69	b	34.50 − 57.10	6C	5.9	6.3	2	20		(22)
1919 May 12	2230	36.19 − 44.01	b			(5.7)		3	10		(23)
Oct. 24	2032	26.11 − 62.05	b	27.50 − 63.60	6−	5.6		3	10		(24)
1923 May 25	2221	35.19 − 59.11	a	34.63 − 61.49	2C	5.8		2+	10	140	
Sep. 14	0810			28.97 − 59.33	2C	5.6					
Sep. 17	0709	37.63 − 57.21	a	37.97 − 56.50	1D	6.3	6.1	2+	13	260	
Sep. 22	2047	29.51 − 56.63	a	29.20 − 56.90	1B	6.7	6.9	2	19	330	
Nov. 29	0336	33.62 − 59.40	b	31.20 − 61.60	6D	5.6		3	25		(25)
1924 Feb. 19	0701	39.00 − 48.32	c	38.59 − 48.50i	1B	5.9	6.8	3		280	(26)
Jun. 30	0341			27.50 − 53.80	6D	5.8					
1925 Sep. 24	0438	26.60 − 55.40	d	25.51 − 55.38	2D	5.5	6.1				(27)
1927 May 9	1031			27.68 − 56.70	1B	5.8	6.4				
Jul. 7	2006			27.00 − 62.26i	3B	5.7	6.4				
Jul. 22	0355	34.90 − 52.90	c	34.72 − 53.64	1A	6.3	6.9			310	(28)
Nov. 12	1446	32.53 − 47.38	a	32.39 − 46.97	1C	5.6	6.0	3	8	180	(29)
1928 Mar. 8	1814	31.54 − 60.10	b	31.45 − 60.16	2C	5.5		3	25	150	(30)
1929 May 1	1537	37.73 − 57.81	a	37.86 − 57.65	1A	7.3	7.1	1	26	700	
Jul. 15	0744	32.08 − 49.48	b	32.06 − 49.60	1A	6.0	6.3	3	13	300	
Sep. 3	1207			26.59 − 62.07i	3B	5.6	5.9				
1930 May 6	2234	38.24 − 44.60	a	38.15 − 44.65	1A	7.2	7.0	1	23	400	
May 11	2235			27.70 − 55.27i	1B	5.8	5.8				
Aug. 23	1053			27.88 − 55.02	2C	6.1	6.2				
Oct. 2	1532	35.76 − 51.99	a	35.86 − 52.08	2C	5.2		3	7	80	
1931 Apr. 27	1650	39.48 − 46.09	a	39.34 − 45.97	1A	6.4	6.3	3+	23	350	(31)
1932 Sep. 8	0725	31.59 − 58.15	c	30.99 − 58.66	1B	5.6				180	(32)
1933 Oct. 5	1329	34.52 − 57.07	b	34.58 − 57.31	1B	6.0	6.2	3	30	330	(33)
Nov. 28	1109	32.01 − 55.94	a	32.10 − 56.02	2B	6.2	6.4	2+	12	160+	
1934 Jan. 2	2055	29.97 − 57.42	b	30.08 − 57.56	1B	5.6		3	15		(34)
Feb. 4	1327			30.65 − 51.64	1B	6.3					
Feb. 22	0807	38.76 − 45.94	b	38.23 − 45.04i	1B	5.7	6.1	4	48	160	(35)
Jun. 13	2210			27.63 − 62.64i	3B	6.6	6.9		(1000)		(36)
Oct. 29	1615	39.90 − 47.80	b	40.66 − 49.01	1B	5.7		4		200	(37)
1935 Mar. 5	1026	35.94 − 53.06	a	35.91 − 53.21	1B	5.8		3	10	190	

Table 5.2 (*cont.*)

Date	Time (GMT) hrs : mins	Macroseismic epicentre N° E°	q	Instrumental epicentre N° E°	Q	M	m	i_0	r_0	r'	
Apr. 11	2314	36.36 − 53.32	a	36.59 − 53.61	1A	6.3	6.8	2	15	220	
1936 Apr. 21	0214			26.29 − 55.28i	2B	5.5	6.2				
Jun. 10	0329			26.50 − 64.00	6C	5.7					
Jun. 30	1926	33.68 − 60.05	b	33.67 − 60.45	1B	6.0	6.2	2+	15	220	(38)
1938 Jan. 26	0340	33.44 − 45.78	b	33.12 − 45.87	2C	5.4	5.7	3		130	(39)
Feb. 14	0254			40.39 − 53.68	1B	6.2					(40)
Apr. 23	0926			27.22 − 53.28	2D	5.5					
Dec. 19	1856	36.64 − 58.50	c	36.23 − 57.96	2D	5.6				180	(41)
1939 Sep. 19	0323			38.36 − 57.34	1B	5.6	6.0				
Nov. 4	1015	32.40 − 48.52	c	32.60 − 49.02i	1B	5.7	6.0	4		150	(42)
Nov. 8	1721	36.32 − 57.97	b	36.33 − 58.12	2C	5.5	6.3	3	10		(43)
1940 May 4	2101	35.76 − 58.53	b	35.91 − 58.51	1A	6.4	6.2	2+	20	100+	(44)
1941 Feb. 16	1639	33.41 − 58.87	a	33.47 − 58.92	1A	6.1	6.4	2	13		
Jun. 10	2038			33.50 − 46.84	1B	5.5					
1943 Feb. 6	0236			24.89 − 63.25	3B	5.9	6.2				
1944 Jul. 23	1200	29.86 − 56.82	b	30.50 − 55.00	9D	5.5		3	25	60+	(45)
1945 Jan. 15	1721			27.04 − 54.85	2C	5.5					
Nov. 27	2156			25.02 − 63.47	1B	8.0	7.6			700	(46)
1946 Mar. 12	0221	29.80 − 51.45	b	29.72 − 51.72	2C	5.7	5.7	3	20	160	(47)
Jun. 20	0037			29.50 − 66.00	6D	5.8					
Jul. 27	1625	35.60 − 45.83	b	35.76 − 45.97	2C	5.5		3	25		(48)
Nov. 4	2147	39.32 − 55.20	b	39.82 − 54.65	1A	6.9	7.4	2+	32	650	(49)
1947 Aug. 5	1424	25.25 − 63.20	c	25.06 − 63.44	2B	7.0	7.6	3	20	400	(50)
Sep. 23	1228	33.67 − 58.67	a	33.42 − 58.66	2C	6.8	6.4	2	21	310	
Oct. 3	0613			25.99 − 57.41	2D	5.8	5.8				(51)
1948 Jun. 18	1844	37.53 − 57.96	b	37.49 − 57.73	1B	5.5		3	20	150+	(52)
Jul. 5	1353	29.88 − 57.73	b	29.46 − 57.78	2C	6.0	5.9	2+	20	180	
Oct. 5	2012	37.88 − 58.55	a	37.79 − 58.44	1A	7.2	7.3	1+	12	640	(53)
1949 Apr. 24	0422	27.28 − 56.46	a	27.22 − 56.42i	2C	6.3	6.1	3	19		(54)
1950 Jan. 19	1727	27.31 − 52.83	a	27.29 − 53.04	1B	5.5	5.8	2+	12	140	(55)
May 9	1116	37.93 − 58.37	b	38.34 − 58.41	1A	5.8	6.1	3		280	(56)
1951 Dec. 30	1821	27.14 − 57.08	c	28.34 − 57.21i	1A	5.5	5.9	4			(57)
1953 Jan. 15	2006	31.07 − 56.78	b	31.30 − 56.90	8−	5.5	5.6	2+	7		(58)
Feb. 12	0815	35.39 − 54.88	a	35.40 − 55.08	1A	6.5	6.9	2	22	290	(59)
1956 Oct. 31	1403	27.27 − 54.55	a	27.21 − 54.39	1A	6.3	5.9	2+	14		(60)
1957 Jul. 2	0042	36.07 − 52.47	a	36.14 − 52.70	1A	6.8	7.0	2	20	320	
Dec. 13	0145	34.58 − 47.82	a	34.35 − 47.67	1A	6.7	6.5	2	18	380	(61)
1958 Aug. 16	1913	34.30 − 48.17	a	34.36 − 47.86	1A	6.6	6.2	2+	11	280	(62)
1960 Apr. 24	1214	27.70 − 54.29	a	27.72 − 54.44	1A	5.8	6.0	2+	6	100+	(63)
1961 Apr. 6	1812	28.10 − 56.80	a	27.80 − 56.70	1A	5.3	5.7	4	20	170	(64)
Jun. 11	0510	27.78 − 54.51	a	27.90 − 54.53	1A	6.5	6.4	2	17	280	(65)
1962 Apr. 1	0045	33.21 − 58.87	a	33.39 − 58.87	1A	5.5	5.5	3	9	110	
Sep. 1	1920	35.71 − 49.81	a	35.59 − 49.85	1A	7.2	6.9	2	26	650	
Oct. 1	1213	27.76 − 54.00	c	27.91 − 54.77	1A	5.5	5.8			100	(66)
Nov. 6	0009	28.17 − 55.79	a	28.09 − 55.57	1A	5.4	5.6	3	7	160	(67)
1963 Mar. 24	1244	34.50 − 48.02	a	34.37 − 47.80	1A	5.8	5.5	2	4	130	(68)
1964 Aug. 19	0933			28.21 − 52.63	1A	5.5	5.3				
Dec. 22	0436	28.12 − 56.80	a	28.20 − 56.94	1A	6.1	5.6	3	12		(69)
1965 Jun. 21	0021	28.17 − 56.01	a	28.14 − 55.99	1A	5.4	5.7	3	9	120	(70)
1966 Sep. 18	2043	27.81 − 54.21	a	27.86 − 54.30	1A	5.4	5.9	3	17	100+	(71)
1967 Jan. 29	0756	26.56 − 54.89	c	26.49 − 55.24	1A	5.5	5.1			170	(72)
1968 Apr. 29	1701	39.28 − 44.27	a	39.24 − 44.23	4−	5.5	5.3	3	7	170	(73)
Aug. 31	1047	34.02 − 58.96	a	34.04 − 59.02	1A	7.4	6.0	1	25	650	
Sep. 1	0727	34.05 − 58.23	a	34.10 − 58.28	1A	6.4	5.9	2	17	250	
Sep. 14	1348	28.34 − 53.23	a	28.34 − 53.16	1A	5.9	6.1	3	12	140+	(74)
Nov. 15	0625	38.10 − 58.25	a	37.60 − 58.50	4−	5.7	5.4	3	9	170	(75)
1969 Jan. 3	0316	36.92 − 57.78	a	37.12 − 57.90	1A	5.5	5.4	3	7	120	(76)
Nov. 7	1834	27.42 − 60.40	c	27.86 − 60.04i	1A	6.5	6.1			450	(77)
1970 Feb. 23	1122	27.83 − 54.64	a	27.82 − 54.50	1A	5.6	5.2	3	13		(78)
Jul. 30	0052	37.67 − 55.89	a	37.85 − 55.94	4−	6.6	5.8	2+	10	310	(79)

Table 5.2 (*cont.*)

Date	Time (GMT) hrs : mins	Macroseismic epicentre N°	E°	q	Instrumental epicentre N°	E°	Q	M	m	i_0	r_0	r'	
1971 Feb. 14	1627	36.62	55.69	a	36.58	55.63	1A	5.5	5.9	3	14	180	(80)
Apr. 12	1903	28.37	55.75	a	28.39	55.62	1A	5.8	6.0	3	11	170	(81)
May 26	0241	35.50	58.30	a	35.56	58.14	1A	5.6	5.8	3	12	140	(82)
Sep. 8	1253	29.03	60.19	a	29.10	60.04	1A	5.6	5.3	3	16	100+	(83)
Nov. 8	0306				27.07	54.46	1A	5.9	5.6			150	(84)
Dec. 9	0142	27.29	56.38	a	27.26	56.43	1A	5.7	5.3	3	13	170	(85)
1972 Apr. 10	0206	28.38	52.98	a	28.39	52.78	4—	6.9	6.3	2	20	450	
1973 Nov. 11	0714	30.57	53.04	a	30.53	53.00	4—	5.5	5.5	4	10	120	(86)
1975 Mar. 7	0704	27.47	56.44	a	27.47	56.25	4—	6.1	5.9	2+	13	220	(87)
Dec. 24	1148	26.98	55.67	b	27.04	55.50	4—	5.5	5.7	3	10		(88)
1976 Apr. 22	1703	28.80	52.11	b	28.71	52.12	4—	5.7	6.0	3	8	130	
Nov. 7	0400	33.82	59.19	a	33.86	59.23	4—	6.4	5.8	3	10		(90)
Nov. 24	1222	39.12	43.92	a	39.05	44.04	4—	7.3	6.2	1	21	470	(91)
1977 Mar. 21	2118	27.59	56.45	a	27.59	56.38	4—	6.9	6.2	3	9		(92)
Apr. 6	1336	31.90	50.76	a	31.99	50.70	4—	6.1	5.6	3	13	120	(93)
Jun. 5	0445				32.64	48.08i	4—	5.7	5.8			360	
Dec. 19	2334	30.90	56.61	a	30.93	56.48	4—	5.7	5.8	3	9	150+	
1978 Feb. 10	2050				25.30	62.40	7—	5.5	5.1				
Sep. 16	1536	33.40	57.12	b	33.39	57.43	7—	7.3	6.7	1+	27	610	
Nov. 4	1522				37.67	48.90	7—	6.0	6.2				
Dec. 14	0705				32.14	49.65	7—	6.1	5.9				
1979 Jan. 10	1505				26.52	61.01	7—	6.0	5.6				
Jan. 16	0950	33.80	59.50	b	33.90	59.47	7—	6.8	6.0	2+	18		
Nov. 14	0221	33.91	59.81	c	33.92	59.74	7—	6.6	6.0	3	10		
Nov. 27	1710	34.05	59.63	c	33.96	59.73	7—	7.1	6.1	2	15		

References to table 5.2

(1) Gubbins 1944.
(2) *Iran* 1319:995; Gulriz 1961; Willey 1963:228.
(3) *FO* 60 651; *Habl al-Matin* 1320:38,40,41; *Erdbebenwarte* 1902:2.1; 1903:3.162.
(4) *Tsentr.Seism.Kom.* 1905:48.
(5) *Tiflis.Fiz.Obs.Ezhemes.* 1903:6; *Tsentr.Seism.Kom.* 1905:116.
(6) *Tsentr.Seism.Kom.* 1905:184; Oddone 1907:322; *Atlas.*
(7) *Tsentr.Seism.Kom.* 1907:14; cf. Sieberg 1932:816 & figure 442.
(8) *FO* 248 842; *Muzaffari* 1329:43; Monti 1907:211; al-Qusi:249; Lorimer 1915.
(9) *FO* 248 846; *Gazet.Persia* (Nasratabad Sipi).
(10) *Tsentr.Seism.Kom.* 1909:68; *Tiflis.Fiz.Obs.Ezhemes.* 1907:4—6.
(11) *FO* 248 902; *Tsentr.Seism.Kom.* 1909:141.
(12) *FO* 248 939; Patterson 1908:70; Sieberg 1917:89.
(13) *FO* 248 968.
(14) *FO* 248 1030.
(15) *IO* LPS 7 253.
(16) *IO* LPS 10 827.
(17) *IO* LPS 10 814.
(18) Gubbins 1944; *IO* LPS 10 827; Mitchell 1958:137.
(19) *FO* 248 1173; *Iran* 1335:119.
(20) *Iran* 1336:145.
(21) *FO* 248 1173; *Atlas.*
(22) *FO* 248 1206.
(23) Hay 1921:265; Gubbins 1944.
(24) *IO* LPS 10 814.
(25) Stratil-Sauer 1950:annex; Ambraseys & Melville 1977.
(26) *Shafaq-i Surkh* 1302:89,90; *Atlas.*
(27) *FO* 371 11490.
(28) *Iran* 1306:2443—8.
(29) *Times* 1927.11.19.
(30) *FO* 371 13063; *Shafaq-i Surkh* 1306.12.20.
(31) *Atlas*; Shebalin 1974; Abdalian 1935*b*.
(32) *IO* LPS 12 3403, 3413.
(33) In Ambraseys & Melville 1977, read Surkhu for Sur, and Ma'den Ozbaku for Maydan.
(34) *FO* 371 17907; *Bidari* 1312:190.
(35) *Shafaq-i Surkh* 1312.12.6; *Atlas.*
(36) *India Weather Review* 1935; *Bombay Bul.* 1936:D.11.
(37) *Atlas.*
(38) *IO* LPS 12 3406; *Ittila'at* 1315.4.11.
(39) Gubbins 1944.
(40) *Atlas.*
(41) *IO* LPS 12 3406; *Ittila'at* 1317.9.30—10.7; *Iran* 1317.10. 1—7.
(42) *Iran* 1317.8.15—23.
(43) *IO* LPS 12 3406; *Ittila'at* 1317.8.18.
(44) Ambraseys & Melville 1977.
(45) *Ittila'at* 1323.5.3.
(46) See: Case histories, § 3.3
(47) *Pars* 1324.12.23; *Iran-i Ma* 1324.12.23; *Ittila'at* 1324.12. 23—4; Stahl 1962*a*:10.

(48) Hitchen 1946.
(49) *Atlas*; Shebalin 1974.
(50) See: Case histories, 27 November 1945.
(51) *Iran* 1326.7.15.
(52) Rustanovich 1967. '
(53) *Ittila'at* 1327.7.15—19; *Atlas*; Shebalin 1974; Krumbach 1949; Tchalenko 1975.
(54) *Ittila'at* 1328.2.5—6.
(55) *Ittila'at* 1328.12.1—4; James & Ghashghaie 1960.
(56) *Atlas*; Rustanovich 1967.
(57) *Ittila'at* 1330.10.18.
(58) Mislocated by *Bul.Seism.Soc.Amer.* 1953:183, at Rayin, in lieu of Rayhun.
(59) Abdalian 1953; Ambraseys & Moinfar 1977*c*.
(60) *Ittila'at* 1335.8.10—15; Muvahid 1970:19.
(61) Ambraseys *et al.* 1973; Tchalenko & Braud 1974.
(62) Ambraseys & Moinfar 1974*a*; Tchalenko & Braud 1974.
(63) James & Ghashghaie 1960; Stahl 1960; Afshar 1960; Mustafavi 1963:368—9.
(64) Red Lion & Sun statistics.
(65) *Ittila'at* 1340.3.22—31; *Bul.Seism.Stn.Shiraz* 1961.
(66) *Bul.Seism.Stn.Shiraz* 1962.
(67) *Kaihan* 1341.8.17—18; *Bul.Seism.Stn.Shiraz* 1962.
(68) Ambraseys & Moinfar 1974*b*; Tchalenko & Braud 1974.
(69) *Ittila'at* 1343.10.7—8.
(70) *Bul.Seism.Stn.Shiraz* 1965.
(71) *Ittila'at* 1345.6.28—9.
(72) *Ittila'at* 1345:11; *Kaihan* 1345.11.10.
(73) Ambraseys & Moinfar 1977*b*; Nabavi 1970.
(74) Ambraseys & Moinfar 1977*b*.
(75) *Atlas*; Ambraseys & Moinfar 1977*b*.
(76) Ambraseys & Moinfar 1976.
(77) Ambraseys & Moinfar 1976.
(78) Ambraseys *et al.* 1975.
(79) Sobouti & Eshghi 1970; Ambraseys *et al.* 1971; Ambraseys *et al.* 1975.
(80) Goudarzi 1975.
(81) *Ittila'at & Kaihan* 1350.1.25—31; Berberian 1976:378.
(82) Red Lion & Sun statistics.
(83) *Ittila'at* 1350.6.18—20.
(84) Moinfar *et al.* 1973.
(85) Moinfar *et al.* 1973; Berberian 1976:383 gives 19 December.
(86) *Ittila'at & Kaihan* 1352.8.21—4.
(87) Moinfar 1975; Berberian 1976: 387.
(88) Red Lion & Sun statistics; *Ittila'at & Kaihan* 1355.1.21.
(89) *Ittila'at* 1355.2.4.
(90) Goudarzi & Ghaderi-Tafreshi 1976; Khoshbakht-Marvi 1977.
(91) Toksöz *et al.* 1977; Arpat *et al.* 1977; Gülkan *et al.* 1978; Berberian 1977:179.
(92) Berberian & Papastamatiou 1978.
(93) Ambraseys 1977*a*; Berberian 1977:51; Ambraseys 1979.

Notes

Chapter 1

1 Ibn Sina, *al-Shifa* v, ed. 'Abd al-Halim Muntasar, Cairo 1965: 15−19 and Zakariya Qazvini, *'Aja'ib al-makhluqat*, ed. Wüsten-feld, Göttingen 1849: 149. See also the Introduction in Taher 1979.

2 Further works of the eighteenth century are introduced by Taher 1975, but with the exception of Hamid al-'Imadi's *al-Hauqala fi 'l-zalzala*, they are merely portions of larger works, such as Ibn al-Jauzi's *Shudhur al-'uqud fi dhikr al-'uhud*, ms. in Cairo National library, no. 95. Similar occidental catalogues were published by Coronelli 1693, Seyfart 1756, in the *Dressdnische Gelehrte Anzeigen* 1756: ii−xi, by Berryat 1761 and Walther 1805. Other works containing catalogues of earthquakes in general but very little on Persia are those of Huot 1837, Hoff 1840, Perrey 1845−75, Tholozan 1879, Schmidt 1879, Oldham 1883, Fuchs 1886 and Milne 1911. Most of these catalogues do not give their sources of information and some of them contain a rich variety of errors. Exceptions are the catalogues of Musketoff 1891, 1899, Musketoff & Orloff 1893 and to some extent Mallet 1850−8. For a critical review of Perrey's and earlier work, see Vogt 1979.

3 See for example accounts of events in Baghdad in 1094, 1118 and 1130 discussed by Melville 1978: 96−7 and of events in Nishapur in the thirteenth century in Melville 1980: 109−10.

4 Meteorite falls have been noted by Ibn al-Jauzi, *Mukhtasar*: fol. 85, in Tabaristan in 238/853; in Gurgan in 375/985 by Sibt ibn al-'Ajami: viii, 8; in the late 1870s near Gwarko in Baluchistan (28.36°N-60.69°E), Dyer 1921: 85; and on 16 October 1883 in southwest Fars (27.76°N-52.19°E), *IO* R 15 1 193. Durand 1902:14 mentions a meteorite fall at Ardal some years before his visit in 1899, but in 1977 no-one in the region knew anything about it. The impact of the meteorite at Khash (28.20°N-61.25°E) on 22 October 1929 was widely felt, *Ittila'at*: 1308, no. 871, but again there is no evidence of cratering. Gojkovic 1973: 310, mentions the presence of at least fourteen roughly circular features scattered over an area of some 150 square kilo-metres southeast of Rayin, near Qal'eh Hasan 'Ali (29.39°N-57.56°E), which he attributes to early meteorite falls.

5 See chapter three, events in 912, 1008 and 1304. A similar juxta-position of events is often found in later Persian and European sources.

6 For a review of political history in the Caliphate period, the reader is referred to volumes four and five of the *Cambridge History of Iran* (*Camb. Hist. Iran*), which contain maps of the areas controlled by different dynasties. The natural barrier of the central desert played its role, confining some rulers to the south

and west (e.g. the Buyids) and others to Khurasan in the north and east (e.g. the Tahirids, Samanids and Ghaznavids); Azarbaijan to the northwest was generally a separate sphere of events.

7 Mustaufi, *Nuzhat*: 75–6, says that in the 300 years after the 1042 earthquake, shocks were frequent in Tabriz, but were not serious and caused little or no damage. See also Melville 1981.

8 Al-Mas'udi, *Tanbih*: 49, says frequent and violent shocks were well known there; for the later decline, see Aubin 1959: 296 and Lewis 1966: 114.

9 Minorsky 1964: 57; Le Strange 1905: 299 contrasts the grouping of settlements in Fars with the isolation of towns in Kirman, separated by broad stretches of uncultivated land.

10 It was found to be a large and prosperous city by Nasir-i Khusrau: text 91, trans. 249, who passed through in April–May 1052, evidently shortly before the earthquake. For the Buyids, see Busse 1975: 262ff.

11 Abu Dulaf: text 27, trans. 60.

12 Not, at least, in the local histories of Zarkub or later, Fasa'i.

13 Ibn al-Faqih: 228, 257; al-Muqaddasi: 384.

14 Mosul is outside the area covered by this book, but it is represented on the maps in this chapter to illustrate the situation north of Baghdad, on the basis of data in Melville 1978: 91ff. and Alsinawi & Ghalib 1975. The latter give the impression that the Mosul and Baghdad districts were of similar seismicity, which is not borne out by the sources: damage to Mosul was consistently far more severe.

15 Some light is thrown on affairs in the province by Bosworth 1977b and Aubin 1979. The caravan trade through Kirman seems to have been at the mercy of continuous depredations by the Baluchis.

16 The mountains on either side of Kuhbanan are said to have been ruptured, evidence of which is still visible at a locality called Gaud-i Zalzaleh-zadeh ('the hollow shaken by an earthquake'), twelve kilometres from Kuhbanan on the road to Asfich (31.32°N-56.27°E and at 31.25°N-56.37°E).

17 We may note that a local Kirman newspaper, ironically named *Bidari* (The awakening), made no mention of the strong earthquake on 28 November 1933, which caused destruction in Kuhbanan and Buhabad and was felt in Kirman and Yazd. It seems unlikely that an event of similar magnitude in the eleventh–twelfth century would have been recorded.

18 The work gives a detailed history of Sistan up to 1056, and a more superficial coverage of the period 1072–1324.

19 Mustaufi, *Nuzhat*: 201–2 says the gold mine was ruined in the time of the later Ghaznavids; they were supplanted by the Saljuqs in Sistan in 1053, see Bosworth 1977a: 44. Although Mustaufi admits the story is merely for the amusement of ignorant people, it is worth noting that Ibn al-Jauzi: vii, 207 reports the discovery of a gold mine in Sistan in 390/1000, authenticating part of Mustaufi's story. Tate 1910: 110 refers to legends associated with earth movements in localities near areas of volcanic activity in Sistan.

20 Ambraseys & Melville 1977. An alternative interpretation would be that they occurred near a main route (hence the survival of information) but not near an important city (hence no places mentioned). The first two shocks are given by Syriac sources, however, and a lack of precision is not surprising in this early period.

21 Abu Dulaf apparently describes the effect of the shock in the Samalqan valley. Gardizi, himself of Khurasanian origin, probably derived his notice of the earthquake from the work of Jaihani, a contemporary Samanid writer, which is now lost but was much quoted. Ibn al-Athir probably owes the same debt.

22 Al-Mas'udi, *Tanbih*: 49; he refers to the proximity of Mount Damavand.

23 Planhol 1968: 414–15; a more favourable picture of the state of affairs in Azarbaijan is given by Aubin 1977a.

24 Minorsky 1964: 46–52; for a political history of the period, see the relevant chapters in *Camb. Hist. Iran*: v and in the *Cambridge History of Islam* (*Camb. Hist. Islam*): i.

25 Mustaufi, *Nuzhat*: 164ff. describes highways to the south (Baghdad via Hamadan); east (Transoxania via Khurasan); north (Darband and Ganja); west (Qonya via Tabriz and Erzerum) and southeast (Qais and Hurmuz via Isfahan and Shiraz). He also has a sixth highway, southwest to Qal'eh Bira (Birecik on the upper

Euphrates), but he does not give the route there (across Kurdistan and via Mosul?). The itineraries of various contemporary travellers are found in several volumes of the *Hakluyt* series and in *Purchas his pilgrimes* (1905).

26 Only the first two of these events are recorded by Persian sources, the rest by al-'Umari.

27 Cf. Lockhart 1960: 69; the establishment of Sultaniyyeh had the effect of removing Qazvin from the main flow of traffic for most of the period. Tabriz recovered its position much more quickly. The few European descriptions of Sultaniyyeh at this time are sparse and no local sources have been seen.

28 The 1384 earthquake should not be dismissed purely because it is reported in an unreliable guise, almost as a legend. It may be that al-'Umari, who mentions the 1495 shock, would also have recorded anything more serious occurring in the region.

29 Clavijo left a valuable account of his journeys in 1404–5; much later, in 1556–7, Sidi 'Ali Re'is gives few details of interest.

30 All three events are mentioned by al-'Umari, so it is irrelevant to view this in terms of the completeness of coverage of the area in Persian sources; there is no way of estimating al-'Umari's comprehensiveness.

31 The accounts are repetitive, based on that of Hasan-i Rumlu: this is one of the rare occasions when Persian historians behave like Arabic annalists. Qayin does not seem to have been a particularly important area and it is not clear why contemporary authors should mention this shock rather than a number of others that they ignore, but it was accompanied by a good story, which may have helped.

32 Mustaufi, *Nuzhat*: 143, Le Strange 1905: 355, Ambraseys & Moinfar 1975.

33 The earthquakes are variously mentioned by Egyptian and European authors (Hurmuz and Lar) and Persian sources; of the latter, Nimdihi was a local, writing in the Deccan and the other, Natanzi, was a Safavid chronicler.

34 *EI*[1] : 'Lur-i Buzurg'; Planhol 1968.

35 An earthquake in Mosul in 661/1263 is given by al-Suyuti: 50, and shocks in Kurdistan in 1310 and 1503 are given by al-'Umari, see chapter 3.

36 Le Strange 1905: 301, Aubin 1977b; the Qutlugh Khans have their own dynastic history.

37 Marco Polo: 88ff.; Oderic calls Yazd ('Geste') the third best city in Persia, *Hakluyt* (ed. Glasgow 1904): iv, 410. The voyages of 'Abd al-Razzaq and Nikitin are both found in *Hakluyt* 22, 1857.

38 For the trips of Jenkinson, Alcocke, Edwards *etc.*, see *Hakluyt* 72 & 73, 1886. The travels of Barbaro are found in *Hakluyt* 49, 1873.

39 Melville 1978: 183, Perry 1979: 312; a useful list of travellers of the period is given by Gabriel 1952. For French and Russian sources see Chaybany 1971 and Petroff 1955 respectively.

40 These have been collected by Hakobyan 1951, 1956. For similar sources of the Caliphate period see Garegin 1951, and for the Mongol period, Khatjikyan 1950, 1955 and Sanjian 1969.

41 We have no accounts of the city by Europeans after the Safavid period; in 1729, Hazin: 169 found it ruined and desolated by the recent Turkish invasion. Olson 1975 has a useful summary of Ottoman–Persian relations and trade rivalries during the period in question. See also Melville 1981.

42 Olivier 1807: iii, 49, 94; cf. Hambly 1964 for the state of Persia at the end of the eighteenth century.

43 See Alsinawi & Ghalib 1975, whose catalogue is particularly inaccurate for the eighteenth century.

44 Thevenot made the journey in 1664, Lucas in 1701, Otter in 1737 and Olivier in 1796.

45 Rabino 1917 records two further events in Rasht in 1709 and 1713, for which no original source has yet been found, see table 5.1.

46 Forster, travelling from Qandahar to Sari in 1783–4, comments on the depopulation of the regions he crossed. Trade in the area, though busy, was small in volume and local in extent.

47 Hambly 1964, Perry 1979: 246–7. Kirman was almost wiped out in 1794 by Agha Muhammad Qajar.

48 Planhol 1968: 434, 445–7; see also the useful chapter on the Qajars by Lambton 1970, esp. pp. 432, 449.

49 Records of the Indo–European Telegraph Department are kept at the India Office and numerous items of scattered information

are contained in British diplomatic correspondence files. A useful account of the early development of the network is given by Preece 1879. A great bulk of material exists that could permit identification of precise periods when various sections of the line were out of operation, a phenomenon particularly common in the lines managed by Persians as opposed to British officials, and some of these intervals, often very long, could doubtless be correlated with gaps in the record of events; but such painful analysis was not considered likely to be rewarding and it has been assumed that periodic failures in the service have not significantly affected the availability of news. It is anyway not easy to identify any systematic bias.

50 Approximately 430 shocks are recorded for the nineteenth century, including aftershocks and minor tremors, which have been pruned from the catalogue (see table 5.1).

51 See Rabino 1913: 304 and compare with Browne 1914: 10–11; issue 471, under the new name, should be dated 16 August 1860, i.e. 28 Muharram 1277. Hashimi maintains that the change came with the next issue. After 1283/1866, newspaper production was reorganised and the illustrated *Ruznameh-yi daulat-i 'aliyyeh-yi Iran* became the *Ruznameh-yi daulati*; an unillustrated paper of the same name appeared at the same time and would probably be a more useful source of data, but this has not been seen. See Hashimi 1948: ii, 306–7.

52 Unfortunately only very inadequate notes were taken when this set was read in 1975; it is possible the collection was incomplete. We should note that three of the earthquakes that occurred during this decade but which have not been found in the press (1864 west of Turbat-i Haidariyyeh, 1865 in Kirman and in Shiraz) are mentioned by Sani' al-Dauleh, who certainly used newspapers as a source of information, as is clear from his identical reports of earthquakes later in the Qajar period.

53 The actual date of the disappearance of the first series of *Iran* is obscure. In 1903 the name was changed to *Iran-i sultani*, and this paper is said to have continued only till 1906, Browne 1914: 89. The first year of this paper has been read from the Tehran University library collection, i.e. to March 1904, after which, according to the catalogue, its name reverted to *Iran*. Reference continues to be made to a paper called *Iran* in contemporary works of the constitutional period (e.g. in *FO* 251 72, for December 1908), but no copies from these years seem to have been preserved in any of the Tehran libraries we used, nor are details given in any of the works available on the Persian press.

54 Nineteen shocks reported in the 1870s, rising to forty-one in both the following decades. The change of emphasis coincides with the appearance of a twin paper, *Ittila'a*, in 1298/1881. Coverage of the years 1881–6 has been read at the Tehran University library, but the paper is concerned exclusively with foreign affairs: this aspect of the contents of *Iran* thus makes way for more room on home news. Browne 1914: 43 incorrectly puts the establishment of the paper in 1295 H., cf. Hashimi 1948: i, 194–8.

55 This was the first daily paper to appear in Persia, running from 1898–1903; issues for this period have been read in the Majlis library.

56 Published in Calcutta in 1893, this paper covered the whole Islamic world. News from Persia was thus not the special interest of the paper, although it became particularly influential through its coverage of the Persian revolution. Issues for 1897–1902 have been read in the Tehran University and Majlis libraries, and Browne's collection at Cambridge for the constitution period.

57 Selection of these particular papers was largely arbitrary, depending on their availability in sufficient numbers to cover the period of interest. Browne's very fragmentary collection of *Muzaffari*, published in Bushire, has been read at Cambridge, covering mainly 1907–11. *Iran-i nau*, from August 1909 to August 1911, but with several gaps, has been read in the same collection. The total output of *Istiqlal-i Iran*, May 1910–August 1911, has been read at the Majlis.

58 *Aftab* appeared irregularly and the Melli collection is missing many issues for 1912–13. *Irshad* was also examined at the Melli, for the period March 1914–October 1915, but yielded no earthquake data, failing to mention even those few events reported by *Aftab*: it was therefore considered to be of no use as a source.

59 This second run, to issue 1489, has been read at the Melli. Spor-

adic reference has also been made to *Shafaq-i surkh*, founded in 1922; year two, March 1923–March 1924, was read at the Majlis, but adds nothing to the information found in *Iran*.

60 For the same period, some sixty shocks are reported in diplomatic sources, of which only ten are also found in the press; similarly, fifty-four events, excluding aftershocks, have been identified from instrumental data, for twenty of which no macroseismic information is available.

61 This statement includes files covering the eighteenth century and earlier East India Co. correspondence, which has not been very fully preserved.

62 Of the reports read, the most important are: 1819–20 Muraviev, 1834 Khodzko, 1834–5 Korff, 1832–6 Karelin, 1836 Blaremberg, Khodzko; 1838–9 Lemm, 1842 Beresin, 1843–4 Voskoboinikoff, 1848 Gamazoff, 1849–52 Chirikoff, 1852 Khanikoff, Gamazoff; 1856 Seidlitz, 1858–9 Bunge *et al.*, 1875 Ogorodnikoff, 1879–81 Zinoviev, 1881–2 Lessar, 1894 Baumgarten, 1898–1901 Zarudnoi and 1910 Trubitzkoi. Full references can be found to these reports in Gabriel 1952 or in A.T. Wilson's *Bibliography of Persia*, Oxford 1930. An annotated list of Russian travellers in Persia is given by Petroff 1955.

63 The Neutral zone was assigned to Britain in 1915.

64 *FO* 248 968. Almost immediately, there was an amusing response to these instructions. Nasrullah Khan, left in charge of the Kirman consulate, reported on 28 October 1909 the occurrence of a fifteen-second tremor felt in Kirman: 'as it is not stated in your circular to what class refer the earthquakes which do not damage buildings, I give it class 4.' Class 0 might have been appropriate. Barclay sarcastically noted on the margin of this letter, 'an unclassified earthquake! Dreadful!! The circular told the consuls not to report unimportant ones. Mr Nasrullah probably measured this one's importance by his own *état d'âme*.' The reporting of small tremors is, nonetheless, valuable and frequently helps to establish the radius of perceptibility of more distant and destructive shocks. The total response to Milne's circular can be seen in the BAAS bulletin for 1911, pp. 18–26.

65 The register of correspondence for 1911 (*FO* 566 1179) has the tantalising entry 'Earthquakes', but the relevant file seems to be missing. Similarly for 1932, a file entitled 'Persia: earthquake in Tabriz and Eran districts; question of all earthquake shocks being reported by consuls for information of F.O.: minute by Mr Gaselee' has also not been located. Its existence suggests that systematic reporting of events went on at least till 1932, which does indeed appear to be the case.

66 In the latter case, *Iran* no. 1375 does not report the date or time of the earthquake, nor in later issues are there details of the villages affected or number of casualties. The other paper consulted, *Shafaq-i surkh* for 16 June 1923, merely refers to aftershocks in the Turbat region: its previous issue appeared on 24 May, two days before the earthquake, which is therefore not even recorded in this paper. By contrast, Mashhad consular diaries for 26 May and following weeks give a detailed report, with a list of all the villages affected, and an account of the event was included in the *Annual Report for Persia*, 1923, see *FO* 371 9035 & 10153. This is a typical example of the tendency of the press to concentrate on the most important town in an affected area rather than the real epicentral region.

67 Collections of diaries from Mashhad and Bushire have been read up to 1940, Kirman diaries to 1939 and various despatches from Tehran up to 1935. Some of the files at the Public Record Office are subject to a fifty-year rule; see *The records of the Foreign Office 1782–1939*, HMSO 1969: 93–4.

68 Stepanian 1942, Byus 1948–55, Rothé 1969, Razani & Lee 1973a, *Atlas* 1977, Buniyatov 1977.

69 Pinar & Lahn 1952, Abdalian 1935a, Ambraseys 1968, 1974b, Ergin *et al.* 1967, Rustanovich 1967, Berberian 1976, 1977, Alsinawi & Ghalib 1975, Heuckroth & Karim 1970 and, for earlier events, Ambraseys 1961.

70 Wilson 1930, Sieberg 1932; these highly inaccurate works have for many years been standard references on the subject for both orientalists and seismologists. Together with Willis's (1928) catalogue they epitomise the twentieth-century trend towards undiscriminating cataloguing; cf. for instance Ambraseys 1962b.

71 See for instance Melville 1978 for Persia, Taher 1979 for the Middle East, Gouin 1979 for Ethiopia and Vogt 1979 for

France. Notable examples of proper cataloguing in earlier works are Baratta 1901 for Italy and, to a lesser extent, Davison 1924 for Great Britain.

Chapter 2

1 These places were badly shaken in pre-historic times, in 1177, 1721?, 1876; 1238, 1678; and 1696 respectively.

2 The Farsi title is *Farhang-i abadiha-yi kishvar.*

3 In the region northwest of Birjand, an elder of Muhammadabad gave us valuable information about the 1941–7 and earlier earthquakes of the region. His story, partly heard from older people, strongly suggested that these events were associated with faulting. This he not only described in great detail, but also showed us, still discernable about ten kilometres from his village. A few years later we returned for more data and to install instruments near the fault zone, but in the meantime our informant had died. However, his son, whom we had not met before, to our surprise volunteered exactly the same information as his father and guided us along almost the entire length of the zone, about which he had heard as a child when he helped his father collect bulbs in the spring along the margin of the salt flats (*daqq*).

4 Ambraseys & Moinfar 1975, 1977c. One other recorded instance of this is perhaps in the 819 earthquake near Balkh, see chapter 3.

5 Some of these locations are in Sistan ($32.35°$N-$60.27°$E), Khurasan ($37.27°$N-$59.10°$E; $37.62°$N-$58.86°$E) and Mazandaran ($36.43°$N-$54.22°$E; $34.43°$N-$50.18°$E; $36.35°$N-$53.33°$E).

6 For example, the Bastak–Lar earthquake of 1911. Shocks that began in June culminated on 13 September with an earthquake that destroyed 150 houses in the Jadid quarter as well as the bazaar of Fath 'Ali Khan, killing ten people. Considerable damage was done to all houses and water *birkehs* (pools), see *IO* 7 251. The shock was recorded by a few stations and was remembered in Jahrum as being responsible for the loss of a number of camels (!) near Juyum.

7 Lumsden 1822: 140; Tancoigne 1820: 73; Burgess 1942: 60; Saint John 1876: 92. There is evidence, however, that there are other reasons why houses were restricted to one storey. One of them, perhaps not too serious, is that people disliked rooms raised above the ground floor, because these would enable neighbours to observe their domestic arrangements, and many tales have been recorded of the fierce opposition to this breach of privacy, including that involving the Ottoman ambassador who built a second storey from which it was possible to see something of the ladies of the British Legation, Arnold 1877: i, 242. Even minarets, it is said, had to be pulled down in Kirmanshah because they commanded a view of the interior of houses, Mitford 1884: ii, 342, Serena 1883: 308.

    Because of the absence of timber, fire was not a threat to Persian towns except for those in the Caspian provinces and a few other mountain regions, Binning 1857: ii, 151. The palace at Ashraf was burnt down after 1627, Ouseley 1819: iii, 271, and Rasht was seriously damaged by fire in 1757, 1777, 1779, 1886, 1899 and in October 1902. Lahijan was burnt down *c.* 1505 and again in 1648, Rabino 1917. Fires are known to have caused some damage in Quchan, in the Kopet Dagh as well as in Azarbaijan, but this was generally minor; (see, e.g. reports in *Vaqa'i'-yi ittifaqiyyeh*: 1298, no. 56). In the south, after the earthquake of 10 January 1897 in Qishm, the people replaced their ruined adobe houses with permanent timber huts. However, a year later these caught fire and were all burnt down, Kababi 1963: 129.

8 Wulff 1966; Planhol 1968, with references.

9 For a critical paper on the misuse of modern materials in Iran, see Moinfar 1969. For the misuse of materials and methods of construction in rural areas, see Razani & Lee 1973a, b, Razani 1974, Ambraseys & Tchalenko 1969a, 1969b, Ambraseys *et al.* 1972.

10 Binning 1857: ii, 237, is of the opinion that the custom of allowing all buildings to fall into wrack and ruin is universal in the East. No one, he says, will repair what his predecessor has built; he prefers, if he can afford it, to build something new, which may bear his own name and flatter his vanity. In this way, all great works of utility and beauty are allowed to tumble in ruins.

This opinion was shared by most of the nineteenth century and earlier travellers, e.g. Eastwick 1864.

11 See for instance, Godard 1937: 115, 1941: 11.

12 *Gazette de France*: 1769.11.3; *IO* R 15 1 194.

13 Fryer 1698: 256, 336. He notices that the caravanserai at Kushk-i Zard, the ruins of which are near the Pulvar-rud south of Abadeh, had suffered foundation failure. See also Price 1832: i, 31, Brugsch 1862: i, 326, Mounsey 1872: 307, Feuvrier 1899: 343, Mitford 1884: ii, 6 and Brayley-Hodgetts 1916: 219. In December 1668 Shiraz lost one third of its houses in a freak storm and a terrible flood, Chardin 1811: viii, 435.

14 *Shafaq-i surkh*: 1320, no. 76; *Habl al-matin*: 1320, no. 40; *Muzaffari*: 1329, no. 32.

15 Morier 1812: 278, Freygang 1816: 279, Johnson 1818: 212, Ouseley 1819: iii, 406 and Porter 1821: ii, 502.

16 After the 1871 earthquake in Quchan a new kind of emergency shelter came into vogue; it made an ingenious use of old timbers pulled out of the ruins, which were set up A-shaped against a ridge-pole. The interstices were filled with bushes and the whole then plastered with mud. When a more pretentious house was needed, a second A-shaped structure was erected parallel to the first, and the intervening space was walled in and bridged with a flat roof. Shelters of this type of one to three rooms were tested by the earthquake of 1893, then improved and tested again successfully by the 1895 shock. They were still in use in 1904. For a description of the shelter, see MacGregor 1879: ii, 83, 85, Huntington 1905: 236 (see plate 27), as well as the report of Khan Bahadur Maula Bakhsh in *FO* 248 611 and 612. See also Curzon 1966: 1, 109.

17 Unintentional damage to early structures was also caused by extensive burrowing by treasure hunters. Much of the destruction of the Qubbeh-yi Sabz in Kirman was the handiwork of a former governor, who heard a rumour that treasures existed under the great dome, Sykes 1897: 581. For a description of the construction methods used to build *qanats*, see Beckett 1953, and *EI*. The importance of this type of water supply for the survival of a Persian village is discussed by Landor 1902: i, 374. Quite often, lines of *qanats* would collapse when flooded by rainstorms, their remains occasionally resembling throughgoing features that could look like normal faulting. Petermann 1861 describes one of these features between Bideh and Aghda (*c.* $32.42°$N-$53.80°$E), which proved on examination to be a collapsed line of *qanats*. The existence of blind fish in the *qanats* was noticed by Stewart 1881: 132 and Seymour 1951: 152. They were examined by Smith 1953 and Greenwood 1976.

18 For this type of dwelling, see Andrews 1973.

19 Tchalenko & Ambraseys 1973. One of the seven investigators assessed Intensities from a helicopter and another from a fourday long trip, during which Intensity mapping should have been carried out at a rate of 1200 square kilometres per day!

20 For the first half of this century, information about earthquake damage, casualties, and the geographic co-ordinates of the areas affected in the territory of the USSR was not allowed to be published, Vladimirov 1972: 38.

21 It is amazing how few seismologists in Europe and the Middle East have had the opportunity of studying earthquake effects and structural damage in detail in the field. This they consider to be the job of the engineer and yet they quite happily assess Intensities from a distance, a few hours after an earthquake.

22 The frequence coincidence between earthquakes and plague occurrences, presumably as a result of contamination or breakdown or the water supply systems, is discussed by Kremer 1880.

23 The first national census was not made available until late in the 1950s. Before that it was difficult to assess the number of inhabitants in any rural, particularly tribal area, or in most urban areas. As a matter of fact, no census was taken even within a *shahristan* before 1930. The people not only disliked it, but also considered it wrong, and connected census with either a poll tax or conscription. As late as 1885 they looked upon it as a wicked proceeding, Tancoigne 1820: 96; Morier 1818: 110; Brugsch 1862: i, 228; Stewart 1911: 198; Huntington 1905. Our own field estimates of casualties for recent earthquakes have uncertainties not less than 35%. Quite often official estimates were made by subtracting the number of survivors in villages from the census number of the population.

24 The figures seventy and particularly forty are often mentioned in connection with aftershock duration. For the use of these numbers, see Roscher 1909.

25 For casualties and behaviour among people during earthquakes, see Lomnitz 1970.

26 See Ambraseys 1977b. The earthquake rocking response of monolithic columns was tested in the laboratory by Aslam *et al.* 1980; their results are not applicable to articulated columns. See also Housner 1963.

27 Harrison 1936: 35; Harrison & Falcon 1937; Stein 1940: 206, 229; Harrison 1946. The Shimbar slide has a volume of about 300 million cubic metres and it is located at 32.39°N-49.64°E, Busk 1926. The Irene slide is south of Durud, Ambraseys 1974a. Historical examples of such events occurred in 872 and 913. Al-Ghazzi, in Taher 1975: 71, transmits a story that in Dinavar there was a shrine at which sheep were sacrificed every year. Whenever the sacrifice was forgotten, the village moved. One year, there was a great earthquake and the shrine and all its surroundings were moved bodily to the top of a nearby hill and nothing was altered; but the village was engulfed.

28 The earthquake of 1641 triggered landslides that overwhelmed villages in the region south of Usku, and in 1910 the Jaraghil cave dwellings were destroyed by similar events, *Iran-i nau*: 1329, no. 79.

29 Abu Dulaf: 58; *FO* 248 842–3.

Chapter 3

1 There is no doubt about the old form of the name of Ray. It is in Old Persian Raga, in Avestan Ragha, known to the Greeks as Rhagae, mentioned in the Apocryphal Old Testament: *Tobit.* 4.1 as Ragau. As a matter of fact the Old Persian reference to Raga belongs to the sixth century B.C. (Bihistan Inscription). On the other hand, in Greek the root of the verb 'rent' is *rhag*. However, the attempt by the Greeks to explain the name as that of a place rent by earthquakes should be regarded simply as popular etymologising, although of course no less interesting from our point of view because of that.

2 Al-Suyuti: 12.

3 Mustaufi, *Tarikh*: 276. He says the shock occurred in 95 H. after a fatal outbreak of plague in Basra; aftershocks lasted for forty days. This is probably the earthquake mentioned by Ya'qubi, *Tarikh*: ii, 349 as occurring in 94 H. and it is to be identified with the widespread earthquakes in Syria that year, which were also said to have lasted forty days, see Ibn al-Athir: iv, 460.

4 *Tarikh-i Sistan*: 127. This local history is the sole source for the event and the description is inadequate to give a precise idea of the magnitude of the disaster. The fact that it is mentioned at all suggests, however, that it was serious, for the historical narrative of this period is telegraphic, being little more than a date list of the most outstanding events. The survival of the information over such a long period also suggests the significance of the earthquake. Zarang, as capital, was also called Shahr-i Sistan, cf. Le Strange 1905: 21.

5 Orbelean: i, 51/186. It was after this earthquake that the upper Arpa-chai was renamed 'Vayots-dzor' i.e. Valley of lament, Kirakos Gandzaktsi: 434, cf. Hübschmann 1904: 348, 469. In the *Atlas* this earthquake is wrongly dated 427 A.D.

6 Theophanis: 843, puts the event in 6235 of the Alexandrian era/ 25 March 743 to 24 March 744, before al-Walid became the Arab Caliph on Thursday 16 April. The same information is given by Anastasius: 140 and Cedrinos: 461, who date this in the 3rd year of Copronymous/19 June 742 to 18 June 743. These details suggest the earthquake occurred between 25 March and 18 June 743, but it should be noted that Muslim sources put the accession of al-Walid on Wednesday 6 Rabi' II, 125/6 February 743, i.e. the previous Alexandrian year, al-Tabari: ii/3, 1740. The formal accession of al-Walid followed immediately on the death of Hisham. It is hard to decide whether the dating of the earthquake is prejudiced by this inaccuracy.

The survival of this notice in Byzantine sources suggests that the earthquake, which occurred on a major trade route of great antiquity through northern Persia, was a large magnitude event.

See figure 3.1 and a recent discussion of the location of the Caspian Gates, between the plains of Varamin and Khuvar, by Hansman 1968. Hoff 1840 and Mallet 1852 wrongly identify the Caspian Gates (Caspiae Portae) with the pass of Dariel in the Caucasus.

Plate 27. A-shaped shelters put up in Quchan after the earthquakes of 1893 and 1895. Other views of this type of construction are shown in Huntington (1905: 236) and MacGregor (1879: 85).

7 Agapius: viii/3, 544; Michel: ii, 524; Bar Hebraeus: 114/124. Agapius places the event in the tenth year of al-Mansur and Michel puts it in 1076 Sel. which creates some problems. Al-Mansur's tenth year was 146/March 763 to March 764; Michel's date is about a year and a half later, 1076 Sel. = October 764 to October 765. Michel puts in May the same year the appearance of a comet which went behind the sun and stayed there forty days, an event which Agapius: 542 appears to place much earlier and which must have been the comet reported for 15 May 760, Grumel 1958: 471. Agapius's date for the earthquake is also suspect, however, as it is not consistent with the other event he mentions that year, namely the fall of Kabul to the Muslims. This did not happen until 152/769, al-Tabari: iii/1, 369. A comparison of Agapius with al-Tabari's account of al-Mansur's reign does not show a consistent disagreement between them. Although some doubt must remain as to the correct place of the earthquake in the sequence of events, Agapius's account must derive from an independent source and his date be accepted.

It is to be noted that details of the earthquake have survived in Syriac but not Arabic sources.

Khurasan extended far beyond its present limits in north-east Iran, designating all the east up to the Oxus and including the mountainous district of Kuhistan.

8 *Tarikh-i Sistan*: 174. As with the previous events, the fact that this earthquake is mentioned at all suggests its importance. No later shocks are reported by this local source, implying that a period of relative seismic quiescence followed the paroxysm, see p. 6. The Hirmand (Helmund) river dried up at Bust twenty years later, the ensuing famine and plague also contributing to the decline of the province; cf. Bosworth 1968: 105n.

9 The correct date of the event, Dhu'l-Hijja 203, is given by Abu'l-Fida: ii, 26, and Ibn al-Shihna: vii, 59, whose accounts follow Ibn al-Athir. Sani' al-Dauleh, *Muntazam*: i, 19 incorrectly puts the earthquake lasting seventy days in Khurasan in 23/644, thus omitting a diacritic point. Hajji Khalifeh mentions a great earthquake that lasted a long time in Khurasan in 201/817 and this inaccurate date is repeated by Sani' al-Dauleh, who also refers to the earthquake of 203/819, thus duplicating the same event under two different dates.

10 Ibn al-Jauzi, quoted by al-Suyuti: 24.

11 Ibn al-Athir: vi, 252; that damage occurred in both Transoxania and the towns of Faryab and Taliqan to the west seems unlikely, and it is possible that *Ma vara' al-nahr* (Transoxania) should stand for Marv al-Rud, or lesser Marv, i.e. Bala Murghab. Qudama says that the regions of Marv and Tukharistan were affected. A shock of this magnitude would certainly have been felt across the Oxus, but probably without damage.

12 Qudama: 210. Similar effects of large-scale liquefaction of desert flats have been reported during the earthquakes of 1838 and 12 February 1953, Ambraseys & Moinfar 1977c.

13 Hamza: 187; Ibn al-Jauzi, *Mukhtasar*: fol. 82r; Ibn al-Athir: vi, 371; al-Suyuti: 25; al-Jazzar in Taher 1974: 152; Shushtari 1952: 30. The correct year is given by Hamza.

14 Ibn al-Jauzi, *Mukhtasar*: fol. 83v; al-Suyuti: 26. It is most likely that the reference is to Herat in Afghanistan, but it should be borne in mind that there was also a town of this name in Fars, near Istakhr, Yaqut: iv, 959.

15 Ibn al-Athir: vii, 53; al-Qusi: 82. Kariman 1970: ii, 244 quotes a poem by Qavam Razi saying that 350 000 people died in Ray. The figure bears a vague resemblance to that given for the casualties of the next event in Qumis.

16 Ya'qubi, *Tarikh*: ii, 600; this major earthquake is mentioned by numerous later authors whose accounts, although sometimes confused, permit a reasonable idea of the main effects of the event. See in particular al-Tabari: iii/3, 1433; Ibn al-Athir: vii, 53; Bar Hebraeus, *Duwal*: 143; Muhammad Tahir: 93.

17 Adle 1971: 85; the ruins of Shahr-i Qumis are east of Qusheh, some thirty kilometres southwest of Damghan, Hansman 1968. There is some archaeological evidence that a much earlier earthquake had destroyed a site near Damghan, Houtum Schindler 1877a. The Tazareh—Damghan fault zone that passes about fifteen kilometres north of Damghan (Geol. Map of Iran, sheet 6862) shows every sign of Quaternary activity, particularly the segment between Taq and Rudbar which is Recent, Krinsley 1972. There is no dating evidence to associate this fault with the Qumis earthquake. Plate 28 shows a segment of the fault zone ten kilometres north of Damghan.

18 Mirkhwand: ii, 478, is the first author to refer to Bustam, which he does on the authority of Ibn al-Jauzi. Bustam was later described as being larger than the capital, Damghan (*Hudud al-'alam*: 135). At the beginning of the tenth century Ibn Rustah: 169, remarks on the ruined houses destroyed by the earthquake at Haddadeh on the road out from Damghan to the east.

19 Ibn al-Jauzi, *Mukhtasar*: 85. His account lists many more places than those mentioned by al-Tabari and forms the model of many later descriptions of the events of this year, such as Ibn al-'Imad and al-'Umari's.

20 Al-Mas'udi, *Tanbih*: 49; al-Suyuti: 26; Mihdi: 86; cf. Tahiriya 1968: 8.

21 Indications of the effects of the earthquake outside the epicentral area are vague. Of the towns of Khurasan, only Nishapur is specifically mentioned as affected, but there is no evidence of serious damage there. Perhaps the Nishapur district, rather than the capital itself is meant. The term 'Khurasan' did not at this time necessarily denote the same area as that called Khurasan today. It should be noted that Ya'qubi, *Buldan*: 276, refers to Damghan as the western-most city of Khurasan, which shows the extent of the area dependent on Nishapur.

22 Ibn al-Jauzi; also al-Suyuti and most later authors. Mirkhwand omits Qum and Kashan, although his account is based on Ibn al-Jauzi. Again, there are no details of damage in these cities, which probably mark the limits of an area of 450 kilometres radius in which the earthquake caused concern. Ray had suffered a major earthquake the previous year 241/855 (see above), possibly also affecting Qum and Kashan, so that there may be some confusion of two separate earthquakes being associated with the same date. Some authors put the Qumis event in 241/855, e.g. Hamza: 189, who inaccurately quotes al-Tabari, and Zamakhshari (see Yaghma'i 1947: 10); Bafqi: iii/2, 891 gives the earthquake in 240 H.

There is no reason to suppose that these towns were not affected in 242/856 even though it seems improbable that they were all destroyed, as Mihdi: 86 describes. Quoting 'most of the histories' he puts the earthquake in Rabi' I, 239/August 853, a date which is only otherwise found in a work quoted by Yaghma'i. Mihdi's date, which is clearly inaccurate, is followed by Houtum Schindler 1896, and in *EI*²: 'Isfahan'. Although Mihdi's account of the Qumis earthquake represents the final exaggeration of the descriptions deriving from earlier sources, it is clear that the attention given to the earthquake in historical works reflects the importance of the event, which must have been strongly felt throughout much of Persia. Al-Suyuti's reading of Tunis for Qumis, however, clearly stretches the size of the earthquake beyond the limits of the possible.

As for the simultaneity of the shocks in the various places mentioned, this is not specifically stated by the earliest authors, who refer to several places being shaken by earthquakes in that year. Ibn al-Jauzi, and the later authors on whose information we largely rely for an adequate account of the event, state clearly that the effects of the shock were felt everywhere at the same time.

23 Mustaufi, *Nuzhat*: 75, Karbala'i: 16, Qazvini: fol. 148r. Both the latter authors refer to this event being predicted, cf. Karbala'i: 275. Chardin 1811: ii, 336, misquotes Mustaufi, putting the shock in 234 H., but more importantly, he puts the next event (434/1042, see below) in 235/849, a date that has been adopted for a genuine earthquake in many catalogues, e.g. Berberian 1976: 406. See further, Melville 1981.

24 Al-Suyuti: 27 reports this event together with the earthquakes in Syria in 245 H.

25 Al-Tabari: iii/3, 1439; Ibn al-Athir: vii, 56; al-Suyuti: 27. Confused with the large earthquake in Syria by Wilson 1930, Sieberg 1932, Alsinawi & Ghalib 1975.

26 Yovhannes Kat'oghikos: 169.95.120; Thomas Artsruni: xxii.184; *Codex Sinaiticus 34*; Moses Kaghankatuatsi *ad an.*; Asoghik: ii.2.80; Samuel Anetsi: 312; Brosset 1874: 185; Garitte 1958. Inaccurate dates for earthquakes in Dvin in 851, 858, 865 and 869 are given by Abich 1882; Stepanian 1942; *Atlas*.

27 Al-Tabari: iii/3, 1515; Ibn al-Athir: vii, 82. Sani' al-Dauleh, *Muntazam*: i, 97 also includes Qazvin (cf. Gulriz 1961: 872)

and Tabriz in the area affected. Both this and the alleged earthquake of 868 in Berberian 1976: 407, may refer to the 858 Tabriz event, see above. In the *Matla' al-shams*: iii,121, Sani' al-Dauleh mentions an earthquake in Nishpur in 249 H. which is also unreliable. The Ray earthquake may be a belated large magnitude aftershock of the 855–6 events, marking the end of the paroxysm.

28 Al-Tabari: iii/3, 1872, Hamza: 190. Ibn al-Jauzi: v/2, 8 has Basra and al-Suyuti: 28, Wasit. Both these seem to be misreadings, but note that Sani' al-Dauleh, *Muntazam*: i, 100 says shocks affected Iraq between 256–70/869–83. Al-Mas'udi, *Tanbih*: 49 mentions the high seismicity of both Mihrjan Qadhaq and Masabadhan, in which Sirvan was situated, particularly referring to Saimareh, cf. Schwarz 1921: 471. Stein 1940: 206, 229. Izad Pinah 1971: 472 gives erroneous dates for the event.

All sources use the term *hidda* to describe the shock, i.e. the rumble and crash of a falling building rather than the force which caused it. The earthquake must have been accompanied by loud noises, presumably from large-scale landslides for which the Saimareh valley is well known, Harrison & Falcon 1937, Harrison 1936: 36, Harrison 1946.

29 According to the Ibn Khallikan: iii, 358, the earthquake occurred shortly after the defeat of Ya'qub ibn Laith Saffari by Hasan ibn Zaid, in Muharram 261/October–November 874. This is the version translated by de Slane, Paris 1868–71: iv, 311; the text edited by Wüstenfeld, Göttingen 1835–43, has a rather less complete account, fasc. 11, Life 838: 61.

30 In Baghdad the Caliph was denouncing Ya'qub publicly to crowds of pilgrims returning from Mecca; the Gurganis would doubtless have added to the clamour, for despite the earthquake, Ya'qub acted with great oppression and rapacity while in Gurgan.

31 This event is not recorded in the other sources that describe Ya'qub's Tabaristan campaign, which was an expensive failure. The unfavourable climate of the Caspian provinces is usually invoked as a major factor in his reverse, see al-Tabari: iii/3, 1883, Ibn al-Athir: vii, 174 and al-Dhahabi: i, 123; the latter says Ya'qub lost 40 000 men in the terrible cold. Ibn Isfandiyar says Ya'qub's camels were decimated by the fly. Ibn Khallikan's full

Plate 28. Aerial view of a segment of the Damghan fault and of the Babahafiz branch. North is top.

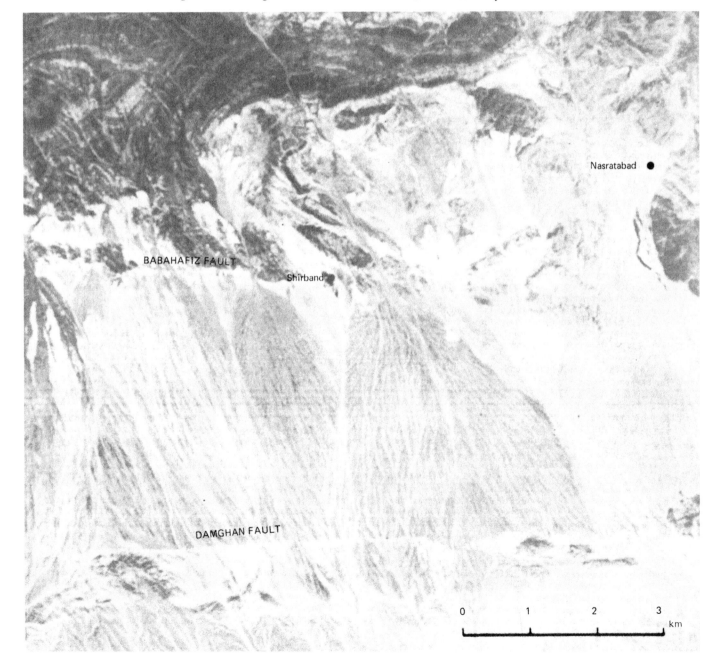

account of these events is valuable and permits the identification of an earthquake in an area for which little information is available at this early period.

32 Thus, the contemporary Yovhannes Kat'oghikos, who also refers to a letter of consolation addressed to the survivors of Dvin by the Catholicos, Mashtots. In neither place does he give the date of the earthquake.

The most precise account comes from Arabic authors of the period, who say that on the night of 14 Shawwal/27 December there was an eclipse of the Moon (cf. Grumel 1958), which was followed the next night by a violent earthquake with aftershocks, the death toll in Dvin reaching 150 000, al-Tabari: iii/4, 2139, Hamza: 123, Ibn al-Jauzi: v/2, 143 and Ibn al-Athir: vii, 323. ·

33 Armenian sources indicate that at least 70 000 people were killed in the region; Thomas Artsruni, a tenth-century author, puts the event in the third year of the reign of King Smbat I. It is mentioned by numerous later authors. Colophons given by Garegin 1951: 89, 99, place the event in Ar. 342 = 17 April 893 to 16 April 894. Tchamtchean puts the earthquake in A.D. 894, cf. Brosset 1874: 185, Macler 1917a: 35, 47. Elias puts it in 1204 Sel., following the eclipse of 27 December.

Although the details of this earthquake are quite clear, many and various errors of date and location have been associated with it. The main problem has been the identification of Dabil, which is the Arabic for Dvin (Duvin) in Armenia, see Minorsky 1930, 1953: 116. Ibn Kathir: xi, 68, puts the event in Ardabil in Azarbaijan, and this is quoted by al-Suyuti: 28. The error must be the result of a corrupt text rather than a misunderstanding of the true place involved, since al-Suyuti's account in *Khulafa*: 370, is correct and he also quotes al-Dhahabi's location of the 280 H. shocks in Daibul or Dabul (*sic.*). Further evidence that al-Suyuti: 28 is a corrupt text comes from his concluding statement that Ibn Kathir puts the Daibul or Dabul event in 288 H., for in fact this author, xi, 84, puts an earthquake this year once again in Ardabil. On both occasions he clearly follows Ibn al-Jauzi's accounts which both read Dabil (= Dvin). The second earthquake also caused 150 000 casualties and followed a solar eclipse (on 23 January 901, Grumel 1958). The similarity between the reported effects of these two earthquakes raises some doubts as to whether they both occurred. Nejjar 1974, translating al-Suyuti, identifies Daibul with a location in India, see below.

The reading Dabil = Ardabil is also followed by al-Jazzar in Taher 1974: 153 and al-Qusi: 86–7. The latter author says that in this year, 280 H., there was a strong earthquake in the city of Ray and its districts, preceded by the drying up of the water in Ray and Tabaristan. This was followed by the Ardabil (*sic.*) earthquake. A number of inaccuracies can be traced to account for this entirely spurious report: Ibn al-Athir: vii, 322 also mentions the drying up of the waters without mentioning the cause, before his account of the *Dabil* earthquake. Later, Hajji Khalifeh: 47 compresses this information into the sentence, 'in 285 (*sic.*) there was a total eclipse of the Sun (*sic.*) in Shawwal and a mighty earthquake and the drying up of the waters of Ray and Tabaristan', a statement which may be responsible for al-Qusi's report, although it is clear that the year referred to should be 280 and the sequence of eclipse and earthquake refer to the Dvin earthquake.

Hajji Khalifeh's text is seriously mis-translated by Carli 1697: 56 as *Nelli regni di Gor* (for Ghar, a district of Ray?; the Persian text reads *ghaur* = drying up, i.e. of the waters) *e Taberistan, successe un estraordinario Terremoto*; this has added to the confusion of later writers. Wilson 1930 lists an earthquake in Ray and Tabaristan for 9 January 878 (i.e. 285 H.) on the authority of Mallet 1852 and Hajji Khalifeh. This spurious event is also given by Ambraseys 1974b.

More exotic locations of the Dvin earthquake stem from the reading of Dabul or Daibul for Dabil and refer to the town of Daibul, whose ruins are differently identified with localities lying southwest of Tatta in the Sind, where the Indus disembogues into the sea, east of Karachi, *Hudud al-'alam*: 123, 372. Thus Bar Hebraeus, who refers to the event without any details beyond the date (280 H.), puts it in outer India, according to Budges's translation (p. 151). This led Oldham 1883 and

Mallet 1852, to place the earthquake in India, between Lahore and Multan. De Ballore 1924: 144, identifies Dabil with Daibul on the Makran coast and so does Sieberg 1932. Mallet, however, has two earthquakes, one in 893 in India and another in 894 in the region of Erivan, at Doun (i.e. Duvin) on the basis of Armenian authors. The *Atlas* gives 891 Ardabil and 893 Dvin.

34 Internal evidence suggests that this was a locally destructive earthquake and that its effects did not extend very far. Most of the survivors removed to near-by towns and the Catholicos moved to Edjmiadsin. The town was rebuilt and its economic life became more active than before, see *Bal'shaya Savetskaya Etsiklopediya*: xiii, 467, and references on recent archaeological findings in *EI²*: 'Dwin'.

35 Ibn al-Jauzi: vi, 31, 33, 37; Ibn al-Athir: vii, 361; al-Suyuti: 29. Storms blew around Basra, destroying palm trees and a place near there was engulfed with the loss of 6000 people.

36 Siunetsi 1885; Stepanian 1942.

37 Al-Mas'udi, *Muruj*: viii, 282, followed by Sani' al-Dauleh, *Muntazam*: i, 112 who duplicates the event, also mentioning an earthquake in Kufa and Baghdad the previous year, 298/910. The earthquake occurred either during or after the storm, which caused much damage.

38 Narshakhi: 59. Although it is not stated that an earthquake was responsible for these events, an alternative natural cause is unlikely, especially if the whole city was affected. The minaret was later restored after further damage to the mosque the following year, perhaps in an aftershock.

39 Gardizi: 33. The event is given in less detail by Ibn al-Athir: viii, 302, who calls it 'a well-known and very great earthquake', and by al-Suyuti: 29. They all refer to the district of Nisa and not to the city of the same name.

40 It seems highly probable that the effects of the earthquake were observed by Abu Dulaf, who was in Nishapur when he heard of a great landslide in the Samalqan valley and went back to look at it. He found that tens of villages had been swallowed up and the ground had sunk more than 600 feet. Streams were cutting their way through the slide material, Abu Dulaf text 26, trans. 58. Minorsky's commentary on pp. 105–6 should be read in conjunction with Aubin 1971: 113, who establishes that the correct reading for Minorsky's Shaqan is Isfinqan = Samalqan, a valley of the Atrak river, downstream and west of Bujnurd and on the northern side of the Ala Dagh range, see figure 3.38. This was counted as one of the districts of Nisa at this time, al-Muqaddasi: 51, 300n.; cf. *Hudud al-'alam*: 325–6.

Although the exact date of Abu Dulaf's journey from Gurgan to Tus and Nishapur is not known, there is no reason to doubt that he could have been in the area at this time. He was in Bukhara before 331/943, Ray some time before 333/944, and Kirmanshah in 340/951. The *risala* refers to other events between 935 and 952 (ed. Minorsky: 5). The location of Samalqan and the the close coincidence of date suggest that what Abu Dulaf describes was the result of the earthquake. The heavy rain that he noted in the district when he first passed through would have made the natural slopes more unstable to shaking.

The *Atlas*: 172, places this event at Nisa in 953 August 20; the ruins of Nisa lie near Bagir, west of Ashkhabad, Gorshkov 1947.

41 Ibn al-Athir: viii, 388 gives Hamadan and Astarabad. The inclusion of Astarabad is most misleading. The nearest place of that name to Hamadan is in Gurgan, about 590 kilometres away. It seems likely that Astarabad should read Asadabad, the town fifty kilometres west of Hamadan. Sani' al-Dauleh, *Muntazam*: i, 131 and al-Suyuti: 29 mention only Hamadan.

42 Ibn Miskawaih: ii, 167; Ibn al-Athir: viii, 390.

43 Al-Dhahabi: i, 168; Ibn al-'Imad: ii, 371. Both authors quote Ibn al-Jauzi's *Muntazam* for this information, which is not, however, to be found in that work; see also al-Suyuti: 30.

44 Al-Biruni: 20, quoting a contemporary for this information.

45 The accounts of this earthquake and of the destructive shock that followed in April are rather confused.

46 The lowering of the water revealed islands and outcrops of rock that had not been seen before. It is not stated whether this was the Persian Gulf or the Caspian. Although later authors, Hajji Khalifeh: 52; Sani' al-Dauleh, *Muntazam*: i, 131, propose the

former, the Caspian seems more likely. Ibn Miskawaih records that the previous winter had been rainless, as though this was a sufficient cause for the drop in sea level. In fact, between the ninth and eleventh centuries, the Caspian fell to its lowest level of −32 metres, Apalloff 1956, see for a summary, Gerasimov 1978.

47 Ibn Miskawaih: ii, 168; Ibn al-Jauzi: vi, 387. At Hulvan, there was much destruction, the ground 'throwing up the bones of the dead and water pouring out of the earth', Ibn al-Athir: viii, 389.

48 This earthquake occurred while damaging aftershocks continued in the Ray−Taliqan region following the catastrophic event of 23 February 958 in the previous Muslim year, 346 H., and the accounts of these two events are confused in the sources. Ibn al-Athir puts the events in the Zagros *before* the Ray−Taliqan earthquake and has the whole sequence of events under the year 346 H. This in turn lends confusion to the sources that follow his account. Jabiri 1943: 82 puts these events all in 346 H. and adds Hamadan and Isfahan to the list of places affected, but no confirmation of this has been found in an early source. His inclusion of Hamadan in the area affected by the February earthquake is reasonable, but the effects of these shocks in Isfahan must have been negligible, cf. Ambraseys 1979.

    The earthquake in Baghdad occurred at the same time as in Hulvan, to which Ibn al-Jauzi adds Qum and Kashan, where he says many people were killed, but this was almost certainly the result of the Ray−Taliqan earthquake. Al-Suyuti: 30, says that the shocks 'returned to the Zagros towns in 347 H.', implying that they were also affected by the shock in February 958, but his account may be muddled; see also al-Suyuti, *Khulafa*: 400, for the information about Hulvan; he omits to mention Baghdad and Kashan in this account.

49 Ibn al-Jauzi: vii, 86, al-Suyuti: 31. Ibn al-Athir: viii, 513 and Sani' al-Dauleh, *Muntazam*: i, 139, put this earthquake in 368 H.

50 Ibn al-Jauzi: vii, 87.

51 Al-Muqaddasi: 426. Siraf, near Bandar Tahiri, was the chief entrepot in the Persian Gulf at this period and the earthquake is generally taken to be responsible for its subsequent decline. This effect is exaggerated, for the port continued to flourish for some time, Whitehouse 1968; Aubin 1959, 1969. It has also been proposed that part of the water-front fell into the sea in the earthquake, making the harbour impracticable, Mustafavi 1963: 520. This is based on Yaqut's statement that in the thirteenth century there was no shelter at Siraf and ships had to go to the Naband penninsula for safe anchorage, Yaqut: iii, 212. Le Strange 1905: 259, takes this to mean that the harbour had silted up by the beginning of the seventh/thirteenth century, but there is no river there. In fact there is some indication of uplift of the land at Siraf, for Stein noted and photographed what he described as a harbour wall there well above sea level, Stein 1937: 204, 237, plate 69. There is no reason, however, to attach this apparent uplift to the 978 earthquake rather than to other episodes, including aseismic uplift.

52 For the ruins of Dinavar, see Morgan 1894 and Strauss 1911. They are situated north of Bihistan (Bisitun), on the confluence of the Gamishan and Dinavari rivers, in a region damaged by the Farsinaj earthquake of 13 December 1957, see Ambraseys *et al.* 1973.

53 This was within a century of the last catastrophic landslide recorded in the region. According to 'Arib: 39, a mountain near Dinavar called al-Till was fissured and collapsed in 300/913. A great deal of water emerged from beneath it and a number of villages were flooded. This account is followed by Ibn al-Jauzi: vi, 115 and al-Suyuti: 29. These authors do not mention an earthquake. The frequency of destructive landslides in the region, probably around the Tang-i Dinavar, is later noted by al-Ghazzi in Taher 1975: 71.

54 Ibn al-Jauzi: vii, 238; he is followed by Bar Hebraeus: 183/204, who mentions that the ground was rent open; al-Dhahabi: i, 186, and Ibn al-'Imad: iii, 150, both give the casualties as 10 000. Al-Antaki: xxiii/3, 475, the earliest source by at least a century, puts the earthquake in 396/1006, but this must be rejected in favour of Ibn al-Jauzi's most precise account. Al-Antaki puts in the same year, 396 H., the appearance of a comet that remained visible for four months: this was sighted on 3 April 1006 according to Grumel 1958. The later sources also record the comet in

396 H. and thus correctly separate it from the occurrence of the earthquake. Al-Qusi: 89, however, records both the comet and the earthquake in 396/1006, but says the latter affected Iraq, not Dinavar. Other incorrect reports are given by Mallet 1852 and Sieberg 1932 who put it in 1007 and by the *Atlas* which puts the earthquake in Kars in 1003.

55 In Ramadan 398/May 1008 news arrived in Baghdad of an earthquake at Siraf and along the coast (*al-saif*). A number of ships were sunk and many people were killed, Ibn al-Jauzi: vii, 238. Later authors have different reading for *al-saif*, which can be disregarded. Al-Suyuti: 31, puts the earthquake in Shiraz (not mentioning the storms there), and the editor suggests the reading Shaizar (Caesarea); Siraf is clearly correct. Al-Suyuti says that because of it (*bi-sababiha*) many ships were sunk. Ibn al-'Imad: iii, 150 says there was an earthquake in Siraf and *al-Sabab*.

    Al-'Umari: fol. 48v says the earthquake affected Siraf (and *al-Bast*). Houses were destroyed and about 10 000 people perished beneath the ruins. The sea formed a wave and sank a number of ships, from which no one survived. Al-'Umari also gives 10 000 for the death toll in the earthquake at Dinavar which happened shortly beforehand, but there is no confirmation in earlier sources that the shock in the Persian Gulf killed so many. His account clearly suggests that a tsunami was responsible for the sinking of the ships, which is a significant interpretation of events. It should be noted, however, that it may simply have been storms that were responsible. High winds affected the Tigris valley and southern Iran, especially Shiraz, during the same month, and Bar Hebraeus: 184/204 refers to the sinkings without mentioning the earthquake.

    Although the details are not satisfactorily recorded, the earthquake must have been as destructive in Siraf as the event of 978, so making a similar contribution to the slow decline of the port. Lack of data for later periods reflects the removal of Siraf from the main routes of trade and communication after the eleventh century, see p. 5.

56 Nasir-i Khusrau: 6. He was in Tabriz in 438/1046.

57 According to Karbaba'i: 16−18, the people escaped to the foothills of the Surkhab mountain to the north of the city and witnessed the destruction of their homes. This suggests that destructive aftershocks continued for some time.

58 Arabic sources, referring to a document from Mosul dated 4 January 1043, describe the total destruction of Tabriz with 50 000 casualties: Ibn al-Jauzi: viii, 114, Ibn al-Athir: ix, 351; al-'Umari: fol. 54r gives 100 000 casualties.

    Later Persian accounts greatly confuse the date of this event without adding any further information. Thus Mustaufi: *Nuzhat*: 75, dates the earthquake 14 Safar 434/3 October 1042, relating a story to the effect that the earthquake was predicted by the astrologer Abu Tahir Shirazi but the prediction was ignored. This account, of which Wilson 1930: 103 gives an incredibly confused version, forms the basis of later reports; Karbala'i dates it 4 Safar 433/3 October 1041. Fasih: ii, 146 gives 423/1032 for the earthquake and 433/1041 for the reconstruction of Tabriz. Hajji Khalifeh: 59 gives 432/1040 and Sani' al-Dauleh, *Muntazam*: i, 160, covers all possibilities of being wrong by putting earthquakes in Tabriz in 432/1040, 433/1041 *and* 434/1042. Little attention need be paid to these alternative dates, beyond suggesting their removal from modern earthquake catalogues. At best they indicate a prolonged sequence of shocks in Tabriz at this time, with the main shock on the date given by Nasir-i Khusrau.

    The different accounts of the event are discussed at some length by Daulatabadi 1964: 3−12 and Mashkur 1973: 398−404. Chardin 1811: ii, 337, following Mustaufi, is in a terrible muddle, putting the earthquake in 235/849 and this has misled Berberian 1976: 406−7; cf. Melville 1981.

59 Four years after the earthquake, Nasir-i Khusrau found Tabriz small in extent but prosperous and populous. Reconstruction was ordered by the ruler, Vahsudan, who had escaped the destruction of his palace. The astrologer who had predicted the earthquake watched over the proceedings, and so auspicious was the birthchart that he arranged for the new Tabriz that, according to Mustaufi, writing *c.* 1340, the city had not been affected by a severe earthquake since (but see A.D. 1273). Mustaufi attributes this freedom from damaging earthquakes to the wells

under the city, which released the powerful vapours below the ground before they could build up sufficiently to cause an earthquake. This Aristotelian view of the generic cause of earthquakes has not gone out of fashion in Iran: an article in *Kayhan International* for 28 Murdad 1349/19 August 1970 states that the two million cesspools underneath Tehran act as buffers against earthquakes, as was confirmed by the then director of the Institute of Geophysics at Tehran University! It remains an improvement, however, on belief in the influence of heavenly ascendencies on such events.

The contemporary poet Qatran composed two odes on the earthquake and in praise of his patron. They are discussed by Kasravi 1929: 89, who suggests that more than one earthquake occurred at this time.

60 Ibn Funduq: 52, 267; Ibn al-Athir: ix, 405; Abu 'l-Fida: ii, 181. Several poems were written about the event.

61 Ibn al-Jauzi: viii, 154, followed by al-Suyuti: 32, quoting Ibn Kathir; see also al-Qusi: 92.

The importance of Arrajan at this period is reflected in the sudden increase of information available. Abu Dulaf, echoed by Yaqut: i, 416 and by Zakariya Qazvini: 201, says that earthquakes were frequent in Izeh a century earlier, although the only event previously recorded in the region affected Ahvaz in 225/840. Nasir-i Khusrau estimated the population of Arrajan to be 20 000 males in May 1052, evidently shortly before the earthquake, see p. 5.

According to Ibn al-Athir: ix, 405, a large mountain near Arrajan was cleft open and cracked apart, to reveal in the middle a flight of steps built of bricks and gypsum. The story, from an eye-witness account, probably refers to the slumping of the slopes of a *tepe* and is repeated by Abu 'l-Fida: ii, 181. The geographical co-ordinates for Arrajan given in *EI*[2] are in error by more than 100 kilometres.

An inscription on the *masjid-i jami'* at Shushtar dating repairs to the building in 445/1053 may refer to the effects of this earthquake.

62 Ibn al-Jauzi: viii, 190; Ibn al-Athir: ix, 449; Abu 'l-Fida: ii, 188; al-Suyuti: 33. It is said that mills were set rotating in the earthquake, which was felt over a large area in the Mesopotamian plains.

63 Ibn al-Jauzi: viii, 323. Other authors give Wasit as among the places affected by the earthquake that occurred in Syria in Sha'ban/August this year, e.g. al-Suyuti: 33, illustrating the tendency of the chroniclers to mention all the events of one year in a single description.

64 Ghaffari, *Nigaristan*, says the fall of Ani to the Saljuq Malikshah (*sic.*) in 456/1064 was made possible by the sudden occurrence of an earthquake in the middle of the siege. The shock caused the eastern side of the citadel to collapse and fill in the ditch with rubble from the walls, Brosset 1849: 148, 150. As other Muslim authors mention the same collapse of the walls without any apparent cause, Ghaffari's statement is superficially very plausible. Canard 1965, however, reviews the various accounts of the siege and casts great doubt on Ghaffari's story, which is based on the *Vasaya* of Nizam al-Mulk, who was supposed to have been assisting Malikshah in the operations. This work is of dubious authenticity and there are several inaccuracies of detail, not least that the siege was conducted by Alp Arslan and not Malikshah. Adequate military reasons can be put forward for the fall of Ani, in particular the mining of the walls referred to by one author.

This earthquake and the event of 1069 are dubious, see below. There are great similarities between them and the veracity of the original source is disputable.

65 Ibn al-Jauzi: viii, 241. According to Ibn al-Athir: x, 35, the Jibal was also affected, but al-Dhahabi: i, 207, quoting Ibn al-Athir, omits the Jibal, as does al-Suyuti: 33. Sani' al-Dauleh, *Muntazam*: i, 168, however, merely gives an earthquake in the Jibal this year, while Jabiri 1943: 104 highlights the confusion by stating that the earthquake affected Khurasan, Hamadan (*sic.*, in the Jibal province), *and* the towns of Kuhistan. References to Jibal reflect what happened in the mountains (*jibal*), rather than in western Persia. In that Jibal and Kuhistan are synonymous (Arabic and Persian) terms to describe the mountain regions in western and eastern Persia, it seems probable that the earthquake occurred in

the Kuhistan district of Khurasan, where villages were swept away by landslides. The event may tentatively be located in the Qayin–Birjand highlands, Ambraseys & Melville 1977.

66 Husaini: 253 quotes the *Vasaya* for the information that while Nizam al-Mulk was laying siege to the rebel Fazlavaih in the castle of Istakhr in Fars, an earthquake during the night made the water spill out of the cisterns of the fort. Another source, Sadiq Isfahani, quoting Nizam al-Mulk, puts the incident in 467/1075, in the reign of Malikshah, but does not mention the earthquake, Ouseley 1819: ii, 405–7.

Considerable confusion surrounds these events. Ibn al-Athir: x, 48, puts the campaign in 464/1072, the last year of Alp Arslan's reign and mentions the very unexpected surrender of the garrison, saying the water had dried up in the cisterns but not referring to an earthquake. The correct date, as given by Sibt ibn al-Jauzi, is 461/1069, see Cahen 1978. It is also clear that the castle involved was not Istakhr, which had been besieged two years earlier by Nizam al-Mulk on a separate campaign, Ibn al-Athir: x, 36. The campaign against Fazlavaih ended with a siege at Khurshah, near Jahrum, where he was captured and then sent to confinement in Istakhr, Ibn al-Balkhi: 131, 166. Ghaffari, *Tarikh*: 128, follows Ibn al-Athir's date, 1072.

The accounts derived from the *Vasaya* of the events in 1064 and 1069 are thus an amalgam of confused details from separate incidents and great suspicion must be attached to the stories of earthquakes during these sieges. Had an earthquake at Jahrum been sufficiently strong to throw water out of cisterns or crack them, it is likely that it would have been felt by the besieging army who were on the contrary astonished at the capitulation of the garrison the following morning. Similarly, had an earthquake at Ani permitted its capture, it is likely that Armenian sources would have said so, as this would be a good excuse for their failure. The element of mystery involved in both cases is artificial and more realistic accounts are available: Sibt ibn al-Jauzi's version of the capture of Fazlavaih is quite different, see Cahen 1978: 114.

Although some of the material in the *Vasaya* may be authentic, these two incidents must be apocryphal. See also Bowen 1931, *EI*[2] 'Istakhr' and Matheson 1976: 213.

67 Ibn al-Jauzi: ix, 14; Ibn al-Athir: x, 94. Al-Suyuti: 35, says that many Byzantines (*kathir min al-Rûm*) perished along with their flocks, which would appear to put a second event in Asia Minor this year, leading Ambraseys 1961: 25 to put an earthquake in 478/1085 which caused the deaths of many Greeks in Erzerum. *Rûm*, however, is clearly an error for *radm*, wreckage, under which men and beasts perished.

68 Ibn al-Jauzi: ix, 38.

69 Ibn al-Jauzi: ix, 81; he quotes a first-hand authority, who says this occurred shortly before the death of al-Muqtadi (Caliph, d. 15 Muharram/4 February 1094).

70 According to Sani' al-Dauleh, *Muntazam*: i, 176, 181, there was an earthquake in Girdkuh in 485/1092 and again in 495/1102. No details are given of either event.

The year 485 H. saw an attack by the Saljuqs on the Assassin castle at Alamut, of the murder of the minister, Nizam al-Mulk, and the death of Sultan Malikshah, leaving the kingdom in confusion, Juvaini: iii, 201. The castle of Girdkuh, overlooking Damghan, did not fall to the Assassins until later, however, and the fact that information about an earthquake before then is unlikely to have been recorded casts doubts on the authenticity of the report.

There is an indication that Girdkuh experienced an earthquake in the twelfth century, but it should have occurred several years after 495/1102. Some years after the death of Ra'is Muzaffar, who held Girdkuh for forty years as the agent of Hasan-i Sabbah (head of the Assassins), there was an earthquake which caused a spring of water to gush out in the well of the castle. This well had been dug by Muzaffar, but abandoned because he had not struck water after going down 300 metres, Juvaini: iii, 208; Rashid al-Din (II): 32. All the dates relevant to establishing the date of the earthquake, however, are uncertain. Rashid al-Din says Girdkuh was taken by the Assassins in 489/1096 and Muzaffar built the well in 493/1100; both dates are disputed, but this is not relevant to the present argument. Assuming Muzaffar ruled Girdkuh for some time, the date Rashid al-

Din: 34 gives for his death, 498/1105, is highly improbable; one variation puts it in 520/1126, but this is only marginally more acceptable. But whenever Ra'is Muzaffar died, it is evident that the earthquake that followed in Girdkuh happened some time after it had been an Assassin stronghold and not within the very first few years.

It seems unlikely that Sani' al-Dauleh could have read the accounts of these events so uncritically as to misplace the earthquake they mention to 495/1102; but on the other hand, it is only at this period that Girdkuh enjoys a brief prominence in the histories. Possibly some independent account of the Girdkuh earthquakes exist in a source that has not yet come to light, to vindicate his statements.

71 According to Isfizari: ii, 55, on the night of Friday 8 Jumada I, 495, there was a calamity (*nazileh*) in Herat; the western side of the mosque and most of the northern and southern sides were destroyed. A rather different version specifies that a strong shock, with a north—south motion, caused the collapse of many buildings in Herat, the *masjid-i jami'* suffering particularly heavy damage, Fami in Barbier 1860: 519. The closeness of *nazileh* to *zalzaleh* (earthquake), and the apparent support of this reading in a variant text, confirms that an earthquake was responsible for the damage.

72 Ibn al-Jauzi: ix, 193 quoting an eye-witness. The shock in Baghdad caused screens and walls to sway from side to side. See also Ibn al-Athir: x, 373 and Sani' al-Dauleh, *Muntazam*: i, 186.

73 Al-Rafi'i, quoted by Taher 1979: 214/68, al-Suyuti: 36 and Gulriz 1961: 872. This is probably the earthquake described by Chardin 1811: ii, 398, which he wrongly puts in 1067−8, see Melville 1978: 141.

74 Mahjuri 1966: i, 181. The exact date of this event is not mentioned, but it occurred in the same year as the death of Khatun Saljuqi, the wife of 'Ala' al-Dauleh, in the second decade of the sixth century H. The last date mentioned before this notice is 521/1127. The wife of 'Ala' al-Dauleh (Ispahbad 1118−39), was the sister of Sultan Muhammad ibn Malikshah. Although there is no other record of the death of the princess or of the earthquake, and Mahjuri does not name his source, the information appears to be reliable. None of the sources for this period used by Rabino 1936 mention these events, but clearly some local information must exist. For the Hazar Jarib district see Rabino 1928: 56, 124. Details of Kunim (Konim) and Zarim are in *Farhang-i jughrafiya-yi Iran*, Tehran 1329/1950: iii, 140, 246. The location of Daulat is not known.

All available topographic maps of the region of the Zalim-rud and Tijan-rud east of about 53.5° were found to be useless and our first field study of the region was carried out by dead reckoning. This placed Zarim (locally pronounced Zulum), northeast of Sika on the banks of what was thought to be a tributary of the Nika. The correct location of Zarim was made possible later thanks to a map of the Nika−Damghan sector prepared by Dr I. Stöcklin in 1954, which was made available to us. Zalim was in fact found to be on the Zalim-rud which turns to the southeast past Lay.

75 Ibn al-Athir: x, 469. Ibn al-Jauzi: x, 14, on eye-witness authority, says that in Baghdad the earthquake caused the ground to swell like the sea. Had this motion continued, he adds, casualties would have been very heavy (implying that as it was, they were not).

A different date, 8 Adhar 1441 Sel./8 March 1130, is given by Bar Hebraeus: 255/289, but Ibn al-Jauzi's source specifically states that the earthquake was at the end of February (*shabat*). See also al-Suyuti: 37.

76 According to Ibn al-Jauzi: x, 46, numerous shocks occurred in Baghdad, beginning on Thursday 11 Shawwal 529. Earthquakes continued, five or six daily, until the night of Friday 27 Shawwal/10 August (a Saturday?). On the following Tuesday, at midnight (i.e. on the night of 1−2 Dhu'l-Qa'da/13−14 August) a strong shock caused ceilings to crack and the collapse of walls. Tremors continued until dawn, causing much distress. The author was a young man at the time and experienced the event; his account is followed by al-Suyuti: 37, who nevertheless gets the dates wrong. He ends the first sequence on the 17, not the 27 Shawwal. Al-Qusi: 95 puts this sequence of shocks in 525/1131. It is certainly to this event that Ibn al-Athir: xi, 22, refers when he says that a strong earthquake affected Iraq and other

places in the Zagros, where many people perished. He puts this in Sha'ban 529/May 1135, followed by al-Qusi: 95, but this is two months too early.

One indication that Ibn al-Athir's date is too early is that the earthquakes seem to have started after the capture of the Caliph by the Saljuq Sultan Mas'ud, which occurred in Ramadan/June this year. Contemporary observers associated the sudden sequence of earthquakes and other phenomena with this action, which caused Mas'ud's uncle, Sanjar, to write to him from Khurasan, suggesting he crave the Caliph's pardon. Sanjar's letter, written at the beginning of Dhu'l-Qa'da, remarks that unusual phenomena had been going on for twenty days, Ibn al-Jauzi: x, 45−7; Juvain: iii, 218, which tallies with the information given above. The Caliph was captured at Hamadan and taken into confinement in Maragheh, where he was later murdered by the Assassins. The earthquake must have been very widely felt throughout this region to have caused such concern and to have become popularly associated with the political events with which it coincided. News of it clearly spread to Sanjar in Marv, although he could not have experienced it for himself, so far away.

77 Ibn al-Jauzi: x, 46, himself experienced this shock in Baghdad. Al-Qusi: 95 wrongly gives 525/1131.

78 Ibn al-Jauzi: x, 108, himself felt the shaking in Baghdad. Sibt ibn al-Jauzi: 111 says it was a great earthquake.

79 Mustaufi, *Tarikh*: 849; Barbier 1857: 294. The earthquake must have affected the houses of the wealthier citizens as well as the poor. The emigration of leading members of society is an unusual result of an earthquake, and other factors were probably involved, see Melville 1980. It is also mentioned by Le Strange 1905: 385 on incorrect authority.

80 Ibn al-Jauzi: x, 138; Ibn al-Athir: xi, 96; Sibt ibn al-Jauzi: 121; Ibn al-Furat (in Kremer 1880: 131). Sani' al-Dauleh, *Muntazam*: i, 194, incorrectly puts this earthquake in 542/1147.

It is probably to this event that Lockhart, in *EI²*: 'Hulwan', refers when he says that the destruction of Hulvan was completed by an earthquake three years after its fall to the Saljuqs, which in fact occurred a century earlier in 437 H., when it was burnt. But there is no mention in the sources of an earthquake there three years later, Ibn al-Athir: ix, 360.

The *ribat* of al-Bahruzi cannot be positively located, but was presumably a stage along the Khurasan road. The post might have been named after Mujahid al-Din Bahruz (d. 540/1145) who enjoyed great power in Iraq during the first half of the twelfth century. His name is particularly associated with the district of Takrit, north of Baghdad, and it is possible that the *ribat* was in this area, Ibn al-Athir: x, 373, 460; xi, 70. Ibn al-Athir: xi, 51 says that a castle belonging to Bahruz was destroyed in the Ganja earthquake of 534/1140 (*sic.* for 533/1139), a detail which he probably mentions in the wrong place.

81 Ibn al-Athir: xi, 133, al-Suyuti: 38, Sani' al-Dauleh, *Muntazam*: i, 197.

82 Ibn al-Jauzi: x, 266. Ibn al-Athir: xi, 287, puts this event in 571/1176, followed by Sani' al-Dauleh, *Muntazam*: i, 205, who gives the wrong location, i.e. he has Iran and Transoxania (*ma vara' al-rud*) instead of the region beyond Ray (*ma vara' al-Ray*). Al-Qusi: 99, says an earthquake recurred the next year in Ray, possibly in an attempt to reconcile the conflicting dates.

Apparent confirmation of Ibn al-Jauzi's date comes from Mustaufi, *Nuzhat*: 58, who says that in 572/1177 the walls of Qazvin were restored. Al-Rafi'i, quoted by Gulriz 1961: 873, whose father was involved in the work, also mentions the restoration of the walls, although neither author says why it had become necessary. Al-Rafi'i's father was a friend of the Vizir, Sadr al-Din, who was responsible for the repairs. Chardin, 1811: ii, 398, however, specifically states that an earthquake causing some damage occurred in 562 H. (*sic.*), after which the walls of Qazvin were repaired. his narrative clearly being derived from Mustaufi. Chardin incorrectly puts the circumference of the new walls at 100 300 instead of 10 300 paces. Although Chardin's figures are inaccurate, they appear to confirm the occurrence of the earthquake and the subsequent restoration of the walls of Qazvin in 572 H. He previously gives the foundation of Tabriz as 165 instead of 175 H. and an earlier restoration of the walls of Qazvin as 364 instead of 373 H. as compared to Mustaufi, so his giving a 6 for a 7 is a common fault, Chardin 1811: ii, 333, 397.

It is strange that the Qazvini authors do not mention the earthquake and it seems probable that the Ray area was the worst affected.

The *Atlas*: 494, following Sani' al-Dauleh, puts the earthquake on the Oxus at Kerki, 1320 kilometres away from where it happened.

83 Ibn al-Jauzi, *Mukhtasar*: fol. 150v says that the earthquake occurred above (*fauq*) Irbil. In the *Muntazam*: x, 287, he merely refers to a sequence of four light tremors in Baghdad, presumably from the same earthquake. See also Sibt ibn al-Jauzi: 224; al-Suyuti: 45.

84 Abu Hamid: 89. This was a local event, causing panic but probably little damage in the city. Hamadan was at the centre of affairs at this period and it is likely that any more serious earthquake would have been more widely reported. The context in which the event is recorded dates it during 587–8/November 1191 to June 1192.

85 Ibn al-Athir: xii, 72 says that the cemetery (*jabbana, sic.*?) near the shrine of 'Ali at Najaf, about eight kilometres west of Kufa, collapsed. Sani' al-Dauleh, *Muntazam*: i, 212, says the whole Zagros region was affected.

86 Ibn al-Athir: xii, 187; Bar Hebraeus: 365. A number of late sources and European authors apparently refer to this event occurring a century earlier, on the basis of a poem (by 'Azizi) which mentions an earthquake in Nishapur in the early 500s H., see Daulatshah: 187; Khalifeh: 147; Zain al-'Abidin: 374. Yate 1900: 413 and Wilkinson 1975: xxviii put an earthquake in 1115 A.D. on the basis of the poem. Khalifeh associates the date in the 500s with 555 H. (see below): both should be read as 605 H.; see further, Melville 1980.

87 Juvaini: ii, 71. Of the villages destroyed, Daneh still exists today in the Zabarkhan district southeast of Nishapur, in which the shrine and village of Qadamgah also stand, see *Farhang-i abadiha-yi kishvar*: v, 71. Another village destroyed, Banask, has not been identified and the reading Banask is not certain. Other authors mention the earthquake but add no details.

88 Sibt ibn al-Jauzi: 351, says aftershocks lasted ten days. Presumably he refers to the period of damaging aftershocks. The alternative figure of two months is given by Mustaufi, *Tarikh*: 494, who says no trace of the town was left and it was rebuilt on a new site nearby. In *Nuzhat*: 148, he identifies the new location as Shadyakh. Contrary to Mustaufi's statement, the position of the city remained the same and it was again flourishing in 613/1216, Yaqut: iii, 230. Shadyakh had become the centre of the population fifty years before this earthquake, following the sack of Nishapur by the Turkish Ghuzz, in 556/1161. Khalifeh: 145, however, puts an earthquake in 555/1160 as the cause of the move to Shadyakh. This is an echo of Mustaufi's misleading statement and the earthquake referred to is the 605/1209 event.

89 Ibn al-Athir: xii, 305; al-Dhahabi: ii, 296; al-Suyuti: 49; al-'Umari: fol. 98v. The entry in Alsinawi & Ghalib 1975: 544 referring to al-'Umari, is most inaccurate. Al-'Umari has '623 H. – an earthquake in Mosul and Shahrizur which continued (*takarrarat*, not Takrit!) thirty days; many villages were destroyed.' There is no mention of fences being destroyed. Shahrizur was on the site now known as Yasin Tepe, *EI*[1]: 'Shehrizur'.

90 Tabandeh 1969: 30; Ambraseys & Melville 1977. This was probably not a very serious event.

91 Fasih: ii, 319. The exact date of this event is not certain, see Melville 1980.

92 Rashid al-Din (ed. Karimi): i, 665, ii, 779. His two accounts are not entirely compatible with each other. The date 669 H. is also in Mustaufi, *Tarikh*: 494.

In *Nuzhat*: 148, Mustaufi gives 679/1280, which is obviously a slip for 669. Nevertheless, numerous secondary sources give 1280 for the earthquake, e.g. Sani' al-Dauleh, *Matla'*: iii, 68, Le Strange 1905 and later authors. In his translation of *Nuzhat al-qulub*, 1919: 147, Le Strange has 629/1232 for 679, and this is also found in later works, e.g. Bosworth 1973: 159.

Later, Timurid authors put the earthquake in 666/1267–8 with no apparent justification, see Fasih: ii, 337 and Khalifeh: 147. The latter quotes the poem by 'Azizi (referred to in note 86), on which basis 1267/8 is given by Yate 1900, Wilson 1930, Lockhart 1960 and Bulliet 1976. Daulatshah: 187, who also

quoted 'Azizi's poem, reads 630 for 666 H., which finds no confirmation elsewhere.

93 Some dispute surrounds the location of the new city, which was founded in Ramadan 669/May 1271 by order of the Mongol Il-Khan Abaqa. This date is given by Pur-i Baha Jami, a poet attached to the retinue of the Vizir responsible for the reconstruction. His verses are quoted by Hafiz-i Abru: 63 and Fasih: ii, 340, also to describe the effects of the earthquake. Kaihan 1932 says the city was rebuilt by Shahrukh ibn Timur, i.e. referring to the 1405 event (see below).

Several authors have suggested that the site occupied before the move to Shadyakh in the mid-twelfth century was again developed, Jackson 1911: 256, Sykes 1911: 156. However, there is no archaeological evidence to support this and it seems more likely that the town approached its present position at this period. Wilkinson 1937: 8, suggests that the town was rebuilt where it stands today, as none of the sites he excavated in the ruins of Nishapur seem to have been occupied later than the end of the thirteenth century; see further, Melville 1980.

94 Bar Hebraeus: 450/527; Rashid al-Din: i, 665, ii, 767; Fasih: ii, 342; Karbala'i: 18.

95 See Mustaufi, *Nuzhat*: 76 and above, pp. 176–7.

96 Sami 1968: 325 is the only reference for this event, which is surprising as the period is well documented. Fasa'i refers to restoration work in the Masjid-i Nau in the 990s H., misquoted or misprinted by Sami as having occurred in the 890s. It is likely that a similar error is responsible for the date 690/1291; see under A.D. 1591.

97 This statement is made by Vahid Mazandarani 1965: 97 and Mahjuri 1966: ii, 1 n. 1, on the authority of one Sayyid Husain Banafti. Although the evidence for the earthquake is tenuous, there is no reason for doubting the substance of the account, which must have survived in local tradition. The account suggests that among the villages affected were Bula, Isas, Kuhnehdeh and Kakulu (= Kangalu?); also possibly Khishtistan (the place of bricks), see figure 3.6. There is no evidence that the earthquake caused any serious damage to the eleventh-century tower of Raskat, plate 15.

An inscription dated 700/1301 records repairs carried out on the *sauma'eh* of Bayazid at Bustam, 160 kilometres from the meizoseismal region.

98 Kashani: 41 says that on this Friday night in 704 H., a thunderstorm and a very fearful earthquake occurred in Tabriz. Lightning fell and there was much damage. Sani' al-Dauleh, *Muntazam*: ii, 20, mentions an earthquake in Tabriz in 714 H., certainly an inaccurate reference to the same event, ten years out.

99 Ibn Bazzaz: 220 relates that a strong shock occurred one night in Sarab, causing the inhabitants to rush from their houses out into the streets. The walls of the mosque appeared to sway to and fro, in unison with the lamps hanging from the ceilings. No damage is reported to have been done.

This anecdote is undated. It concerns Shaikh Safi al-Din of Ardabil who died in 735/1334, and it could refer to any date between 700/1301 and 735/1334. Sandy Morton, to whom we are indebted for the reference, suggests a date in the latter part of this period. It is possible, however, that the event referred to is the same as the shock experienced in Tabriz, both occurring at night. The ground movements at Sarab described by Ibn Bazzaz suggest a rather large magnitude event occurring at some distance from the town.

100 Kashani: 43. This would appear to have been a strong aftershock of the preceding event. Had it affected Tabriz or Sarab, it is likely the author would have said so.

101 Al-'Umari: fol. 121r.

102 Kashani: 179. Sani' al-Dauleh, *Muntazam*: ii, 23 mentions an earthquake in Hamadan in 725/1325 which is probably a reference to the same event, ten years out. The shock followed a week of heavy rain.

103 From a contemporary document in Ra'in 1970: 89, and the Continuator of Samuel Anetsi. Inscriptions in the monastery date the earthquake in 1319 and the reconstruction of the church in 1329, Kleiss 1968, 1969. There is no evidence that the damage extended beyond St Taddeus.

Abich 1882, quoting the Continuator of Samuel Anetsi, puts the earthquake ten years earlier in Ar. 757/1308 and identifies the monastery of St Taddeus with that at Tatev in the Karabagh; he is wrong on both counts. Abich is followed by many modern writers, including the *Atlas*, which gives the earthquake a magnitude of 6.1, locating it 150 kilometres away. Berberian 1977: 155, 158, gives one earthquake in Zangezur in 1308 and one in 1319, both on the authority of O.Yu. Schmit 1974 (*sic.*, = *Atlas*).

104 The exact date of this event is not satisfactorily established. Tchamtchean: ii, 281, who gives the fullest account of the earthquake, puts it in Ar. 769/1320. Hakobyan 1951: 392 notes a contemporary statement by an anonymous writer that an earthquake took place in 1321, with no locality being specified and Sanjian 1969: 64 notes a colophon written by a native of Ani in 1321 which refers to recent 'tremblings of the ground'. It is only in the works of nineteenth-century authors that the year 1319 becomes associated with the event, see Saint-Martin 1818: i, 112, Bjeshdsian 1830, Chakhathuno 1842: ii, 19, Brosset 1861 and Macler 1917*b*.

105 Internal evidence suggests that this was a locally destructive shock, see Tchamtchean and references in *EI*[1]: 'Ani'. Its effects have been grossly exaggerated by a number of early writers, such as the author of the *Edjmiadsin Manuscript* no. 60 in Brosset 1849: 70, who confused Ani with Ani = Kemakh west of Erzincan, which was destroyed in Ar. 494 and again in the great Eastern Anatolian earthquake of 1318–19, which also ruined Amasya; see Husameddin: *sub anno*.

106 Tchamtchean refers to a dispersal of the inhabitants of Ani following this earthquake and it is generally believed by later authors that this marked the end of the city, which was never rebuilt. This seems to be an exaggeration of the effects of the shock and other causes can be adduced for Ani's gradual decline. Several structures dating before 1320 were left undamaged, and new coins were struck after the earthquake, signifying among other things that life in the city did not come to an abrupt halt; see Khanikoff in Brosset 1849: 145–6, Brosset 1860, Manandian 1965. Orsolle 1885 notices that some of the early buildings of Ani were destroyed by an earthquake as late as 1880.

107 Majd al-Din: 117; Fashi: iii, 52. Jird no longer exists and its location is rather uncertain, but it was probably situated to the northwest of Zauzan, to the south of Akbarabad. Hafiz-i Abru names the three main centres of Khwaf as Jird, Salumeh and Zauzan, while the earlier author, Mustaufi, calls them Sanjan, Salameh and Zauzan, which would seem to equate the districts of Jird and Sanjan.

Aubin 1967 draws attention to the fluctuations in the importance of the various local centres in the district. It remains strange that the town of Khwaf is not itself mentioned as being affected by this earthquake, and that Jird must have replaced Sanjan as the main town in the northwest after the earthquake. Mustaufi, describing Zauzan, mentions the great palace there, so that although he wrote *c.* 1340, his information must apply to the period before the earthquake.

There are two places in Khwaf called Sanjan, one in the northwest and the other called Sanjan-i Burabad or Sanjan-i Khwaf in the southeast, but although it is tempting to associate Jird with a location in this area, near to the town of Khwaf itself, the reference to the epidemic in the northwest makes the location of Jird in that area preferable. The occurrence of the epidemic as a result of the earthquake suggests that maximum damage was done there, in particular a disruption of the *qanat* water supplies on which the region depended.

108 In the thirteenth century both Khwaf and Zauzan, then separate districts, seem to have been more populous than they are now: Khwaf had 200 villages and Zauzan 124, Yaqut: ii, 486, 958.

109 From a cursory field survey of the region between Rushtkhwar and Farahabad it appears that there is no throughgoing Quaternary fault zone in the region. Recent tectonic features are short and discontinuous, devoid of any evidence of major tectonic activity, Ambraseys & Melville 1977.

110 The length of the duration of shaking can be inferred from Majd al-Din's account of the death of the ruler of Zauzan, Ghiyath al-Din Firuz, who, according to one of his intimates who survived the shock, rushed back and forth a number of times, when the earthquake struck, between the centre of his palace and the veranda, before the whole place was overturned on top of him.

111 Al-'Umari: fol. 130v. The earthquake seems actually to have caused some damage in the city but it was quickly repaired; Ibn Battuta, who was in Isfahan in 1347 does not mention the event; Ambraseys 1979: 67.

112 Al-'Umari: fol. 130v. Grigor Kemakhetsi, in Hakobyan 1956: 264, mentions a shock felt this year in Erzincan, to the west; if these shocks are connected, it suggests they may have originated from the Malazgirt region in Turkey.

113 This event is mentioned by Razani & Lee 1973*a*, on the authority of Professor A. Sami. We have not seen Sami's statement in print in a work of his own composition, and unfortunately he himself cannot remember the source of his information (conversation, Shiraz 3 April 1974). In Berberian 1976: 393, Sami's statement is published on the authority of Razani, and 762 H. is given as 1930–31 A.D.!

114 Hafiz-i Abru: xxxvii; Hafiz-i Abru, *Majmu'*: 48; Fasih: iii, 96.

115 The Falak al-Din mosque is referred to in 712/1312 and 844/ 1441. On the first occasion it is mentioned in connection with the construction of the Suq al-Sultan nearby, by Saifi: 595. On the second occasion, its minaret again fell down, landing on the Nizamiyyeh madraseh which was thereby partly destroyed, burying several people, Fasih: iii, 289.

116 Whether or not it is the same building, or a building on the same site as the mosque damaged in the 1102 earthquake is not clear. It was situated in between the Khush gate to the east and the Qipchaq gate to the northeast of the city. The mosque was built, or heavily restored, by the Ghurid Ghiyath al-Din in 597/1201, Isfizari: i, 33; Hafiz-i Abru, *Majmu'* says it was completed in 602/1205 by Ghiyath al-Din's son, Mahmud. The mosque was destroyed or badly damaged by the Mongols twenty years later. It was probably repaired by Fakhr al-Din Kart in 699/1300 and certainly restored by Ghiyath al-Din Kart in 720/1320, by which time the eastern and southern sides were in ruins, Saifi: 439, 746; Fasih: iii, 31. After the 1364 earthquake the mosque was repaired by Mu'izz al-Din Kart.

117 This statement on the basis of an autobiography of Timur, which is of dubious authenticity, see Kariman 1970: ii, 247. The histories of Timur speak metaphorically of earthquakes shaking the world every time he set out on an expedition with his army.

118 Hafiz-i Abru: 64, who does not give a date; al-Maqrizi: iii/2, 682; al-Suyuti: 56. The silence of the Persian sources about the 791/ 1389 earthquake contrasts strangely with their full coverage of the most highly destructive event that followed in 808/1405.

Some authors consider that it was only after the 1405 earthquake that Nishapur was built on its present site, although al-Maqrizi's account of the aftermath of the 1389 event suggests that this could have occurred equally well after that shock. Yate 1900, Jackson 1911, Lockhart 1960 and Bulliet 1976, all consider the modern site to have been developed after the 1405 earthquake; there is no indication of this in the Persian historical sources, cf. Melville 1980. It may be, however, that the move had already occurred after the 1270 earthquake, as proposed above (p. 179). The recovery of the city after 1389 seems to have been comparatively rapid, for Clavijo 1928: 181–3, who was there in 1404, describes Nishapur as a great and flourishing city, in a heavily populated district.

119 Herbert 1634: 53, 1638: 120. No confirmation of this in Muslim sources has yet been found.

120 Hafiz-i Abru: 64, Fasih: iii, 160, quoting some verses by Luftallah, cf. Schefer 1881: 283. Khalifeh: 147 and Daulatshah: 187 quote the quatrain by 'Azizi mentioned above (p. 179). Clavijo 1928: 181 who passed through the region late in July 1404, describes Nishapur as an impressive city on the eve of the earthquake. Its later importance was greatly reduced.

121 Khatjikyan 1950: 57; he refers to an anonymous colophon.

122 Al-'Umari: fol. 153v; the text reads *bilad al-aghwan*.

123 Al-'Umari: fol. 159v. In the absence of further information it is hard to determine whether the Taliqan of Qazvin or of Marv is meant; both regions are subject to violent earthquakes (cf. 819, 958, 1608). Information on events in Khurasan is marginally more likely to have been recorded at this period than for central Persia and al-'Umari's source may have been the same as for the 1410 event around Balkh; this suggests that a tentative location of the earthquake in northern Afghanistan is to be preferred.

Taliqan is to be identified with Qal'eh Vali, see *Hudud al-'alam*: 335; Le Strange 1905: 424n considers Chechaktu to mark the site, because Timur is said to have gone there, but this is not conclusive. The name Taliqan certainly survived beyond the Timurid period.

124 According to al-'Umari: fol. 160r, in 833/1430 there was a destructive earthquake in Hamadan, the town of Wasit and Audar (?). Sani' al-Dauleh, *Muntazam*: ii, 57, mentions only Hamadan and Wasit.

The location of Audar is obscure. It may be suggested that the area involved was on the lower Tigris, between Amara and Basra, where not only Wasit is situated, but also several villages called Hamdan and Auda or Audah. However, the correct Arabic spelling of these places does not show the same similarities as the English (i.e. Ḥamdān for Hamadān and 'Audah for Audar), and it is unlikely that al-'Umari, an Iraqi author, would have misspelt them. In view of this, the events should probably be located around Hamadan in Persia and a separate earthquake occurring the same year in Wasit and Audar (?).

125 See previous note.

126 About fifty people were killed at Gunbad-i Kavus; al-'Umari: fol. 161v.

127 No major town is mentioned by al-'Umari: fol. 161v; it may be supposed that the shock was only locally destructive.

128 Nimdihi, *Tabaqat*: *sub anno*. The following year destructive floods swept away a bridge on the Sakkan, which had possibly been weakened by the earthquake; Aubin 1969: 35.

129 Al-'Umari: fol. 167r; Melville 1978: 101 wrongly converts 861 = 1471.

130 Al-'Umari: fol. 167v; his notice of the events of this year seems in general to be confused and unreliable, cf. Ambraseys 1979.

131 Al-'Umari: fol. 167v, reported as part of the preceding event. The earthquake would appear to have been felt widely in Azarbaijan, causing no damage in Tabriz.

132 Al-'Umari: fol. 169v.

133 Nimdihi, *Tabaqat*: *sub anno*; Nimdihi, *Kanz*: fol. 169r ff.; his accounts of the earthquake give conflicting dates. The less detailed account apparently of the same event, puts the earthquake in 881/1476, but this must be inaccurate, cf. Aubin 1973: 89.

134 The Oman earthquake, which is undated, is discussed by Aubin 1973: 111, who considers that it must have occurred after 1442. His main source is de Barros, *Decadas da Asia*, Lisbon 1945ff., ii/2, i, 44, whose account is also referred to by Wilkinson 1969: ii/vi, 11, who, however, puts the earthquake 'apparently earlier than the mid-fourteenth century'. The earthquake assisted the decline of Qalhat, on the Oman coast, to the advantage of Muscat during the fifteenth century. In short, the date of the Oman event is disputed and one cannot make a positive association of these two earthquakes. However, there is no doubt that the seismic activity in the Strait of Hurmuz shows a paroxysm during the late fifteenth and early sixteenth centuries, see earthquake of 1497.

135 The author of this notice, Mar'ashi: 453–4, 458, himself experienced the earthquake and was sent to supervise relief and restoration work in Gurjiyan. Sutudeh's edition used here gives the earthquake as 889 H., but the 890 H. in Rabino's edition (Rasht, 1911) is far more likely. The notice comes between sections on 889 and 891 H. This unique account is quoted by Rabino 1917, and Sutudeh 1970–2, the latter putting the earthquake in 889 on ii, 17–18 and 890 H. on ii, 420–2. The event is also mentioned, incompletely, by Gulriz 1961: 873–4.

136 In recent usage Tanikabun is a large district containing a number of sub-districts and occupying the western end of Mazandaran. Mar'ashi appears to use it in a more limited sense for he regularly mentions Gulijan and Gurjiyan, which are now considered parts of Tanikabun, as separate districts. Probably, for Mar'ashi, Tanikabun meant the area around modern Shahsavar, the hinterland connected to it by the Chashmeh Kileh river system and possibly areas further southeast.

137 The name appears to have disappeared and it is not even certain how it was pronounced, Gurjiyan, Karjiyan. Rabino's suggestion that Karjikuh may have something to do with it may be correct. It is, however, possible to locate Gurjiyan approximately from Mar'ashi's mention of Jandeh Rudbar. He says 'In the Jandeh Rudbar in Gurjiyan, a pig, terrified by the earthquake, jumped off the top of a mountain, fell into the *rudkhaneh* (river bed) and died.' Sutudeh's identification of Jandeh Rudbar as modern Jinnat Rudbar is very attractive. Jinnat Rudbar is a small district

on the upper reaches of the Chalak-rud, to the west of Gulijan. This puts it in the upper valley of the Chalak-rud. Since Mar'ashi frequently mentions Gurjiyan and Gulijan together we can safely assume that Gurjiyan included part or all of the upper Chalak-rud system. It could of course have extended further; see *Farhang-i Abadiha-yi Iran*, Tehran 1339/1960.

138 Gulijan is the district round the mouth of Chalak-rud, between the districts of modern Ramsar and Khurramabad, *Farhang-i Jughrafiya-yi Iran*, iii, Tehran 1329/1950.

139 Shakur or Shakavar is identified by Sutudeh with modern Ishkavar. The main part of Ishkavar is in Gilan and consists of the upper part of the valley of the Pul-i Rud, above its confluence with the Chalk-rud. There is also a small neighbouring district of modern Tanikabun which is called Ishkavar.

140 Since Taliqan was certainly affected by the earthquake, it is reasonable to assume that the area between it and Gilan, i.e. the remainder of the upper Shah-rud river system, was also damaged. Mar'ashi only uses Rudbarat this once, but he frequently mentions Rudbar-i Lamasar in the Shah-rud valley below the confluence of the Alamut and Taliqan rivers. Presumably Mar'ashi's Rudbarat refers to both the Lamasar and Alamut regions; see Willey 1963, for the area under discussion.

141 The district of Taliqan is meant, i.e. the district of the valley of the Taliqan river or upper Shah-rud. This is made clear by another mention of the same castle in the account of the course of a campaign: after taking Palisan (also spelt Falisan) the army proceeds to take the castle of Amameh, which is in the Jaj-rud valley, see Mar'ashi: 157. Exactly where Palisan was, is not known.

142 The district of Dailaman is basically the valley of the Chalk-rud, the eastern tributary of the Pul-i Rud. The town of Dailam, where Mar'ashi and Sultan Mirza 'Ali experienced the earthquake, is towards the western end of the valley.

143 Raniku included a coastal district in which Rudsar and the former town of Timajan are to be placed and part of the valley of the Pul-i Rud, Rabino 1917: 335. The story of the destruction of Sayyid Razi Kiya's palace is a conjecture of Rabino's (p. 339). The building of this palace is described by Mar'ashi: 140–1.

144 For the location of these districts, see Ambraseys 1974b: 58 and cf. figure 3.8, for correct location of Gurjiyan = Karjiyan. For comments on Mar'ashi's account, see Melville 1978.

145 The author, Isfizari: ii, 99, is exactly contemporary with the event. He says that 'for two *farsakhs* between Nauzad and Mask the ground was fissured to such a depth that the bottom of the crack was invisible.' The fault break was noticed by Tate 1910: i, 29, in the early 1900s. He adds that it was several miles in length, and traces were still visible in a narrow long glen or ravine, most probably referring to the segment of the break between Kalateh Mazar and Khunik. Tate gives the date of the event in the Old Style.

146 See also Aubin 1967: 201; Ambraseys & Melville 1977; Melville 1978; Stöcklin *et al.* 1972.

147 Dilley & Dimand 1931: 118; see below, 1903.

148 Al-'Umari: fol. 174v. He says that a mountain near Hamadan was fissured by the shock into two pieces, implying landsliding rather than faulting.

149 News of this destructive event reached Cairo in Sha'ban 902/April 1497, Ibn Iyas: ii, 320. It is not clear, however, whether the town affected was Jarun, on Hurmuz Island, or a town on the mainland of the Gulf such as Gambrun (modern Bandar 'Abbas). Hieronimo di Santo Stefano 1857, who passed through Hurmuz, probably shortly before the earthquake, does not mention the event, nor does Varthema 1863, who was in Hurmuz and Gambrun in 1504. Cf. Aubin 1973: 111n, who quotes an anonymous source referring to a destructive cyclone and earthquake shortly before 1507.

150 Al-'Umari: fol. 175v.

151 Al-'Umari: fol. 176v. It is probably to this same event that Sa'igh: i, 262 refers under the year 980/1572; he mentions only Mosul, but Longrigg 1925: 37 says that in 1572 there were severe earthquakes (*sic*.) in Azarbaijan, felt as far south as Mosul, which again suggests that these references are for the same event.

Longrigg's information is apparently based on the Mosul *vilayat* calendar, which in turn incorporates the information in the *Manhal al-auliya*, a source which is specifically quoted by Sa'igh for his account of the earthquake. He further records a solar eclipse and a famine caused by a long drought in the same year (980/1572). No reference to these events has been found in the source quoted, which was written by al-'Umari's elder brother, in the edition by Sa'id Diwahchi, Mosul 1386/1967; see Longrigg: 330 on these sources. In that both this source and ours (*Athar al-jaliya*) were almost exactly contemporary and written by brothers, the initial confusion over the correct date may have originated at source rather than in the secondary works that mention the event.

Al-'Umari mentions all three natural occurrences under 908 H. which has a strong resemblance to 980 H., and in the absence of a confirmed reference to the latter in an original source, 908/1503 is to be preferred. Unfortunately, there does not seem to have been a solar eclipse in either 1503 or 1572, so this independent possibility of resolving the issue cannot be applied. The only contemporary traveller's account of this region does not refer to the earthquake, see the Anonymous Merchant 1873. He was travelling between Erzincan—Malatya—Aleppo—Tabriz between 1507 and 1510, not, as his editor believes, in Tabriz 1511—20.

152 Mustafavi 1963: 64, Karimi 1965: 83 and Sami 1968: 337 refer to a *vaqfnameh* for a description of the repairs, which Stahl 1962a: 9 dates in 1505.

153 Rumlu: 342, Qummi: 254, Munshi: 117, Riza Quli Khan: viii, 110.

154 The villages are unfortunately not named. However, the large number of casualties, even if exaggerated, suggests that a part of the region with large villages must have been involved. The absence of any information from Gunabad (Tabandeh 1969), the lack of large settlements in the Nimbluk valley, and its immediate proximity to Qayin, suggest the possibility that the shock originated in the fertile Zirkuh district, forty-five kilometres east of Qayin, in which the large villages of Isfad, Istind and Shahrakht are located, see figure 3.7.

155 Sani' al-Dauleh, *Muntazam*: ii, 121, gives an earthquake in five villages belonging to *Qazvin* in Muharram this year, which is clearly inaccurate. He duplicates the event by also recording an earthquake in Khurasan this year. This mistake is taken up by Gulriz 1961: 873, where we find an extraordinarily muddled account of the so-called Qazvin earthquake, see Ambraseys & Melville 1977. Jabiri 1943: 180 mentions a severe earthquake in Fars this year which is either again inaccurate, or else an event about which we have no further information.

156 Al-'Umari: fol. 190r says that a mountain near Tabriz split into four pieces and a great deal of smoke (dust?) poured out, blotting out the horizon. Alsinawi & Ghalib 1975: 544, on the authority of al-'Umari, put an earthquake in Baghdad and Mosul this year. No notice of this was found in the manuscript consulted, where reference was only made to the Tabriz event.

157 Tarbiyat: 173, quoting the contemporary *Takmilat al-akhbar*, cf. Storey 1972: ii, 798. The death of these princes in the castle is confirmed by Munshi: 75.

158 Khan Ahmad, the rebel governor of Gilan, was sent there the following year, see Rumlu: 440. The castle is frequently referred to in the course of later Safavid history.

159 Al-'Umari: fol. 199v says that the ground moved in waves destroying many houses. The event is mentioned without a precise date by Fasa'i: ii, 154, 160. Modern authors, apparently following Fasa'i, put the earthquake in 997/1589: thus Mustafavi 1963: 64, Karimi 1965: 85 and Sami 1968: 337, who all add that half of the dome and the buildings of the Shah Chiragh were ruined. Fasa'i merely puts the earthquake and subsequent repairs to the ageing structure in the 990s H. Mustafavi: 62 puts the restoration of the Masjid-i Nau (formerly the Atabeg mosque, founded at the end of the sixth/twelfth century by Atabeg Sa'd ibn Zangi), without authority in 995/1587, but it is obvious that the buildings were restored at the same time, following this earthquake. Nevertheless, 995/1587 is also given by Stahl 1962a: 9 and Meshkati 1974: 125. Sami: 326 mentions an earthquake in Shiraz in 690/1291 after which the Masjid-i Nau was repaired, and also refers to the later restoration under 895/1490. Both these dates are almost certainly misprints for the event of the 990s.

160 There is no evidence that this earthquake caused any damage along the main routes from Shiraz to Isfahan and Lar, or that damage in Shiraz itself was serious. Fernberger 1898 passed this

way at the end of 1591, but he does not mention anything unusual in Shiraz or along his route. Travellers who passed through the region a few years later do not mention an earthquake in Shiraz, although they notice the effects of the 1593 earthquake in Lar.

161 Natanzi: 528, 530; he adds that tall houses and lofty buildings became a heap of dust, to the extent that houses and districts were indistinguishable from each other.

162 Natanzi: 528. Teixeira 1902: 241, a contemporary traveller, says that the earthquake occurred in September. Herbert visited Lar in 1628. In the first edition of his book (1634: 52) he says that the earthquake happened 'thirty years ago'; in the second (1638: 120) he puts the earthquake in 1593, of 'their account, 973', which is incorrect. He consistently converts dates with a difference of 620 years between the Christian and Muslim eras. Herbert reports that out of 5000 houses of the town, 3000 were ruined and 3000 people perished. Teixeira's figures are to be preferred; Stevens 1715: 362.

163 The shock had been predicted by an astrologer and the people of the town had moved out before the disaster occurred, which suggests that there were some foreshocks, Natanzi: 529. There is no evidence that damage extended beyond the immediate vicinity of Lar. Travellers do not mention any damage to the caravanserais along the main routes, and Lar must have been rebuilt quickly. Gouvea 1646: i, vii who passed through Lar in 1602 did not notice the effects of the earthquake, nor did Gaspar 1820 or Figueroa 1667.

164 Godard 1941: 12, on the evidence of a *kashi* (title) commemorating the repair of the cupola in 1008 H., kept in the repository of the sanctuary. Not confirmed by other sources.

165 Na'ima: i, 202 calls this a tremendous earthquake.

166 Teixeira 1902: 28—30 arrived in Basra on 6 August 1604. He saw many houses in ruins within and without the walls, which were being rebuilt very hastily. The reason was that about ten days before, the arsenal had caught fire and 5000-odd leather sacks of powder exploded, with such uproar that men thought the end of the world had come. There was great damage in most of the city.

167 Munajjim: fol. 57r. The contemporary author of this account incorrectly dates the earthquake on 2 Muharram, which is not consistent with the narrative of events surrounding the episode nor the astrological data that he also provides. He says that the hunt started on 1 Muharram and ended on the 3rd, the day after the earthquake, having continued for five days. He also says the Sun was 27° plus in Aries on 1 Muharram, and 1° in Taurus on the 3rd. The sun would not have entered Taurus until 5 Muharram.

168 The shock was felt by members of a hunting party in the Miyan Kaleh peninsula, held by Shah 'Abbas. Munajjim Yazdi was the Shah's astrologer. The peninsula of Miyan Kaleh lies due north of Ashraf. The shock was possibly felt as far as Shamakha, Alishan 1881: 446.

The restoration and repair work carried out of the *masjid-i jami'* in Damavand as recorded by an inscription dated 1024/1615, may be associated with this event.

169 Munshi: 947. Oldham 1883 and Wilson 1930 date this event 1611, as does Berberian 1977: 86. In Ambraseys & Melville 1977: 187, *Mahvalat* is wrongly translated as *mahulat* = barren (regions).

170 Valle 1677: 481, who was a witness of this event. Although the destruction in Bandar 'Abbas was extensive, Della Valle notes that the houses were poorly made of mud bricks and the castle was not strongly constructed.

171 In the first week of January 1624, Poser 1675 passed through an area he describes as having been destroyed by an earthquake, in the region around Qal'eh-yi Shikasteh and Istakhr. He gives no indication of the time that had elapsed between the earthquake and his visit to the area. Della Valle passed by the two castles shortly before in 1621, and does not refer to any earthquake damage. This limits the earthquake to the years 1622—3, unless Poser is referring to a more distant literary tradition (cf. above 1069). However, Ouseley 1819: ii, 236 notices that between 1621 and 1627 the number of free-standing columns in Persepolis was reduced from more than twenty to nineteen, see also § 5.2. Damage in the region is also alluded to by Kaempfer 1712: 324.

172 Kasravi 1956: 161, followed by Daulatabadi 1964: 12, probably
refers to the Tabriz earthquake of 1133/1721, which is well
documented. Daulatabadi incorrectly gives 1033 = 1633, which
is followed by Berberian 1976: 408. Kasravi is also one hundred
years out when he says, p. 162, that there was an earthquake in
Tabriz in 1090 H., quoting Sani' al-Dauleh, *Mir'at*, whose basis
for this information is unknown. In fact, however, Sani' al-
Dauleh, *Mir'at*: i, 352, gives 1190/1780 for the event. The correct date is 1194 H., see below.
　　The *Atlas* also has a spurious event in Tabriz in 1623.
173 Sani' al-Dauleh, *Muntazam*: ii, 190. There seems to be no record
of this earthquake in contemporary sources, for instance
Philippe 1669. Sani' al-Dauleh may refer to the event of 1608
which affected the Qazvin region, or alternatively he may be
inaccurately following Hajji Khalifeh: 134, who puts an earthquake in Tabriz this year, 1049/1639, which is also incorrect
(see under 1641).
　　This spurious event is mentioned by Gulriz 1961,
Ambraseys 1974b: 56, 78 and Berberian 1977: 86.
174 The date of the earthquake, Friday at about eight on the night
of Ar. 22 Ovtan 1090/22 Shawwal 1050 H./5 February 1641, is
given in full or in part by several contemporary or near-
contemporary sources, see Ar'aqel and others in Hakobyan
1951, 1956. It is confirmed by Rycaut 1680:6, Lotichius 1646:
ii, 747, Coronelli 1693: 322. Numerous other references of varying accuracy are available for this earthquake: e.g. Walther 1805;
Hammer-Purgstall 1844: ii, 504, 1829: v, 307, on the authority
of Morier 1818, puts the earthquake a year earlier at the time of
the ascension to the throne of the Ottoman Sultan Ibrahim
(1049/1639). The earthquake is also put in 1049/1639 by Hajji
Khalifeh: 134, who says that the Shami-Ghazan was destroyed.
Tavernier 1681: i, 51 puts this ten years later in 1651, a date
followed by Godard 1934: 13. Tavernier's inaccuracy is unfortunate for he visited Tabriz in January 1655. An earlier visitor to
the city, Evliya Chelebi, there in 1647, refers to the earthquake
destruction of the Sham-i Ghazan, but does not give a precise
date for the event. Other incorrect dates are given by Andreas
Evdoketsi, who describes the event as occurring in the same year
as the shock that ruined Van in 1648, putting both earthquakes
in 1646, and also by the *Anonymous Armenian Chronicle*, which
puts the event in 1091/1642, causing 1200 casualties, see
Hakobyan 1951: 326. Wilson 1930: 115 puts earthquakes in
Tabriz in 1640, also affecting Damascus and in 1641, also felt at
Baghdad, which appears to be without foundation. Wilson quotes
Hajji Khalifeh, who mentions neither Damascus nor Baghdad,
and he also quotes Mallet. The earthquake in Baghdad is also
given by Alsinawi & Ghalib 1975.
175 According to the *Theatrum Europaeum* 1644: iv, 484, the earthquake of 1641 in Persia was followed by shocks for three days
during which rockfalls overwhelmed a town in which 200 000
(*sic.*) people perished. Then another shock, in the province of
Anziron (= Azarbaijan), almost completely engulfed Riangasan
and Rikan where 100 000 (*sic.*) people were killed. *Dresdnische
Gelehrte Anzeigen* 1756: 117, gives Ricky for Rikan and Seyfart
1756: i, 36, Riken for Rikan. Hoff 1840 and Mallet 1852 give
Rikan. It seems very probable that the information in the
Theatrum Europaeum refers to the same earthquake and that the
second source should have Armenian origin, for it explains the
corruption of Usku (i.e. Urkan in Arm.) into Rikan and of
Sham-i Ghazan (Arqghazan) into Riangazan.
176 Melville 1981.
177 A story of a shepherd and part of his flock being swallowed up
by a ground fissure during an aftershock is related by Ar'aqel:
497. Internal evidence suggests that this was the result of a landslide. Berberian 1976: 183, maintains that this was the result of
opening and closing of an earthquake fault fracture (*sic.*).
178 Flash floods and large-scale landslides continue to occur in this
part of Mount Sahand, causing great damage, Rawlinson 1840:
3. In 1910 the village of Jaraghil was destroyed by a landslide
which was attributed to the weakening of the mountainside by
previous earthquakes, which had caused deformations on the
valleyside, *Iran-i nau*: 1329.1.26.
179 Described in great detail by Ar'aqel: 37/499; also in contemporary or near-contemporary notices: Andreas Evdoketsi, Grigor
Varagetsi, Ananun Vanetsi, Martiros Khalifa in Hakobyan 1951:
161, 356; 1956: 284, 483. Cuinet 1890: ii, 695, says that 2000

people were killed in Van which seems to be an exaggeration; see
also Lynch 1901, Tchalenko 1977. Sani' al-Dauleh, *Muntazam*:
ii, 194 dates the event two years earlier in 1056/1646 and so
does Abdalian 1935a. Abich 1882: 440, 446 gives the correct
year but wrong dates (2 and 12 April).
　　In 1075/1664 another earthquake destroyed part of the
citadel of Van, one of the towers and some of the walls, Rashid,
Tarikh: i, 102.
180 A poem by Mir Baqa-yi Badakhshi describes an earthquake that
did much damage in Tabriz in 1060/1650, Kasravi 1956: 161,
followed by Daulatabadi 1964: 13. The poem is also given by
Nakhjuvani 1964: 450. The chronogram yields the date 1060 H.,
but the poem is not by itself an adequately reliable source for
this event, which seems rather unlikely to be a genuine occurrence. Tavernier 1681: i, 51, who was in Tabriz in 1655, mentions an earthquake as having destroyed many houses and the
Sham-i Ghazan in 1651, but this almost certainly refers to the
shock of 1641. The entry in Berberian 1976: 410 on this event
is heavily disguised by a series of misprints and that in the *Atlas*
is based on a secondary source.
181 Hanway 1753: iii, 92n. There is no supporting evidence for his
statement and Hanway cannot be regarded as a reliable source
for events of this period. Contemporary travellers who passed
through Tabriz shortly afterwards do not refer to any earthquake
damage, Tavernier 1681, Poullet 1668.
182 Ar'aqel: 234/446, *Anonymous Armenian Chronicle* in Hakobyan
1951; Stepanian 1942 gives 1658 and Ergin *et al.* 1967 place this
event in Erzerum.
183 Mentioned only by Muteferriqa, the continuator of Hajji
Khalifeh: 137. Once again, it is not clear whether many places
were ruined in Tabriz itself or whether the earthquake was
destructive outside the city. Travellers passing through the region
late in 1664 do not mention any earthquake damage, Tavernier
1681, Daulier-Deslandes 1926.
184 Sani' al-Dauleh, *Muntazam*: ii, 204, who adds that the damage
was repaired by Shah 'Abbas II (d. 1077/1666). An alternative
date of the end of 1074/June–July 1664 is also given for this
event, *ibid*: 210–11. These rather contradictory statements by
Sani' al-Dauleh indicate two dates for the restorations, which is
quite probable, but it is clear that only one earthquake was
responsible for the damage. Of the two dates he gives, 1075/
1665 is perhaps to be preferred, if only because 1074/1664 is
the date given for an earthquake in Tabriz (see text) with which
he might have been confused. Sani' al-Dauleh frequently follows
Hajji Khalifeh and both authors are unreliable (cf. pp. 15, 173).
It is clearly desirable to emphasise the separation of the Tabriz
and Damavand events.
　　The inscription, which was uncovered by a painter in
1296/1879, is on a beam on the wall to the left of the *mihrab*.
While in Damavand in April 1974 we were told that the inscription is still there, although extremely difficult to reach; cf.
Smith 1935.
　　It is probable that this event was the most destructive of
a series of shocks at this period, for Chardin, who was in Iran
between 1665 and 1677, remarks that earthquakes in Gilan and
Mazandaran are frequent and furious, although they usually pass
without fatal results, Chardin 1811: iii, 285; iv, 162. No other
contemporary visitor to Persia mentions this earthquake.
185 Al-'Umari: fol. 216v. We are indebted to Bibi Azam Bakhtiyari
of Ardal and to her household for providing information from
unpublished family documents; see Ambraseys 1979.
186 Chardin 1811: iii, 133, Isahak Vardapet in Hakobyan 1951,
Sani' al-Dauleh, *Muntazam*: ii, 211. The Shah sent officials from
Isfahan to study the damage and carry out repairs. Inscriptions
in various buildings in Mashhad testify to the widespread damage
caused in the city. The dome of the shrine was repaired in
1086 H.; the Dar al-Sayyadeh in the same year; the Masjid-i
Gauhar Shad in 1084, 1087 and 1089; the Parizad madraseh in
the bazaar in 1091; the Dau Dar madraseh in 1088; the Balasar
madraseh in 1091 and the Pa'in Pay madraseh in 1087 H., Sani'
al-Dauleh, *Matla*': ii, 44, 97, 140, 148, 151, 245, 256, 258;
Khanikoff 1861: 104; Jabiri 1943: 203; Godard 1941; Meshkati
1974, cf. Sykes 1910: 1114, 1137.
　　The small town also affected may have been Qadamgah,
see Melville 1980.
187 The shocks must have occurred between July 1677 and February

1678, i.e. between Fryer's two journeys from Bandar 'Abbas to Isfahan and back, see Fryer 1698: 240, 301, 309.

188 Quoted by Rabino 1917: 282, on the authority of the *Mahboub oul-kouloub* of Koutb ed-Din Mohammed Lahiji. A careful examination of this book, however, yielded no trace of this information, which must be contained elsewhere; see Melville 1978.

189 This account is based on local tradition, Tabandeh 1969: 31, and is subject to certain reservations, as the author himself implies. The old Friday mosque of Gunabad (thirteenth century) bears no inscription relative to repairs about this time. However, an inscription in the *masjid-i jami'* of nearby Qayin records restoration work in 1086/1675, possibly associated with this earthquake, which Tabandeh only vaguely puts in 1089/1678, Meshkati 1974: 76; Ambraseys & Melville 1977.

190 Well covered by contemporary Armenian sources in Hakobyan 1951: 303, 311, 327, 395; 1956: 272, 414, 522; Khatjikyan 1955: 215–16 footnote, and the references in Stepanian 1941: 71, no. 21, 28 and 59; also in Monier 1723: 228, Tchamtchean 1781; Chakhathuno 1842. Villotte 1730: 65, who was in Erivan in 1688, puts the earthquake in 1676. Hoff 1840: 330, dates the event correctly in 1090 H. but writes Ravan for Rewan = Erivan. Sieberg 1932 and Alsinawi & Ghalib 1975, follow Hoff's reading for Erivan and put the earthquake at Rawa on the Euphrates near 'Ana, 660 kilometres from where the earthquake actually happened.

Kasravi 1956: 162 mentions an earthquake in 1090 H. as having been felt in Tabriz, but this is in error for 1190 H., see p. 183. The location of some of these places is shown in figure 3.2.

191 Mahjuri 1966: ii, 108; it remains to find confirmation of this event in earlier sources.

192 *Dressdnische Gelehrte Anzeigen* 1756: xviii, 292; no record of this event has yet been found in Persian sources.

193 This information is preserved in a letter from Kuran to a native of the town, living in Damascus, and given in Arabic by the Damascus author, al-Ghazzi; see Taher 1975: 69.

The account derives from a local eye-witness report. It is the first recorded earthquake for the Isfara'in valley, but it is probable that earlier references to earthquakes in unspecified locations in Khurasan apply to this area. There are no major towns with which news of earthquake damage can be associated, but the region has always been an important corridor of through traffic from east to west. For the routes in the thirteenth century see Aubin 1971.

Yaqut: iv, 319, mentions Kuran as one of the villages of Isfara'in and Yate 1900: 382, remarks on the ruins of former sites at Adkan and Kuran, where two of the ancient towns of the valley were supposed to have existed.

194 Hakobyan 1951: 309, 312, 359 and 284 (*Divan Hayots Patmut'ean*: x, 115). For St Taddeus see Ep'rikean 1903: ii, 1; Ra'in 1970: 78 mentions only one earthquake which caused the collapse of the walls and ceiling of the church in 712 H., see above under 1319.

195 Razani & Lee 1973a on the authority of A. Sami; not confirmed by other sources.

196 Gollancz 1927 (*Bassora Manuscript* no. 285/510). Alsinawi & Ghalib 1975, refer to an earthquake in Baghdad in 20 Safar 1114, on the authority of al-'Umari, but we can find no mention of this event by that author.

197 Ananun Vanetsi in Hakobyan 1951: 367, says that the country of Mahmatun was particularly affected and that the dome of Bardughimeos Ar'aqeal and fortified towers of Kara Sarai collapsed. The country of Mahmatun may be identified with the region of Mahmatan or Mehmedik, around Hoshap', Hübschmann 1904. The church of St Bartholomeus should be at Deir (Albayrak or Aghbak, mod. Sikefti); however, Stepanian 1942, suggests that this church was in Van. Moreover, his sources give Kara Hisar Sarai instead of Kara Sarai which implies that both Karahisar and Sarai were affected. It is probable that the repair of public buildings and bridges in the region of Van contributed to by Durri Efendi in 1720 was necessitated by this earthquake, Durri 1810. No local tradition of the event has survived in the Zap-suyu valley, which was visited late in 1966, and no Armenian monuments were found.

198 Extract from a letter dated Tabriz, 12 July 1717, from the Capuchin missionary Pierre d'Issoudun in Correspondence politique, Perse, vol. 5, fol. 139r, *Archives du Ministère des Affaires Étrangères*, Paris; cf. Kroell 1977: 64. Nothing more has been found about this event which marks the beginning of a series of destructive earthquakes in the Tabriz region, culminating in a devastating shock in 1721. Unfortunately the correspondent died before the end of the year.

Passing through Tabriz in January 1717, Bell 1788: 90, found 'several well built mosques with stately minarets'. At the time of the earthquake he was travelling between Daulatabad and Isfahan. He does not mention the shock being felt.

It is possible that Porter 1821: i, 227, writing a century later, refers to this event when he says that 70 000 people were killed by an earthquake in Tabriz in 1727, but this more probably describes the 1721 disaster, see below.

199 The date is given in full by Zunuzi: 163, *Journal Historique*, Oct. 1721: 276 and Hanway 1753: iii, 92. Isahak Vardapet and Grigor Urakh Karintsi in Hakobyan 1951: 294, give only the year and Anonymous 1721 in Dzhanashvili 1902: 321, gives 16 April. Later sources, cf. Daulatabadi 1964: 14, give 1134 (which begins in October 1721) for 1133 H. and it is also probably to this event that Kasravi 1956 refers under 1033 H. (see above p. 183).

Porter 1821: ii, 227, mentions an earthquake in Tabriz in 1727 which killed 70 000 people, followed sixty years later by another which can only be the 1780 event. That his 1727 is in error for 1721 (a far more destructive and widely reported shock than any other at this period, cf. earthquake of 1717) seems beyond doubt, but the mistake has been repeated by a number of authors, see Berberian 1976: 411. The only apparently independent indication that there might have been an earthquake in 1727 is in a poem by Sayyid Najib Urdubadi, kindly brought to our notice by Riza Razani. It has a chronogram dating a disaster in Tabriz in 1139 H. (August 1726 to August 1727), but it is not clear how reliable this is. Whether or not Urdubadi was contemporary with the earthquake and therefore presumably accurate, the notice on its own cannot be considered sufficient evidence for an earthquake of this gravity, even though there may have been a small shock at this time. Wilson 1930, puts the event on 18 November 1727, which is the date of an earthquake in England and the preceding entry in the list of Mallet 1852.

200 This is the lowest estimate, given by Gardane who was writing from Isfahan in June 1721, see Correspondence politique, Perse, vol. 6, fol. 13r, *Archives du Ministère des Affaires Étrangères*, Paris.

Hanway and an entry in the *Mariastein manuscript* dated 20 August 1721, fol. 71, put the casualties at nearly 100 000. Other estimates range between 80 000 given by Du Cerceau 1728: i, 272 and 250 000 given by Sani' al-Dauleh, *Muntazam*: ii, 229. Some later sources, following Mallet, give only 8000. This wide range in the figures and their general exaggeration is found in similar accounts of the 1780 event. See also Anonymous 1783: 814.

201 Gardane says the shock ruined three-quarters of the houses, while the *Mariastein manuscript* adds that only one-third of them collapsed and that none of the many Christian merchants were killed. Moreover, Zunuzi confirms that 'most of the tall buildings, such as mosques, madrasehs and shrines were excessively fissured but not completely destroyed.' That the heaviest destruction occurred at some distance from the city is suggested by Anonymous 1721 in Dzhanashvili 1902: 321, who says that 'certain localities in the [region] of Tabriz were tested.'

202 Brydges 1834, who passed through the region in the summer of 1809, says 'between the camp [at Ujan] and Bosmeech [Basminj], we passed over ground which some years before had been rent by a succession of earthquakes in the most extraordinary manner, and on the left hand [W.] of the road I was shown a mountain, riven at that time from top to bottom. This dreadful calamity took place in the year 1724 [*sic*] . . . ' Wilson 1930, misreads this 1774, cf. Berberian 1976: 411. The succession of fault breaks in the vicinity of Shibli, unequivocally associated with the earthquake of 1780, would on this evidence appear to have been previously associated with the 1721 event. Ground deformations are also mentioned by the *Marienstein manuscript*; see following note and Berberian 1977: 229.

203 Gardane: fol. 14r, also refers to an earthquake in Qazvin at this

time, which did not do so much damage. It is not clear whether the two events were connected, but news of them evidently reached Gardane (in Isfahan) at the same time. It is probable that the shock felt in Qazvin was from this large earthquake on the southeast extension of the Tabriz fault. In a separate communication, fol. 55v, written a few days earlier (6 June), Gardane also notes a shock in Qazvin, which ruined several houses, apparently a reference to the same event. That there was, however, some distinct activity in the Qazvin region after the Tabriz earthquake is indicated by the *Marienstein manuscript*: fol. 72, which says that one Monday in July 1721 the settlements of Kiareh (Khiyarij), Dang (Dadkan or Dukhan?) and Kiesal (unidentified) were ruined and many houses were damaged as far as Qazvin. The region most affected would thus appear to be around Buyin Zahra. No record of this in the Persian sources has yet been found.

Gardane goes on to say that the mullas had predicted an earthquake for Isfahan on 5 June, because of the extraordinary pallor of the Sun and Moon over the previous few days, but this had proved false. Krusinski, in Du Cerceau, also notes a phenomenon in the skies over Isfahan, the Sun appearing blood red through thick clouds for two months during the summer. This was again seen as an evil portent, in the period shortly before the Afghan siege of the Safavid capital, cf. Mihdi: 191–2.

204 *Gazette de France* 1755: 529; *Journal Historique* 1755: 462, followed by Seyfart 1756: 139 and Berryat 1761: 627, who says that 6000 houses were ruined. Walther 1805: 114, gives 7 July.

Later authors misreport this event. Huot 1837: 110 puts an earthquake in Tabriz in July 1755. Hoff 1840: iv, 423, says that on 7 June or July, an earthquake in North Persia was felt at Tabriz and in Kashan, Hann (*sic.*, = Qum?) and Isfahan, on the authority of Seyfart and Porter. The former mentions only Kashan, as above, the latter only the Tabriz events of 1727 and 1780 (see text). Despite this, Wilson 1930 gives earthquakes in 1755 in Tabriz, Kashan and Isfahan, with 40 000 casualties, a figure Porter gives for the 1780 Tabriz event. All the many similar references to an earthquake in Tabriz in 1755 are totally spurious and are a further reflection of the confusion surrounding the 1721 earthquake. There is, on the other hand, every reason to suppose that the Kashan shock would be felt in Qum and Isfahan, as in 1778 (see text).

205 Riza Quli Khan: viii, 147 describes this apparently similar event, with 1200 casualties, under 982/1574; Sani' al-Dauleh, *Muntazam*: ii, 128, puts it in 962/1555. The same notice is given by more recent authors, such as Naraqi 1969: 19.

It may be that the strong resemblance between the 1574 or 1555 and the 1755 earthquakes in Kashan is purely coincidental. However, the suspicion that one of them is spurious can be strongly argued. It is most improbable that an earthquake of this severity in Kashan in 1574 would not have been recorded in the contemporary Persian sources. Hasan Rumlu, a native of Qum, completed his chronicle in 1577 and two other historians, Qummi and Natanzi, from the same area, in 1590 and 1599 respectively. All three authors mention earthquakes elsewhere in Persia and are unlikely not to have done so for a local event. Riza Quli Khan is a late and not entirely reliable author and his statement on its own requires confirmation from earlier sources.

On the other hand, contemporary European accounts of the 1755 earthquake are available and it is inconceivable that they are referring to an event two centuries before. The fact that no contemporary Persian sources mention this earthquake is not so significant, for the coverage of this confused period is less good than for the Safavid period. Gulistaneh's work covering the interregnum between Nadir Shah and Karim Khan was written in India. Historians of the Zand period have little information on this early part of the reign, although one might have hoped that the authors who give the later 1778 event would also refer to this earlier one. The question remains, however, of how European sources acquired the information; presumably through missionaries in Isfahan, since we are not aware of a contemporary traveller who passed through Kashan during that period. In contrast, Newberie found Kashan thriving in October 1581.

If these negative considerations are sufficient, the account of the 1574 earthquake in Kashan should be taken to apply to the 1755 event and represent the Persian tradition of the occur-

rence. This would make it necessary to reject the 1574 event in Ambraseys 1974*b*, Berberian 1977: 86 and Melville 1978: 76.

206 Experienced by Niebuhr 1776: ii, 139; see also Karimi 1965: 89.

207 Walton 1865: 126, from local sources.

208 Without quoting an authority, *Ittila'at*: 1960.5.2. mentions a destructive earthquake in Lar in 1180/1766. It is possible that the account by Qazvini: fol. 147v. of an earthquake at an unspecified location in Fars during the reign of Karim Khan of Zand (*Vakil* from 1765 to 1779) refers to this event.

209 Contemporary sources such as *Gazette de France*: 1769.11.3 and *Journal Historique* 1769: 474, state quite clearly that torrential rains and a hailstorm in Baghdad destroyed 4000 houses and killed a number of people. They add that during the hurricane earthquake shocks were felt. Without any good reason, Richard 1771: viii, 504, misquotes these sources and says that it was the earthquake that destroyed 4000 houses which was then followed by torrential rain. Sieberg 1932, Alsinawi & Ghalib 1975, as well as other later writers, repeat this misleading information. As a matter of fact, it is questionable whether the shocks alluded to in the sources were genuine earthquakes. See above under 902, 912.

210 *IO* R 15 1 3, pp. 11, 13; Abu 'l-Hasan: 213; Qazvin: fol. 147v; see also Naraqi 1966: 155, for an unpublished family document on the earthquake, and Perry 1979: 241. The shock was preceded by three or four days of continuous heavy rain. A later author puts the casualties at 30 000, Zarrabi: 207.

211 He was the brother-in-law of the historian Abu 'l-Hasan. Poems by the contemporary writers Azar Shamlu, Hatif Isfahani and Sabahi Bidguli, both on the earthquake and the subsequent restorations, give dates for the completion of different buildings in Kashan, showing the progress of the reconstruction work. Hatif describes building by 'Abd al-Razzaq in 1196, the repair of his own house in 1194, and the restoration of the *masjid-i jami'* in 1196 H., *Divan*: 30, 99, 116, 119. Chronogrammatic verses by Bidguli are quoted by Afshar in the notes to his edition of Zarrabi: 497 and by Naraqi: 157; the poem refers to the repair of the bazaar in 1195 H. Another poem by Bidguli on the earthquake and reconstruction in Kashan is given by Nakhjuvani 1964: 452. The poem yields the date 1198 H., which Nakhjuvani incorrectly takes to be the date of the earthquake. Naraqi: 158, mentions repairs to the *masjid-i jami'* and the adjoining *ab-anbar*, dated by another verse of Bidguli's to the year 1193 H. Inscriptions above the *mihrab* and the plaster-work on the inside of the dome of the *masjid-i jami'* are still visible, referring to repairs in 1194, 1196 and 1207 H. (1793 A.D.).

Olivier 1807: iii, 94, who passed through the region in October 1796, found Kashan in very good condition, with only one-fifth of its houses still in ruins. But he notes the population was only about 30 000 (compared with an estimated 150 000 in the Safavid period), and that from Abibeh (Bidashk) to Quhrud most villages were destroyed or badly damaged. Dupré 1819 and Flandin 1851 also note the desolation of the country. Allah-yar Salih notices that in the graveyard above the ruins of Saruq most headstones had moved from their place and toppled over 'either at the hands of man or as the result of an earthquake.' Almost all the dates on the stones are well before the earthquake, see Afshar, *loc. cit.* Kashan never regained its former prosperity, Watson 1866.

European sources do not mention this earthquake. Mallet 1854 gives an earthquake on 14 March 1794 (1208 H.) which, according to the *Memorial de chronologie*: ii, 932 and Perrey 1850, ruined Kashan, but this is not mentioned by local writers. The area around Saruq was destroyed by local earthquakes on 7 February 1890 and again in 1895, Windt 1891, Stürken 1906.

212 Armenian authors, e.g. Martiros Khalifa, put the event on a Friday night under the Old Style date 27 December 1779; see others in Hakobyan 1951, 1956. The event is reported by Abu 'l-Hasan: 281, Zunuzi: 163, al-'Umari: fol. 260r and Maftun: ii, 211. They all say it was a Saturday, but are equally divided on whether it occurred on the last day of 1193 or the first of 1194 H. The Muslim day ends at sunset and it is probable that the date 1194 is to be preferred, with a strong foreshock in 1193 and a strong aftershock in 1194. This scheme accommodates the chronogrammatic inscription quoted by several authors, e.g. Daulatabadi 1964: 20, which distinguishes three destructive shocks.

213 Kasravi 1956: 162, in addition to the sources referred to in the previous note.

214 Zunuzi gives twelve *farsakhs*, Tabataba'i gives twenty; Jaubert 1821 includes Sufiyan.

215 See the Armenian sources in Hakobyan 1951, 1956.

216 Beaumont, writing in July 1780, gives the figure as 50–60 000, *IO* R 15 1 3, p. 50; the lowest estimate (50 000) is made by Qazvini: fol. 147v. The highest estimate is 205 000 by Abu'l-Hasan. Probably all these figures are exaggerated.

217 Drouville 1825: i, 54 says that as a result of the earthquake, between Tabriz and the mountains and to the northwest of the city, a terrace of grey material was thrown up which contrasted clearly with the red of the mountain and the green of the lower slopes. The terrace, he says, was up to ten metres high, 100 metres wide and two (geographical) miles long. The same is indicated, less precisely, by Ouseley 1819: iii, 406, who nonetheless says that it occurred to the northeast of the city. He adds the delightful story of the French gentlemen in Tabriz who 'acquired a bad name among the lower classes, having made artificial earthquakes by burying under ground a composition of steel-filings and other ingredients, which, after a certain time, fermented and exploded with a violent concussion; on this account, the old women of Tabriz accused them of having set the mountains on fire, and attributed to those experiments the several shocks which have alarmed them since the French departed' (1808–9).

218 The extension of the fault break to the southeast of Tabriz for about seven *farsakhs* towards Shibli, is mentioned by Zunuzi. Maftun adds that the Surkhab mountain split apart and a deep fissure formed. Its width was two metres and its length was about two *farsakhs*, heading southeasterly. According to al-'Umari the throw was four yards. Splitting of the ground to the north of Tabriz and in Marand is mentioned by Qazvini, Tabataba'i, and Kotzebue 1819: 145.

219 Maftun; Tabataba'i. Similar phenomena occurred in the city, Freygang 1816: 279, 284.

220 Maftun says that about two *farsakhs* east of Tabriz a meadow (of the size requiring more than twenty *mann* of seed for sowing) was carried off for a distance of a quarter of a *farsakh* or more. This was probably the low land between the ruins near Arpadarrehsi and the Talkheh-rud.

221 Martiros Khalifa and Ananun Vanetsi say that the shock was felt in Van where it did not cause any damage. Tabataba'i adds that the shaking of the ground was felt for a distance of a month's journey, and Hakob Tivriktsi says that it was also felt on the east side of the Furat River in the region of Malatya. Zunuzi gives the details of the two aftershocks; Tabataba'i and Nadir Mirza also record the long sequence of shocks, the latter saying that they continued for twelve years, see Daulatabadi 1964.

222 Those affected are listed by Zunuzi and Tabataba'i; many of these are hard to identify, see Melville 1981.

223 See Zunuzi and Nakhjuvani 1964: 523.

224 A valuable account of many of these monuments is given by Karang 1972, who mentions some of the inscriptions dating restoration work, e.g. on the Dal-o-Zal mosque, p. 667.

225 According to Berberian 1976: 403, the Tazeh Kand–Zabarlu section of the fault shows a throw to the northeast. We could find no convincing field evidence for this reversal. They also suggest a minor right-lateral horizontal component of the fault movement which we have been unable to substantiate.

226 See statements by Freygang, Ouseley and Porter 1821: ii, 502. Morier 1812: 276 notices that domed buildings withstood the earthquake better than other types of masonry construction. Tancoigne 1820: 73, puts the earthquake in 1559. Sani' al-Dauleh gives 1190 for 1194 H. (see note 172), and Johnson 1818, puts it in 1814. Several other authors describe Tabriz and its surrounding over the next thirty years, possibly including the effects of a later earthquake, see text under 1786.

227 Qazvini: fol. 148r. The exact location of the event is not known.

228 Given by Reinegg 1796: i, 28. The fact is denied by Porter 1821: i, 184 while Ritter 1840: vii, 507 and Anonymous 1845, on the basis of local information, maintain that no such eruption could have taken place. However, the *Atlas* lists an earthquake in Ararat on this date.

229 Michaux 1911: 381, who experienced the event.

230 Nadir Mirza: 56, recounts an eye-witness report of the shock in Tabriz which he dates in 1201/begins October 1786, towards the end of autumn. Brosset 1849: 25 records an earthquake in Erivan in Ar. 1235/begins September 1786. Jaubert 1821: 136, Tancoigne 1820: 67, and other authors passing through the area from Khuy to Tabriz in the first decade of the nineteenth century describe the effects of earthquakes in the region and refer to a destructive shock some years beforehand to account for the numerous ruins still visible. It is not entirely clear whether these accounts refer to the 1780 event or to subsequent shocks, such as the one at the end of 1786. The fact that Porter 1821: i, 227, mentions another earthquake sixty years after that of 1727, although clearly intending the 1780 earthquake (he says the shock killed 40 000 people), may contribute to the adoption of the date 1786 or 1787 in some catalogues, such as Berberian 1976: 411 (quoting Nadir Mirza) although it is not in fact well supported. Abdalian 1935a, 1964 and Ergin *et al.* 1967 put an earthquake in Tabriz in 1791, apparently confusing this with a separate event in Armenia in 1784, cf. Berberian 1977: 160.

231 Rich 1836: i, 387; Longrigg 1925: 208.

232 The date of the event is uncertain. Dupré 1819: ii, 209 and Morier 1812: 259, mention the event as having occurred shortly before their visit in 1809, see Kaihan 1932: ii, 378; Barthold 1930: 265. Manestey 1812, Jaubert 1821, and Tancoigne 1820 who visited the site in 1804, 1806 and 1808 respectively, do not mention the event.

233 Truilhier 1838: 256, 8.

234 Morier 1818: 355; Rabino 1928: 46 refers to a pertinent inscription in the *masjid-i jami'* in Babul; also, local information. Ritter 1840: viii, 426, 540 and 563 confuses the dates of the earthquakes in the first decade of the century; see also Stahl 1911.

235 Bontemps in Ittihadiyyeh: 71; Gardane 1809: 34; Tancoigne 1820: 71; Morier 1812: 297.

236 This event is described as being on a serious and destructive scale, Mahjuri 1966: ii, 369. Local tradition has it that this earthquake ruined the settlements along the Turud–Shahrud caravan route and caused widespread damage along the Mashhad–Tehran route. This is consistent with Mahjuri's narrative although he seems to exaggerate the effects of the earthquake (probably amalgamating it with the event which followed); see also Gansser 1969: 449. Between 1805 and 1810 there were a series of damaging shocks in the central Alburz, the effects of which are now difficult to separate.

237 Dupré 1819: ii, 187, 198; Morier 1812: 254; Riza Quli Khan: ix, 289; Rabino 1917: 69. Dupré says that in Tehran the shocks caused great consternation and the Shah supplicated for divine mercy. The people attributed to the French the several tremors which had alarmed them since their arrival, and predicted the total destruction of the French quarters by an impending shock. Sani' al-Dauleh, *Muntazam*: iii, 93 and *Mir'at*: i, 533, reports the effects of this event together with those of the next earthquake, putting them all in Shawwal 1224. An inscription referring to repairs in 1227/1812 of the Imamzadeh Ja'far at Pishva, southeast of Varamin, may be associated with this event.

238 Ouseley 1819: iii, 270, 295; Rabino 1928: 37, 40, 54, puts the event in 1225/1810; Qal'eh Bandi 1969: 158. Unpublished family documents in Babul and Sari mention the effects of this earthquake on the roads to the south, particularly in the Ganj-i rud and Julab region, where all settlements and summer resorts were destroyed. This, they say, happened when Mirza Safi Mazandarani took office. Rabino, p. 44, puts the destruction of the bridge in 1820, and he is followed by Wilson 1930.

239 Yate 1900: 205. The date of the event is questionable. Local information suggests that this was a large shock but it is not certain that the information refers to the same event.

240 Monteith 1857: 118; Karimi 1965: 54; Perry 1979: 275. Al-Qusi: 147 and *EI*[1]: 'Shiraz', put the event in 1813. See also text, § 5.2.

241 Ouseley 1819: iii, 430. An inscription in the church of St Taddeus dated 1229/1813 refers to repairs of the structure.

242 Morier 1818: 355; Murray 1859: 199; Ritter 1840: viii, 563.

243 Porter 1821: ii, 501; Anonymous 1839; later authors, Wilson 1930, Stahl 1911, give confusing accounts of the event, the latter putting the earthquake in 1818.

244 Of the contemporary sources, we know only Willock *IO* 9 83 and the correspondent of the *Bombay Gazette*: 1824.9.7, who happened to be at Kunar Takht at the time of the earthquake, give the exact date of the event. Fasa'i: i, 267 dates the event in

Shawwal 1239 saying that it preceded the earthquake that ruined Shiraz on 27 Shawwal. Later authors confuse the two events and attribute the destruction of Kazirun to the Shiraz earthquake, see text.

245 Willock says the minister of Shiraz reported an exaggerated picture of the damage in which allegedly 2000 people were killed. His object was apparently to obtain a remission of the revenue. Alexander 1827: 100, who passed through the region in June 1826 found Kazirun, among other places, still in ruins and the Russian mission years later noticed the damage caused by this earthquake along the Shahpur valley, Chirikoff 1875. However, the bridge on the Shura near Shahpur was not destroyed by the shock; two of its arches had been carried away by floods some time after the event. Although there are no estimates of the casualties for the whole region, Willock in a later report (*loc. cit.*) says only about 150 lives were lost in Kazirun.

246 Anon. 1825; *Journal de Francfort* 1825: 50; Willock in *IO* 9 83; Zain al-'Abidin: 326; Sani' al-Dauleh, *Muntazam*: iii, 130; Fasa'i: i, 269; cf. Arago 1859 and Karimi 1965: 83 (Farsi text). Poetic descriptions of the event are given by Mirza Kuchik in Ethé 1903: 363, and Vassal, quoted by Imdad 1960: 38.

Casualty figures officially reported for Shiraz are grossly exaggerated. Anon. and Fasa'i estimate them between one and several thousand, and Zain al-'Abidin puts the material damage at three crores. Willock's dispatch in *IO* mentions the destruction and distress in Shiraz, but note Fasa'i's statement that after the earthquake all the people stayed on the roofs of their houses! A local tradition has survived that the tree in the *sahn* of the Masjid-i Nau was planted by survivors from Kilistan and Qalat who came to Shiraz. In Guyum and Kilistan we could learn of no local tradition specifically concerning this earthquake, but it was said that sites north of Shul and at Guyum were abandoned as the result of an earthquake that had happened 'one hundred years ago'.

Sani' al-Dauleh, *Muntazam*: iii, 121 and *Mir'at*: i, 540 puts two strong shocks in Shiraz on 27 Dhu'l-Hijja 1236 and in Safar 1237/25 September and mid November 1821, but probably refers to events in 1239/1824. Fraser 1825, who was in Shiraz in September and October 1821, does not mention anything unusual, except the effects of the cholera epidemic in Kazirun and Shiraz. Jabiri 1943 is quite wrong when he includes Isfahan in the places affected by the earthquake of 1239/1824, cf. Ambraseys 1979.

247 Alexander 1827, who was in Shiraz two years after the event, says that ' . . . since the last earthquake the water in the wells has risen very near the surface, where formerly there were ten and fifteen yards of line there are now only three or four feet.' Willock's dispatch in *IO* reports that ' . . . the country in general appears to have rather benefitted by its effects, which have considerably increased the supply of water in the aqueducts (*qanats*).' In this connection, note the remark of Daulier-Deslandes 1926: 35, who was in Shiraz in May 1665 and was told that at Shiraz the water rises in the wells for thirty years, then slowly sinks during another thirty, cf. Chardin 1811: viii, 432, who was told this occurred every twenty years.

248 The exact date of the earthquake is not known. Zain al-'Abidin: 327 puts it in 1239/1825 and Bell 1840 says that it occurred eleven or twelve years before his field trip in the region in the spring of 1837.

249 In the defile of the Harhaz, between Kuhrud and Bul Qalam, Bell 1840: 579, noticed that the piers of a masonry bridge ruined by the earthquake and built on solid rock 'seemed as if they could never have been intended to support the same arch, so different was their parallel . . . and the opposite sides of the ravine had no doubt suffered displacement by a tremendous earthquake, which occurred about eleven or twelve years before . . . ' The construction works of the new road along the Harhaz has obliterated most of the ruins of the older road works, including the piers of the bridge mentioned by Bell which should have been located near Bayjan. Local tradition has it that in 1239 H. and again five years later, earthquakes devastated the region between the Harhaz and the Talar-rud and that mountain passes had been entirely filled in.

250 Mallet 1854: 169; Arago 1859. Sani' al-Dauleh, *Mir'at*: i, 540 puts this event in November 1821, cf. above.

251 The exact date of this event is not known. Contemporary docu-

ments put the earthquake in Davud Pasha's time (1817–31), *Archives du Ministère des Affairs Étrangères*, Dossier de Coupures, Perse 1850–70, Paris.

252 *FO* 248 62; Connolly 1838, was in Tehran at the time of the earthquake and he describes the situation caused by the shock in the city. He also witnessed the damage caused by the aftershock of 6 April at Jaj-rud. Unfortunately the notes of his journey from Tehran to Astarabad are now lost.

253 *IO* 9 91; Watson 1866; Bell 1840, who passed through the region in the spring of 1837, noticed the damage. He describes the ruins of the caravanserai at Jaj-rud and of the mosque at Damavand, as well as of other buildings as far as Sari which he attributes to the 1825 earthquake. See also, *Preussische Staatszeitung* 1830: 1320.

254 Cf. Wright 1977: 26. The damage caused to the British Residency and other buildings in Tehran is described in the dispatches to the *FO* and *IO* (*loc. cit.* above). An inscription on the *masjid-i jami'* in Qum refers to repairs carried out in 1248/1832.

255 Modern authors put this earthquake on 9 May, which is in fact the date of the report in the European Press, Wilson 1930; Sieberg 1932; Rustanovich 1967.

Rabino 1928: 163, says that the villages of Qal'eh Zardavan, Varzan and Kharabdeh, which are situated northwest of Damghan, were destroyed by an earthquake. However, Holmes 1845: 319, who passed through the region early in 1844 points out that Kharabdeh was in ruins and uninhabited, having remained in this state since it was destroyed together with the other three villages by Zaki Khan *c.* 1779. This evidence would make it necessary to reject the April 1830 event in Ambraseys 1974*b*, and Berberian 1977.

256 Local information alludes to the destruction of Quchan in 1248/ 1832–3, confirmed by Radde 1898: 171. But both Quchan and Shirvan had already been demolished by 'Abbas Mirza, who captured these towns in September 1832, Burnes 1834: ii. The collapse of the gallery of the 'Abd al-Razzaq mine in Ma'dan in 1832, may be associated with this event, Khodzko 1853: 238; Fraser 1838.

257 Armenian sources in Stepanian 1942: suppl.; Brant 1841: 424; local information.

258 Wilbraham 1839: 67, Blau 1863: 203, Southgate 1840.

259 According to local tradition, ' . . . it is said that 130 years ago in the time of Muhammad Shah (Qajar) there was a very strong earthquake which caused complete destruction between Chihil Dukhtaran and Durah (Daureh, Durahi?). This earthquake occurred in 1254/1838 at a time when the men had gone north for a trading and plundering expedition which lasted for two years and four months. Only a few of the people were killed, but the blows of the earthquake were so severe that the plain (desert) was turned into water (a small lake) in such a way that for years afterwards it was not possible for the people to go that way. In places where the ground was firm between Shushki, Nasratabad and Gurgaz (Gurgiz), to the south, the ground opened up, and from Haidarabad to Qal'eh Gurg the ground was transformed into a mountain and blocked the narrow defiles. All the villages (*qal'ehs*) within thirty *farsakhs* were obliterated and all the people had to struggle against the (*dusht-id*?) that came from Farah to plunder them. For this reason, nobody went to Herat . . . '

This legend comes from a not altogether reliable dignitary of Duruh who was told that this story appears in a book called *Tarikh-i Durani*, which, however, we have not been able to consult. This sole evidence for such a major earthquake is very tenuous. It is not clear whether 'Shush-i Nasratabad', a stopping place on the route from Shurgaz to Zahidan is meant or 'Shushki and Nasratabad', the latter being a site settled and fortified by the Persians about seventeen years after the earthquake. This fort, together with its sole *qanat* was destroyed by an earthquake in June 1905, *FO* 248 846.

However, there is corroborative evidence to suggest that there is no reason to seriously doubt the substance of the local legend. Smith, for instance, who passed through parts of the region in 1872 was told that the twelfth-century brick tower of Mil-i Nadiri which is situated about fifteen kilometres northeast of Shurgaz, was destroyed by an earthquake about twenty-five (thirty-five?) years earlier, and that the mountain range that separates the Duruh plain from the Kand Ghinau valley was called locally 'Zalzaleh kuh', i.e. mountain of the earthquake,

Smith 1876: 248, 331. Further to the south all legends of saints after whom the highest summits of mountains are named (Kuh-i Sultan, Kuh-i Husain) are connected with earth movements and local tradition has it that the region of Neh is particularly prone to earthquakes, Tate 1910: 110, Gabriel 1935: 226. The significant changes in the direction of flow of the Helmand River and the flooding of the Hamun in 1839 (Tate 1910), may be associated with this major earthquake in Sistan.

These observations become significant when we find them to be associated with localities which are either very close to, or actually show traces of, recent throughgoing faulting. This follows very closely the fault zone of Neh, which runs along the flysch zone that borders the Lut to the east, Stöcklin *et al.* 1972, Stöcklin & Nabavi 1973.

260 Because of the difference in calendars and times kept in the three countries affected by the shock, the dates and times of the earthquake and its aftershocks were thoroughly confused by the European Press. Many earthquakes in the region for the period 1840—1 reported by Mallet 1854, and after him by Musketoff & Orloff 1893, Byus 1948, Ergin *et al.* 1967, and *Atlas*, appear to be without foundation.

261 In the oral tradition Masun is confused with Sasun and the year of the event is given as 1834, 1840 or 1894, the latter year obviously referring to the massacre at Sasun and not to the earthquake destruction of Masun. Chirikoff, who passed through parts of the region in August 1852, remarks that 'in the great earthquake of 2(0) July 1840, the district of Avajik and the whole area around it were obliterated; not a single person or living being survived the shock', Chirikoff 1875. See also Stepanian 1942, Loftus 1855 and Lynch 1869.

262 Most of the villages affected by the earthquake had suffered war damage and they were already in ruins before the earthquake. Armenian villages had been abandoned and some of them were occupied by Persians who had emigrated from Russia; Wagner 1856, Ussher 1865, Monteith 1852, *EI²*: 'Aghri Dagh'.

263 Anonymous 1840; Voskoboinikoff 1841; Pagirev 1909; Musketoff & Orloff 1893; Stepanian 1942; Pinar & Lahn 1952; contemporary Press reports in Mallet 1854.

264 Mount Ararat is 5170 metres high and stands 4300 metres above the Araxes plain. Snow-fields and glaciers descend for 1000 metres from its summit, the snow-line being at 4000 metres. Its northeast slope is cleft downwards by a steep ravine, the valley of St James, the highest part of which is a spacious basin, being enclosed by vertical walls of rock, while the lower part (now a stony desert) was formerly occupied by the village of Arguri (1740 metres) and the monastery of St James. Since 1840 the water from the well of St James emerges at a different spot; Weidenbaum 1884; Lynch 1901.

265 From an eye-witness description of the slide see Abich 1847, 1896, and for a description of the causes of the event with illustrations see Abich 1882: ii, 395. Also, Anonymous 1845; Parrot 1845; Bryce 1877; Buhse 1855. For the monastery of St James and the village of Arguri, see Brosset 1841; Hübschmann 1904. Wagner 1848, considers that there was a volcanic eruption accompanied by an earthquake and not a landslide triggered by an earthquake. His arguments are of course wrong, cf. Abich 1882.

266 At the time of the earthquake, Mitford and Layard were near Kirmanshah and Ainsworth was at Rawandiz. They do not mention an earthquake being felt, Mitford 1884, Layard 1887, Ainsworth 1841. However, when Mitford arrived in Tehran on 15 August he was told by two English copper miners that 'six weeks before there had been an earthquake near Tabriz, which had done much damage' and that a village situated under a mountain was overwhelmed by rockfalls, apparently referring to Arguri. See also references in notes above and *Annalen für Meteorol. & Erdmagnetism* 1840, no. 1: 161.

267 The notice in Ep'rikean 1903: i, 357—60, concerning the complete reconstruction of the walls of the monument in 1862 by Father Superior Hovhannes, probably refers to major repairs necessitated after the earthquake damage sustained by the monument in 1834 and 1840, Brant 1841. The church was demolished *c.* 1917 and when the site was visited in 1966 there was nothing left to be seen.

268 A much earlier earthquake allegedly destroyed the same region, between Dogubayazit and Balik Göl, of Goghovit or Goghod, in

803 A.D., Michel.L: 268; Alishan 1882: 446; Abich 1882: 435. For the location of the district of Goghovit, see Hübschmann 1904. However, the reading of the place name in Michel.L. is uncertain, and it may refer to Claudias on the Euphrates (cf. Michel: iii, 34) a locality destroyed by an earthquake in 817.

269 *FO* 195 224 X/J1343; Hell 1854: i, 525; *L'Institut* 1843.9.20: 244; 28: 19; Wolff 1845: 84, 219; Sani' al-Dauleh, *Muntazam*: iii, 185. The aftershock sequence is reported by Abich 1857: 52. At the time of the earthquake Voskoboinikoff 1847, was on the Araxes. See Kleiss 1969, for the remains of churches in Khuy. Wilson 1930 puts the earthquake on 26 April. *EI¹*: 'Khoi' puts the event in 1842. See figure 3.30.

270 No local information has survived, but ruined settlements around Chihil Dukhtaran are attributed to earthquakes. In a long ode, Shaibani (in Naraqi 1966: 254) laments the loss of life caused by the earthquake, which he puts on a Sunday afternoon at the end of Rabi' II, 1260, which has to be 12 May. See also Zarrabi: 207. The shock is also mentioned by Burgess 1942: 59, in a letter dated Tabriz 6 June 1844 (cf. below). The caravanserai at Quhrud, which was built in the mid-seventeenth century, was in a ruinous condition well before the earthquake, Flandin 1851.

271 The lack of registers in the villages could mean that the number of casualties is exaggerated, particularly in Quhrud where in contrast with other parts of Persia houses were built with timber bracing or stone masonry. It is unlikely that Quhrud was totally destroyed, Binning 1857; Ballantine 1879. The largest numbers of people were killed in the agricultural areas of Jaushaqan and in those east of Qamsar.

272 The dam near Quhrud, the Band-i Quhrud, a stone masonry construction of the Safavid period, is 37 metres high, 31 metres long and 6.1 to 4.6 metres thick; Mustaufi, *Nuzhat al-qulub* 1919: 72; Brugsch 1862: ii, 258; de Sercey 1928: 230; Smith 1971: 72.

273 The thirty-metre high minaret of *masjid-i jami'* settled in its foundations unevenly, causing the structure to lean, Houtum Schindler 1896: 111.

274 Cf. Burgess; this monument, already structurally unsound and in need of strengthening before the earthquake, was cracked by the shock and its two minarets were detached from the main body of the structure. An inscription on the upper frieze of its south *ivan* gives the year 1261 H. (1845) in which superficial repairs of the damage were carried out; Godard 1937: 115, Zander 1972: 246.

275 Perrey 1845: 1448, confuses the effects of this earthquake with those of the shock that happened the following day in the region of Garmrud and Miyaneh, exaggerating the damage in Isfahan, Ambraseys 1979. Wilson 1930, confuses the two events and Ambraseys 1968 wrongly reports damage at Malayir. Ambraseys 1974*b* dates the event one day earlier.

276 Sani' al-Dauleh, *Muntazam*: iii, 186; Perrey 1845: 1448, gives Akkend, Neghian and Armon-Khare. He confuses this event with the earthquake of the previous day in Quhrud, as does Wilson 1930. It is probable that the inscription on the Tash-masjid in Tark, which dates completion of reconstruction of the building in 1282 H. (1865), refers to the effects of this earthquake. Eastwick 1864: i, 202, passing through the region in October 1860, attributes the dilapidation of the caravanserai at Jamalabad to an earthquake.

277 The correct date is given by Burgess 1942: 59, who experienced the shock in Tabriz and by Abich 1857: 51, and Musketoff & Orloff 1893. The date given by Rabino 1917: 69, for a shock felt in Rasht on 4 Rabi' II/23 April, may be a misprint for 24 Rabi' II, 1260, which is the date of the event, for which Ambraseys 1974*b* and after him Berberian 1977, give 23 April 1844. The only pertinent local information perhaps relevant to this earthquake is that during the reign of Muhammad Shah, the fort (?) at Abkashi was destroyed by an earthquake and had to be rebuilt after his death (September 1848).

278 *Vaqa'i'-yi ittifaqiyyeh*: 1267.8.19, 9.3; 1268.4.7; Sani' al-Dauleh, *Muntazam*: iii, 215 and *Mir'at*: ii, 68; Shakiri 1967: 23, 44; Yate 1900: 176. The aftershock sequence lasted for a long time. Local information suggests that after the event, shocks were felt at Ma'dan in 1855 and 1857, Khanikoff 1861: 91. Local sources place the destruction in Sar Vilayat and Bar-Ma'dan a few years before the Russian expedition to Khurasan went there in 1858; other local sources reported in *Kavkaz*: 1893.11.26, place the

destruction of Khabushan sixty years before 1893. The meizoseismal region of the event is similar to that of the earthquake of
17 November 1893, but somewhat larger and centred further
south, see figures 3.5, 3.27. An inscription on the Balasar
madraseh in Mashhad records repairs carried out in 1271/
1854–5, possibly as a result of this earthquake.

279 A vivid and detailed description of the earthquake is given by
Wills's daughter, the only survivor of the family, Wills 1894, and
by the resident Swedish doctor, Fagergren 1853. See also *Vaqa'i'-
yi ittifaqiyyeh*: 1269, nos. 120–43, Sani‘ al-Dauleh, *Muntazam*:
iii, 225 and *Mir'at*: ii, 130 and poems in Imdad 1960: 43–6.

280 The Armenian church built in 1662 was totally destroyed. It
was rebuilt in 1856, Ter Hovhanianc 1880: i, 311, 475, ii, 105.

281 Fasa'i: ii, 160, Mirza Shirazi 1864; Stack 1882: i, 64, considers
that the thirteenth (*sic.*) column at Persepolis was thrown down
by an earthquake in the period 1852–81, see also text § 5.2.

282 The total loss of life was estimated to be 6–14 000 out of a
population of about 30 000, Fagergren, Brugsch 1862, press
reports. Of the Jewish community, 330 were killed, Petermann
1861, and only a few Armenians survived, Ter Hovhanianc. The
British agent in Shiraz reported about 10 000 killed, *FO* 248
153.

283 The situation created by the earthquake is described by Fasa'i:
i, 308–10; see also Gobineau 1859, Pelly 1864, Brittlebank 1873.
In spite of the remission of taxes, British dispatches in *FO* state
that the government's financial oppression in Fars continued.
Moreover, a large number of the inhabitants whose houses were
destroyed removed to the gardens outside the town and encamped in the suburbs, but were attacked by brigands during the
night and robbed of any property that they had saved. They
were obliged to move back into town the next morning: 'a great
number are now living in the mosques', *FO* 60 179.

284 Internal evidence suggests that damage was widespread but confined within a small area around Shiraz where not more than
1000 houses were ruined.

285 Perrey 1854: 468, says that on the night of 1 May 1853, the
earthquake that ruined Shiraz and Kashan (*sic.*) also caused the
drying up of the Zayandeh-rud which supplies Isfahan with water.
Apparently Perrey misunderstood Fagergren's letter, in which
the effects of the Shiraz earthquake are described, and at the end
of which he adds other natural disasters which befell Fars in
1853 (not connected with the earthquake). Among these was the
drying up of the Zayandeh-rud. The date Perrey gives for the
event, 1 May, is in fact the date (Old Style) of the publication of
Fagergren's letter in newspaper *Kavkaz*, see Ambraseys 1979.
Wilson 1930 also dates the event in the Old Style. Sani‘ al-
Dauleh, *Mir'at*: ii, 133, 148, refers to shocks throughout
Dhu 'l-Hijja 1269/September 1853 and again on 15 Shawwal
1270/11 July 1854, on the basis of press reports.

286 Abich 1858. See also *Vaqa'i'-yi ittifaqiyyeh*: 1273, no. 298;
Gobineau 1859: 509.

287 The earliest instances of the construction of isoseismal lines are
for the Dutch earthquake of 23 February 1828 by P. Egen (*Ann.
Phys. Chem.* 1928, vol. 13: 153) and for the Rhenish earthquake
of 29 July 1846 by J. Nöggerath (*Das Erdbeben vom 29 Juli
1846 im Rheingebiet*, Bonn 1847). Khanikoff's map precedes
that constructed by R. Mallet for the Great Neapolitan earthquake of 1857 which was published in 1862.

288 For a vivid description of the effects of the earthquake in Shiraz,
see Vambery 1973, Bernay 1863, and Rochechouart 1867, who
experienced the shock. Vambery points out that before the
earthquake birds were flying about in a restless and wild manner,
which he took for a sure forerunner of a shock. He adds that
after the event the mobs attacked the *firingis* (usually Europeans) whose sojourning in the city they considered had brought
on this calamity (for similar cases see notes 217, 237 (p. 186)).
Damage in the city was widespread, and the Fars government
allocated funds for the restoration of public buildings, Sani‘
al-Dauleh, *Muntazam*: iii, 275, *Mir'at*: ii, 280, who, however,
puts this in 1278/1861; in *Mir'at*: iii, 10, he says that shocks
were continuous in Shiraz for ten days after 22 Jumada II,
1279. For a description of Shiraz after the earthquake, see Pelly
1863a; Mirza Shirazi 1864; Perrey 1864: 176. Wilson 1930, puts
an earthquake in Shiraz in 1865 which is the year in which
Rochechouart's letter reached Perrey 1867: 31.

289 Thielmann 1875: ii, 28, who crossed the northern part of the

meizoseismal region in October 1872, found many old stone
bridges, close to which the streams in the valley had furrowed
out a new bed, leaving the structures on dry ground.

290 According to Stupin 1864: 24, at Bulgavar the ground dropped
by 8.5 metres, presumably the result of a landslide, killing the
owner of a nearby water mill.

291 Ambraseys 1974b, mislocates Kirt, Niyaraq and Aralu; cf. figure
3.22. See also: Perrey 1865: 208, 1866: 44; Abich 1882;
Musketoff & Orloff 1893; Malinovski 1935; Byus 1948.

292 Sani‘ al-Dauleh, *Mir'at*: iii, 25. This is considered to be one of
the celebrated earthquakes in Kirman, the subject of a poem by
Afsari Kirmani, see *Bidari*: 1353, no. 233 and Afsari 1977: 69.
The event is still remembered by some people in the region,
attributing to it the drying up of the *qanats* in Chatrud and,
wrongly, the collapse of the Qubbeh-yi Sabz (Green cupola).
Stack 1882: i, 203, who was in Kirman in 1881, says that the
height of the tower was about forty feet and its diameter about
thirty feet; the dome remained entire but great gaps and fissures
defaced the walls which, although exceedingly thick, were built
of half baked bricks of clay. A modern inscription records the
fact that straw was stored here for the army of Ja‘far ‘Ali Khan
in 1273/1856. Stack adds that 'to those accustomed to oriental
ideas of colour, it will cause no surprise to learn that the colour
of the Green cupola is blue!'

293 According to Napier 1876, no lives were lost in Quchan as there
had been foreshocks that warned the people. *Iran*: 1288.11.21,
reports only two deaths in the town in amusing circumstances.
The earthquake was felt in Tehran, where the British envoy,
Alison, received a report of 2000 killed by the main shock and
another 4000 by the aftershock, while the figure given by Amir
Husain for the whole region was 30 000, see *FO* 248 278,
Anonymous 1872a–c. Bellew 1874 heard the news about the
destruction of Quchan when he passed through Turbat-i
Haidariyyeh in April 1872. Napier 1876, who visited the region
of Quchan in October 1874, found the town half in ruins and the
villages along a distance of fifteen miles to the north-northeast
of it, destroyed. In contrast, he remarks, the villages to the
south of Quchan had suffered little, and one, about a mile east
of the town, had not suffered at all. A similar description of the
effects of the earthquake is given a year later by MacGregor
1879: 83, who adds that by then half of the town was abandoned,
its walls in ruins, and as to its gates, besides those made by man,
it had, as the inhabitants wittily remarked, several which they
termed 'darvazeh-yi zalzaleh' (earthquake gates).

294 Apart from reports left by travellers, local information about the
damage in the Daulat Khaneh region has survived in contemporary or near-contemporary accounts which, however, are not very
precise as to the date of the event. For instance, the simultaneous
destruction of Quchan and Darbadam is placed in 1869, and that
of Quchan, Shamkhal and Ab-Suvaran (Zubaran) is dated 'ten
years ago', reckoned from 1880, or the second destruction of
Quchan with its forts to the northeast, is placed 'twenty-seven
years ago' reckoned from 1893; *Kavkaz*: 1893.11.15–12.20;
Petrusevich 1880; Grodenkoff 1883. There is still a tradition in
the region that whenever Karaul Dagh shakes, Quchan and the
Incheh valley are destroyed. To the northwest, south and southeast of Quchan, damage did not extend more than a few kilometres. Reports of the damage extending to Shirvan by Perrey
1875: 145, Fuchs 1886, Sieberg 1932 and Rustanovich 1967,
are incorrect. These authors have confused the town of Shirvan
with the synonymous district in the Russian Azarbaijan which
was in fact damaged by an earthquake on 28 January 1872, cf.
Iran: 1289.7.24; *Atlas*. Moreover, both Napier and MacGregor,
who visited the town of Shirvan, state quite categorically that
the earthquake caused no damage there.

295 Damaging aftershocks throughout the region continued for an
abnormally long period. When Napier and MacGregor visited
Quchan in 1874 and 1875 they found not only most of the
people, but also the Ilkhan living in tents, earthquakes being
still sufficiently frequent to make houses dangerous. MacGregor
noticed a new type of timber construction designed by the
people of Quchan to resist earthquakes (see plate 27). See also:
Sani‘ al-Dauleh, *Matla‘*: i, 149; Baker 1876; Yate 1900; Pontevès
1890; d'Allemagne 1911; Sykes 1897; Radde 1898; Tchalenko
1975.

296 The temporary abandonment of Tukhrajeh because of the dry-

ing up of the water that drove its *asiyabs* (water mills) in an earthquake about 100 years ago is still remembered locally. It is said that the shock originated from the mines of Zughalsang and also that it ruined Wasit, a locality as yet not identified (site visit in April 1978). This must be the earthquake reported in *Iran*: 1292.4.29, where the destruction of Jur is mentioned.

297 Information based on cursory field survey that yielded no local sources and on Ambraseys 1974*b*, *Iran*: 1296, nos. 382–3, Petrusevich 1880: 188, and Wilson 1896: 143. Gross errors in the date and location of the event are made by Fuchs 1886; Sieberg 1932: 810, 815; Rustanovich 1967; *Atlas*. Musketoff & Orloff 1893: 471 mention the effects in Mishkidzhik, which is identified as Mushtaqin. See also Byus 1948; *Kavkaz*: 1879, nos. 66–72.

298 About two kilometres north-northeast of Sariqamish at its crossing with the Garm-rud, Berberian 1976: 149, found an exposure of a recent fault, striking 170°E for a few hundred metres, which he called the Buzqush Fault. This feature shows that the western Miocene block is thrusted over Quaternary alluvial deposits. However, on the appended seismotectonic map the western block is shown downthrown.

299 No local information about this event has survived. The consensus of opinion among the older people in Takab is that although earthquakes are often felt in the Shahnishin region, they rarely cause any damage. The shock was felt by Houtum Schindler 1881*a*: 189, 1883: 329, at Kavand. He visited the region shortly after the event. See also *Iran*: 1297.10.23; Sani‘ al-Dauleh, *Muntazam*: i, 245. The shock was not reported from Tabriz.

300 For a description of the damage and aftershock sequence given by the Russian consul in Astarabad, see Musketoff 1891; also Hedin 1892*a*, 1892*b*, 1918, who at the time of the earthquake was returning from his ascent of Damavand. He visited the meizoseismal region later. See also *Nature*: 1890.11.13, p. 42. The date of the event in the Old Style is 29 June. This created some confusion in the date of the event, Ambraseys 1974*b*.

301 No field evidence of faulting could be established. Only long scarps of old landslides in shales and marls can be traced east of Tash and at Shahkuh. This refutes the tentative suggestion of Ambraseys 1975*b*, of ground deformations of tectonic origin being associated with this event.

302 Neither Hedin nor Feuvrier 1899: 240, who at the time of the earthquake were in Damavand and Pulur, felt the shock. The former attests that the shock was not felt in Mayamay either.

303 Aftershocks persisted until well after November 1890, *Iran*: 1308, no. 734, 743.

304 For a vivid description of the effects of the earthquake in Quchan, see: Tsimbalenko 1893 & 1899; Anonymous 1894; Quchani 1929; poems composed on the occasion are given by Fani; Shakiri 1897. The Ilkhan of Quchan, Muhammad Nasir Khan, was away at the time of the earthquake which killed eleven of his wives, *Turkm. Vedomosti*: 1893, no. 92. See also *FO* 248 592; *FO* 60 543 and Sani‘ al-Dauleh, *Ruznameh*: 1094, who puts it in Bujnurd.

305 The collapse of the *qanats* not only reduced the water supply considerably, but also allowed the contamination of the water. The Russian medical team in Quchan expressed great concern that the town might become the source of an epidemic in Transcaspia and Ashkhabad if rebuilt on the same site, see Tsimbalenko.

306 Officially estimated figures for the whole region vary between 12 and 18 000. For Quchan alone the original estimates are 12 000 killed out of a population of 25 000; Huntington 1905, gives 5–7000 and Tsimbalenko 1893, gives 5000; he adds, however, that the number of animals killed may be more than 30 000 as this figure does not include losses outside Quchan. The economic losses were enormous for the region. All the wool stock, animal trade and agriculture was lost for many years, *Kavkaz*: 1893.11.23–12.23; *FO* 60 577.

307 Local information from this region is conflicting and it seems to refer to the effects of more than one earthquake. It is certain, however, that at least two localities, Bidkhan and Chakaneh Ulya, were destroyed together with Quchan. Baumgarten 1896: 30–9, who skirted the region in May 1894 from the southeast, notices its desolate state, and Tsimbalenko 1893, attributes the

origin of the earthquake to the Kuh-i Muhammad Beg mountains. For a vivid account of the shock and its effects on the mountains, see Aqanajafi Quchani: 56, who, at the time of the earthquake, was at Khusrauyeh (Khusraviyeh).

308 Today it is fed by a hot spring only. Tchalenko 1975, observed here some evidence of recent fault movements that may or may not be connected with the 1893 earthquake. In an effort to convince the inhabitants to rebuild Quchan on a new site, the Persian authorities attributed the destruction of the town to its proximity to the hot spring of Utrubad and of other localities. They recommended a site for the new town between Nazarabad and Hay Hay, eleven kilometres to the southeast of old Quchan. Their recommendation was ineffectual for various reasons.

309 These features are no longer visible on the ground or on aerial photographs, nor is there any reliable local information available. They are described in various reports in *Kavkaz*, *Turkm. Vedomosti*: 1893, nos. 92–8 and 1894, no. 1, suppl.

310 For the aftershock sequence, see: Shakiri 1897; Musketoff 1899; *Iran*: 1311, nos. 813–33; Lysakovski 1906; Dmitriev-Mamonov 1903; additional data in: Radde 1898; Yate 1900; Wilson 1896; Hale 1920; Rustanovich 1967; references in *Atlas*. There is a detailed list of aftershocks in Baumgarten 1896: 31.

311 Rebeur-Paschwitz 1895, and Italian Station Bulletins.

312 *Iran*: 1311.9.19, refers to an earthquake in Fars which destroyed villages killing a number of people. Wilson 1930 says that the event occurred in Shiraz. However, from *IO* R 15 1 194, we learn that the epicentral region was the *dihistan* of Kirbal (Kurbal) the chief villages of which are Kharameh and Mansurabad. This document adds that in Shiraz the shock lasted for about a minute, but no damage resulted beyond the tumbling down of some old walls.

An earthquake destruction of Kurbal is still remembered by some of the locals. The question is whether they mean this earthquake, or the earlier event of June 1865. Perrey 1867: 74 writing about the June 1865 earthquake (see catalogue) misspells Mansurabad as 'Muserata', and gives 'Kerman' instead of Kharameh. Ballore 1906: 209, gives Mancharageh for Mansurabad but in his Figure 30 he shows the correct location of the place but writes Mucharageh, a local name of the greater district of Shiraz.

313 For a detailed description of the damage in Quchan, see the report made by the Attaché of the Khurasan Agency, Khan Bahadur Maula Bakhsh, and the Mashhad Political Diaries for the period January–June 1895, in *FO* 248 611–12. *Zakaspieskoe Obozrenie*: 1895, nos. 1–16, 27, and *Iran*: 1312, no. 847 complete the picture of the earthquake. See also Fani, Anonymous 1895 and Tchalenko 1975.

314 Agamennone 1896; Maevski 1899, and references in note 313.

315 Official estimates vary between 2000–8000 killed in Quchan alone and 10 000–11 000 in the whole region, which the report of Maula Bakhsh shows to be excessive. The damage and number of casualties was exaggerated for political reasons. Mr Ferguson of the Imperial Bank visited Quchan but mistook the ruins caused by the 1893 event for evidence of new destruction.

After the earthquake the governor of Quchan, Shuja‘ al-Dauleh, tried to move the survivors by force to a new site near Hay Hay. This attempt resulted in riots and bloodshed and his replacement as governor by his cousin Muhammad Nasir Khan. Only then did the people give way and consent to move. About 2000 families perished in the 1893 and 1895 events out of a total of 6000. Of the survivors, some traders, business people, mullas and sayyids, who either had the means or received government assistance, about 1500 families in all, removed themselves to the new town of Quchan which was built between Nazarabad and Hay Hay, on the highroad eleven kilometres southeast of the old town. The remaining 2500 families, mostly farmers, refused to abandon their fields near old Quchan, and rebuilt their homes preferring to risk the chance of a fresh earthquake, see *FO* 248 652 and Yate 1900. For a description of Quchan after the earthquake, see also Huntington 1905 and d'Allemagne 1911, who have photographs of the small triangular wooden shanties that survived the shock; see also Quchani 1929.

316 Local information suggests that Kalukhi, Katlar and settlements in Shahvardi were destroyed once more, shortly after 1893. The effects of the two earthquakes in other localities south of

Quchan are difficult to disentangle. According to the report of the Austrian Consul in Quchan, Rakovski, the large bridge on the road to Quchan was totally destroyed by the earthquake which, however, caused absolutely no damage to nearby villages, Agamennone 1896. Most probably Rakovski refers to the broken bridge near Sayyidabad, half-way between Quchan and Mashhad, the ruinous state of which is attributed to other causes by Yate 1900: 297.

317 Rebeur-Paschwitz 1895.

318 Anonymous 1895, Musketoff 1899. The damage was quickly repaired and a few years later was hardly noticeable, Zarudnoi 1901, 1916; Ronaldshay 1902; Penton 1902 and Hale 1920.

319 In fact, strong foreshocks, widely felt in Azarbaijan, began on 18 December 1895, causing some damage in Dashanli and surroundings, Agemennone 1895.

320 In Strasbourg the trace amplitudes of the foreshock and main shock, recorded by a Rebeur–Ehlert seismograph, were 7.2 and 12.2 millimetres respectively, Rudolf 1903.

321 A story of two villagers being swallowed up by a ground fissure and crushed to death during the aftershock of the 14 January is related in *Iran*: 1313.8.23. For a similar incident see under 1929 May 1, and Howard 1975.

322 *Neologos*: 1896.1.17–26; *Iran*: 1313.8.5, 23; Sani' al-Dauleh, *Ruznameh*: 1198; Anonymous 1899; Musketoff 1899; Lysakovski 1906 & 1910; Byus 1948. Sarre 1899, passed through part of the meizoseismal area two years after the event, but he does not mention the earthquake. In his memorandum, however, he does mention the desolate state of the region and the effects of earthquakes (*Islamisches Museum zu Berlin*). In lieu of Kivi and Sangabad, the foreign press gives Khoi or Goi and Zanjabad or Gandjabad. This led Agamennone 1896 & 1900 to place the epicentre on the foothills of Mount Sahand, about 200 kilometres to the west of Kivi, and Wilson 1930, to extend its destructive effects from Khuy to Gangabad (*sic.*). Ambraseys 1974*b* and Berberian 1976: 414, 1977: 90, correctly place the foreshock at Sangabad but they are wrong in the location of the main shock which they place at Khuy. *Iran*: 1314.5.10. recounts that a chicken was buried at Kivi, but dug out a month later. During this time it had laid twenty-eight eggs but had gone blind. It was still alive in October!

323 *Iran*: 1315.1.25; Ahmadi: 158. The earthquake is well remembered locally and it is not confused with the event of 1864 (see above). The Qubbeh-yi Sabz was still standing before the earthquake; half of it had collapsed already in 1893 when visited by Sykes 1897: 581, who adds that this was not the result of an earthquake, but the handiwork of a former governor, who heard a rumour that treasure existed under the great dome, Sykes 1902*a*: 264. Today its ruins are incorporated in a modern structure, Byron 1937: 205. According to the records of the telegraph office, this was the last shock in Kirman for twelve years, *Iran-i Nau*: 1327.11.25. Le Strange 1905: 307, and Wilson 1930 put this earthquake a year earlier. So does Berberian 1977: 90. See also Kortazzi 1900 and Rudolph 1903.

324 The people in the region of Khuy do not remember this earthquake. In Salmas they maintain that this was the strongest shock felt for a generation before 1900; see Anonymous 1900; *Khulasat al-hawadith*: 1317.12.7; *Ezhemesyachniy Meteorolog. Byul. Tifliss. Fizicheskoy Observ.*, 1900.

325 This earthquake is known only from its instrumental epicentre, which is of very low accuracy, first shown in Plate 2 of the 9th Report of the Seismological Investigations Committee, *BAAS* 1904, at 33.3°N-76.4°E, in Kashmir. The shock was widely recorded and Tams 1908*a*: 525, from 27 station readings, suggested a different epicentre on the borders of Persia with Afghanistan and Baluchistan. A few years later a third location was given in the 17th Report, *BAAS* 1911: 35, at 35°N-60°E, that puts the epicentre between Turbat-i Haydariyyeh and Turbat-i Jam, at 14 hours 35 minutes GMT, or late in the afternoon (local time).

The problem with the location of this earthquake is that we could find no source that mentions such an event, except perhaps Stratil-Sauer 1937: 310 who briefly refers to the destruction of Ravar by an earthquake in 1903. Stratil-Sauer was in the region of Ravar in the early 1930s and it is strange that he should give an earthquake there in 1903 while saying nothing

about the shock that we know destroyed Ravar in 1911. What makes Stratil-Sauer's reference to an earthquake in Ravar in 1903 suspect is that none of the elders we interviewed in Raihan and Ravar in 1975 and 1978 had ever heard about the 1903 event, but all of them vividly remembered and confirmed that it was in 1911 that the place was first destroyed by an earthquake.

It is probable that the March 1903 earthquake occurred further north, in the meizoseismal region of the 1493 Mu'minabad earthquake, at Durukhsh. There, the carpet industry is said to have been terminated about that time by an earthquake that all but destroyed the town, Dilley & Dimand 1931: 118. We have not visited Durukhsh, so Dilley's information cannot be confirmed. There seems, however, to have been no long-term detrimental effect on the carpet industry, as Gabriel 1935: 150 found it thriving in 1933.

Berberian *et al.* 1979*a* cover all possibilities of being wrong, putting earthquakes in the Ravar–Raihan region in 1903, 1911 and 1913, the former two associated with faulting.

326 From information collected on field trips in 1962 and 1975 and the references quoted by Ambraseys & Moinfar 1975. The magnitude quoted in that publication has been over-estimated by 0.6 units.

327 For a detailed description of this event see text pp. 115–17.

328 Because of the political situation prevailing at the time of the event, evidence for the range of the perceptibility of the shock in Persia is scanty; it was felt in Arak, Gulpaigan, Dizful and Zuhab. It was not reported from Tehran, Ahvaz, Bushire and Shiraz.

329 See: *IO* 7 228, *Iran-i Nau*: 1327.10.23.

330 Based on information from field studies and on sources quoted by Ambraseys & Moinfar 1973; Ambraseys 1974*a*. See also Tchalenko & Braud 1974; Berthier *et al.* 1974. Rustanovich 1967, places this earthquake in Bujnurd (*sic.*), about 600 kilometres northeast of Silakhur, and Wilson 1930, not only gives the wrong year, but also the wrong epicentre.

331 Much of this information was collected in the field. See also the reports of Nasrullah Khan in *FO* 248 1030 and *FO* 371 1184. Berberian *et al.* 1979*a*, without quoting their source, give two earthquakes in Ravar, in 1903 and 1911 with which they associate surface ruptures along a geological fault southwest of the town. We could find absolutely no field evidence of faulting or excessive damage southwest of Ravar in 1911. Tipper 1921, refers to the state of Ravar in 1913 and Stratil-Sauer 1937: 310, who visited Ravar in 1932–3, incorrectly dates this event in 1903, see above.

332 Original official estimates for the damage and casualties were grossly exaggerated; see *FO* 371 10153.

333 This crack is still visible and it is often attributed to the Dasht-i Biyaz earthquake of 1968. However, it is also visible on the photographs of the *ivan* we took, late in 1962.

334 On information collected from a number of field trips, the references in Ambraseys & Moinfar 1977*a*; Tchalenko *et al.* 1973; Mohajer-Ashjai *et al.* 1975. Aftershocks are noted in the Mashhad Diaries, see *FO* 371 9035, and in the Persian press, e.g. *Shafaq-i surkh* and *Iran*, up to the middle of August (see *Iran*: 1341, nos. 1402–28). ISS places the epicentre 405 kilometres away, north of Ravar and IGS, 225 kilometres east of Kaj Darakht, near Herat.

335 When we first visited the region southwest of Gifan and Qatlish in connection with the field study of the Kopet Dagh earthquake of 1 May 1929, we were told by local people that in that earthquake ground deformations in the form of open cracks extended for a few kilometres from Qal'eh Jaqq, bearing 240°E. These we did not see, and it is very probable that they were caused by the 1923 rather than the 1929 earthquake.

336 *Shafaq-i surkh*: 1302.7.21; *Iran*: 1302, nos. 1449–69; references in *Atlas* 1962. Sieberg 1932: 815 places the event in Budjurd (*sic.*) see p. 118.

337 Although there is now no doubt that the meizoseismal area of the earthquake did not extend beyond Lalehzar, Gughar, Chinalu and Bustan, it seems that the aftershock of 18 January 1924 caused damage and rockfalls in the region of Mah Hatuni and Gaud-i Ahmar, about sixty kilometres northwest of Lalehzar. A few sedentary local people interviewed during the first field trip confused the two events, which led Ambraseys 1975*a* to believe that the meizoseismal area of the main shock extended to

Gaud-i Ahmar. This erroneous conclusion is repeated by
Berberian 1976: 80, 247.

Besides this local information, see *FO* 371 10150; *IO* 10
977; *Shafaq-i surkh*: 1342.2.15, 3.3; *Iran*: 1302.7.4–21, and
Gabriel 1929.

338 According to local information available in 1963, about 100
years and again 40 years ago, earthquakes destroyed the fort at
Qal'eh Nau as well as the *qanats* further to the southeast
between Hisar Quli and Karim Khan. This may refer to the
earthquakes of 1868 and 1927.

339 The earthquake damage was far less serious in terms of damage
and casualties than reported in the foreign press; see also
Shafaq-i surkh: 1306.5.8–13; *Iran*: 1306, nos. 2443–8.

340 Tchalenko 1975: 14, describes a donkey being swallowed up by
the fault at this locality, and says that this case 'may be added to
the two other known cases of living beings fallen into, and
crushed by, an earthquake ground fracture, i.e. in San Francisco
in 1906, and at Fukui, Japan in 1948', but this seems to be the
result of a misunderstanding. When we interviewed Dr
Tchalenko's informant, Hajji Muhammad Quli Riza'i of Kakili,
it became clear that the old man was asked the wrong question,
that is whether he had heard of the story of the donkey, to
which he replied in the affirmative. In fact what happened is that
during the earthquake the animal panicked and fell into a hollow,
breaking his legs. The following day he was still alive but was
dispatched by Hajji Muhammad's friend. The other two incidents
are equally untrue, Howard 1975.

341 Based on local information, field evidence and data in *FO* 371
13797, 13785, 14546; *IO* 10 1143; *Shafaq-i surkh*: 1308, nos.
1203 *et seq.*; *Ittila'at*: 1308.2.14 *et seq*. See also *Atlas* 1962 &
1977; Tchalenko 1975; Tchalenko *et al.* 1974a; Gorshkov *et al.*
1941, as well as on Rustanovich 1967; Sadiqi 1972, and their
references.

342 Very few people were found to remember the event. See *Ittila'at*:
1308, no. 812–15; *Shafaq-i surkh*: 1308, no. 1260; Shustari
1952: 30, Wilson 1930.

343 Ambraseys 1975b identified these ground deformations as being
of quality (CK), i.e. of doubtful tectonic origin. It was confirmed later that these features are of landslide origin. They are
not very far from the site of the pre-historic slide of the Shimbar
Valley which involved 280 million cubic metres of slide material,
Layard 1846: 83; Layard 1887: ii, 256; Busk 1926. They
occurred at Tang-i Zireh, east of Ajan, on the west banks of the
Karun, outside the meizoseismal region.

344 In fact southeast of Qumchi, where the road crosses the Tang-i
Gil.

345 At the first warning Salmas was evacuated and the troops were
camped outside the barracks at Dilman. They were of great help
in rescuing and evacuating the wounded after the main shock;
FO 371 14538.

346 For the damage wrought to churches, see Kleiss 1969; Pope &
Ackerman 1939; Tchalenko & Berberian 1974.

347 For a description of the internal deformations of Tell Deir
'Alla, caused by an earthquake in the twentieth-century B.C.,
see Franken 1964; Burney 1973.

348 For a detailed description of the effects of the earthquake in the
Salmas region, see Tchalenko & Berberian 1974, and for
additional illustrations, Berberian 1976: 279–336. For the
northwestern region information has been obtained from local
sources and from technical reports for the design of the Qutur
railway line and bridge at Istaran. See also Forbes 1931;
Richards 1931; Reitlinger 1932; Hitchen 1946; Kasravi 1956;
Nakhjuvani 1964; Tchalenko *et al.* 1974a. Details of the relief
funds are found in *IO* 11 290.

349 The district of Ah consists of the following settlements: Sadat
and Qapuz Mahalleh, Sarpulak, Mubarakabad and Chashmeh
'Ala. In 1933 the bottling factory was rebuilt at Ab 'Ali.
Kushish: 1309, nos. 212–15; Kaihan 1932: ii, 352; Tchalenko
1973: 305.

350 For the Musha—Fasham fault zone, see Dellenbach 1964;
Assereto 1966; Allenbach 1966; Tchalenko *et al.* 1974b. A field
study of the zone in April 1974 provided no evidence that the
1930 earthquake was associated with faulting and all cases of
reported ground deformations were found to be due to
incipient sliding, particularly in the Ira-rud (Siyah-rud).

351 The instrumental location of the event, re-calculated by
Nowroozi 1976, shows a focal depth of seventy kilometres
which is inconsistent with the observed macroseismic effects of
the earthquake.

352 This group of settlements is the last inhabited stage of the old
caravan route from Kuhbanan, across the Buhabad desert to
Tabas, via Rizu, followed in 1272 by Marco Polo. The route was
not used again by a European traveller until Gabriel crossed the
Buhabad desert in August 1928; Gabriel 1935, 1952.

353 Among the ruins there are the remains of the foundations of a
small fort, built in the shape of a quadrangle with hard bricks, a
building material totally lacking in the region.

354 With the exception of the 1:50 000 series of 1974, all other
topographic maps of the region proved useless. Not only are
settlements misplaced relative to each other, but also the grid
position of the major villages is shown six to twelve kilometres
too far to the southwest.

355 Our field survey did not extend beyond Bidun in the north. It
was supplemented by the information in *Ittila'at*: 1312.9.29,
Shafaq-i surkh: 1312.9.15, and the unpublished diary of Fried-
rich Kümel (22–9 April 1941), vol. 3, pp. 481–3. See also *FO*
371 17907, Huckriede *et al.* 1962.

356 Because of their considerable length and linearity, the ground
fractures near 'Aliabad Mulla appeared at first sight to be probably of tectonic origin, Ambraseys 1975b. However, a subsequent and less cursory field survey early in 1978 showed that
these fractures are probably not of tectonic origin. They follow
low-lying areas of high water table running along the western
limits of the *daqq*, and they are likely to be due to slumping.
The ground deformations reported from southwest of 'Aliabad
Mulla are more likely to be of tectonic origin. It still remains
uncertain whether the earthquake was associated with faulting,
Ambraseys *et al.* 1979.

357 Figure 3.44 shows a map of the region affected by the sequence
of earthquakes in 1935 and it is based on information collected
during field trips in 1974 and 1975. The spelling of place names
has not been altered to conform with the system adopted for
the rest of the book.

358 Because the meizoseismal region is sparsely populated and the
few sedentary people could not distinguish between the effects
of this earthquake and those of stronger shocks that followed
that year, the information collected in the field is of limited
value. A summary report written by the resident engineer of the
'Campsax' consortium which was in charge of the construction
of the railway line, was found to be useful; also the report given
by Divanbegi 1969. See also: *FO* 317 18996; *Ittila'at*: 1314,
nos. 2431–5. For the railway line see: Anonymous 1935, 1936,
1937, Yarham 1942.

359 The main shock was followed within twenty-four hours by five
large aftershocks that released half as much seismic energy as the
main shock.

360 Figure 3.45 shows the distribution of damage caused by the
main shock and aftershocks. Only the region affected primarily
by the aftershock of 12 April at 12h 44m (GMT) can be identified, the shock causing the final ruin of Amuri, Pahneh and
Ahudasht, which occurred in the afternoon (local time). This
aftershock caused great panic in Sari and damaged the rectangu-
lar Gunbad, seven kilometres south of Sari (plate 29), southeast
of Bala Dizah.

361 Because of the frequent earthquake shocks and landslides the
thickly wooded region of Khalkumeh defined by Sadat, Mazdeh,
Shishak and Qadikula, is called by the local people 'zalzaleh
kuh', i.e. the mountain of the earthquake.

362 In July 1950 Dr J. Stöcklin and A. Erni visited the middle course
of the Tijan-rud in connection with an oil seepage 1.8 kilometres
south-southeast of Azam. They also noticed ground deformations
between Aryam and Sankur, 1.5 kilometres east northeast of the
latter locality, most probably the same as those which we
attributed to landsliding. Similar ground deformations were
noticed in 1950 by Dr Stöcklin about 1.5 kilometres from
Vastakula, where a landslide down a steep, thickly wooded slope
had cut out a broad 'corridor' through the forest in which a
younger generation of trees had grown. By counting tree rings,
Dr Stöcklin concluded that the oldest trees of the second generation were fourteen years old, so that by allowing one year for

the new stems to grow, the date of the landslide would have been 1935 (personal communication 1.5.74). A similar case was observed southwest of Amuri, at Musishah Tepe.

363 The relief team was ordered back and the Army sent a contingent which, however, did not penetrate beyond Qadikula; see: Divanbegi 1969; Mahjuri 1966; Qal'eh Bandi 1969; *Ittila'at*: 1314.1.26–2.13; *Paiman*: 1314, no. 6; Anonymous 1960.

364 The earthquake gave rise to rioting in Babul and Sari, where police were attacked by people who believed that the earthquake was a punishment for the anti-clerical policy of the government and in particular the prohibition of the Muharram processions, *FO* 371 18996. In Babol the clash claimed a number of victims, Divanbegi 1969.

365 Tchalenko 1974: 105 mentions an aftershock during the third week of April that caused damage to buildings in Sari. He refers to the event of 12 April at 12h 44m (GMT) reported in the *Times*, whose coverage of the events preceded that of the local press. For individual aftershocks see *Atlas*, 1962 & 1977, as well as station bulletins of the USSR network.

366 This monument was first noticed by A. Erni in 1931. In 1933 Godard visited the site and photographed the tower which showed some cracks in its brickwork above the doorway, Godard 1936. These were attributed to the earthquake of 1935, Anonymous 1960. However, the photographs of the monument taken by Savage in August 1957, Savage 1957c, as well as our observations in May 1974, show the same cracks in every respect as in Godard's photographs; see also Bivar 1972.

367 The tower of Raskat was first noticed in August 1873 by Baker 1876, and it was photographed in 1933 by Godard 1936. According to an octogenarian from Raskat Ulya, as far as he can remember the cracks of the north side of the tower and the hole in its roof have always existed. The earthquake of 1935, and not that of 1957, simply caused some of the plaster inside the tower to fall off and the cracks to widen 'so that one could see through them'. This is contrary to what local people said to Savage 1957c, i.e. that the tower was badly damaged by the 1957 earthquake.

368 Ambraseys & Melville 1977.

369 Based on information collected during three field trips in 1974

Plate 29. A *gunbad* south of Sari, damaged by the 1935 and earlier shocks.

and 1978. It was found that local people still remembered quite accurately the effects of this earthquake, but they tended to confuse them with the effects of the earthquakes that followed in 1947, 1962 and 1968.

370 There are few published accounts about the effects of this earthquake, mainly press reports summarised in Anonymous 1945, Pendse 1946, Rothé 1949, and no information whatever from Persian territory, the local press quoting only news from Radio London, *Pars*: 1324.9.8; *Iran-i ma*: 1324.9.8–14. Unpublished reports such as the Intelligence Summaries from Gwadur, Muscat and Quetta, *IO* R 15 6 367 & 359, *IO* LPS 12 3226 & 3535, and various studies carried out by engineering firms in connection with the construction of large projects in the region, provide a considerable amount of detailed information. The effects of the earthquake between Gwadur and Ormara were also investigated in the field by Gates *et al.* 1977 and Page *et al.* 1979.

371 After the earthquake a site for new Pasni was selected about eleven kilometres from the coast, but it was never used because of its distance from the sea. Some traders however re-settled at Gwadur.

372 Sondhi 1947, notices that reported Intensities from northern Baluchistan within the region between Kalat, Quetta and Dera Ghazi Khan, were higher than those reported from the south which was nearer to the Makran coast. He thinks that this was the result of a separate shock occurring simultaneously at Rakhni which suffered considerable damage, although removed by more than 850 kilometres from Pasni. Sondhi refers to an account of this event being prepared for publication in the Records of the Geological Survey of India, but unfortunately this was never published. The files of the Meteorological Department of New Delhi, *Seismological Reports*: Nov. 1945–March 1946, however, do not substantiate the alleged occurrence of a separate shock at Rakhni.

373 *FO* Diaries from Bandar 'Abbas, Zahidan and Tehran do not mention the event.

374 Idrisi: i, 157, in describing Julfar, mentions a bar which was sometimes exposed and necessitated ships coming from Basra to Oman to unload their merchandise on its bank and reload after crossing. This bar was almost certainly an early stage in the development of the sand spit which protects the harbour of present-day Ra's al-Khaima. Some years ago a tidal wave breached this spit and formed a direct channel from the open sea to the harbour, Wilkinson 1964: 345. It is probable that the tidal wave referred to was due to the waves from the Makran in 1945 which had diffracted into the Persian Gulf. See also Beer & Stagg 1946, Berninghausen 1966 and Rothé 1946.

375 On the morning after the earthquake the theory that the cause of the shock was British experiments with atomic bombs in the Arabian sea was being discussed in Muscat, while the loss of a Muscat *dhow* in the open sea was attributed to an attack by a submarine, *IO* R 15 6 359.

376 For a description of the islets formed, see Sondhi 1947. They should have emerged some time after the earthquake, otherwise they would have been eroded and scoured by the seismic sea-waves that followed the event.

377 The mud volcanoes in the region of Las Bela and in the Inggol delta are described by Buist 1852: 154. Walton observed that mud volcanoes in the Makran became active, emitting mud and gas during high tides particularly in the spring, not only inland as far as twenty kilometres from the coast, but also off-shore. In 1864 many miles of sea between Gwadur and Karachi was literally covered with dead fish and there was an unpleasant smell and large emission of gas, *Proc. Bombay Geogr. Soc.* 1864: 622. Many other cases have been recorded of eruption of mud volcanoes off-shore Makran. In October 1925 an eruption off the Las Bela coast had thrown up large numbers of dead fish which polluted the air between Sonmiani and Dam, *IO* 10 1084.

378 Walton 1865: 125 refers to a strong earthquake on 25 August 1864 which was felt at Gwadur. There is evidence that about the same time there was an interruption of the telegraph communications from Karachi to the west.

379 Cf. earthquake of 16 February 1942. The question of whether the ground deformations reported between Turshab and Qal'eh Kuhneh and those between Badamuk and Gurab were associated

with the 1941 or 1947 earthquake remains unresolved. Also it is not certain whether the legend about the two earthquakes responsible for the appearance and disappearance of the spring of water at Turshab refers to these two events. It seems that the first of these shocks should have occurred long before 1941, as the water from Turshab was renowned among the nomads for making bread and used by the grandparents of our informants. A particular difficulty in this region is that very few of the inhabitants are sedentary, and even fewer have lived in the same village for more than a generation. Iliats were noticed as the predominant population in the area in 1858–9 by Bunge 1860.

380 This earthquake was preceded by a damaging shock on 6 May at 15h 06m that ruined Nujnur, Amreh, Namar and Kafa, killing a few people, figure 3.49.

381 Based on unpublished reports and site visits. The damage at Ab-i Garm was reported immediately by Vrolyk 1957a, 1957b, and a summary of his report was published by Rothé 1959. A far more extensive study of the damage was prepared by Savage 1957a–c, who visited the Harhaz and Kasiliyan routes in August 1957, see also Anonymous 1960. About a year later Hagiwara & Naito 1959 studied the damage outside the meizoseismal region and early in 1964 the author, with the assistance of 'Ali Riza Gharib, visited the central and eastern parts of the epicentral area. In 1971, Tchalenko 1973 collected additional information during a field trip in the Sangichal region; see also Gansser 1969; Mahjuri 1966: ii, 370; Fookes & Knill 1969; Bout et al. 1961.

382 This is on the site of the bridge of Hajji Ali 'Mushaee', noticed by Bell 1840. In 1837 Bell noticed many ruined bridges in this part of the Harhaz which had been destroyed by an earthquake prior to his visit, cf. p. 187 (note 249).

383 Access to, and movement within, the epicentral region is difficult even under the best conditions. To the north of the Ahansar–Andavar–Alasht alignment, the terrain is so thickly wooded and to the south so inaccessible and infested with landslides, that with the totally inadequate maps available in 1964 only a major fault break could not have escaped our notice. Rothé 1959 and 1969, on the authority of Vrolyk 1957a–b, alludes to a fault break at Ab-i Garm which was not substantiated in the field by Tchalenko 1974, who suggests that the extensive cracking in the mountains north of Sangichal reported to Savage 1957b, might have been due to movements along the Amir fault zone that runs along the Amir-rud. However, Ambraseys 1975b indicates that these features on examination proved to be of (AEddk)-type, i.e. discontinuous ground fractures of non-tectonic origin.

384 According to Savage 1957c, the earthquake caused considerable damage in the Kasiliyan valley as far as Firrim and Raskat. We have not been able to confirm this. It is very probable that he refers to the effects of the 1935 earthquake which at the time of his site visit he was not aware of.

385 The collapse of the reinforced concrete hotel at Amirabad in Ab-i Garm, reported by Vrolyk 1957a, b and Rothé 1959, 1969 is not evidence of the high Intensity of the shock, but rather of the weakness of the structure, which at the time of the earthquake was under construction and still resting on its shuttering, Savage 1957b. Contrary to newspaper reports, there were no casualties at Alasht, Zirab, Pul-i Safid and Pulur.

386 *Khwandaniha*: 1336, nos. 87–8. According to Malikzadeh the full extent of damage could not be assessed for almost one year. The final assessment made by the Red Lion & Son Organisation was never published (personal communication).

387 Bout et al. 1961.

388 Cf. earthquake of 16 February 1941. Musaviyeh was rebuilt on the same site and new settlements were built on the fringe of the *daqq* around pumping stations.

389 This is the first earthquake in Persia which was studied properly in the field. For details of the damage and aftermath see: contemporary local press reports; Omote et al. 1962; Hendricks 1962; Ambraseys 1962a; Despeyroux & Lescuyer 1963; Zareh 1963; Abdalian 1963; Kobayashi 1963.

390 Parts of the fault zone were mapped by Saraby & Foroughi 1962, Mohajer & Pierce 1963 and Mohajer 1964. Immediately after the earthquake and again eighteen months later, the whole zone was mapped by Ambraseys 1962a, 1963, 1965, and certain parts of the zone were visited again 1974, 1975 and 1978. Much more

detailed information was gathered in the field than can be set forth here.

391 For a detailed geological study of the region, see Soder 1959. See also Gansser 1969, Stahl 1962b and Berberian 1976: 419.

392 The fault movements and the mechanism of the earthquake as deduced from field observations is in agreement with the results obtained from source mechanism studies, Petrescu & Purcaru 1964; Wu & Ben-Menahem 1965; McKenzie 1972.

393 Eighteen months after the earthquake, the whole zone was mapped again. It was found that at a number of control points on rock, the throw had increased by about 15% but not the lateral displacement. Although some parts of the fault break were difficult to detect immediately after the event, in early 1964 most of them were easy to distinguish from some distance away. Figure 3.51 shows the fractures observed immediately after the earthquake, cf. Ambraseys 1965, which shows all fractures, including small deformations observed in April 1962 and 1964. For an attempt to measure creep of the Ahangiran segment of the break, see Mohajer-Ashjal 1974.

394 See also Brown 1963; Freville 1964.

395 The damage in the epicentral region has been the subject of numerous field studies by Ambraseys & Tchalenko 1968, 1969a, 1969b; Bayer et al. 1969; Brown 1969; Bubnov 1968; Gansser 1969; Institute of Geophysics Univ. Tehran, *Publ.* no. 46, 1969; Moinfar 1969; Niazi 1968, 1969; Pakdaman 1968; Reinemund 1968; Sobouti 1969; Tabandeh 1969; Tasios 1969. The total number of casualties is not known. Estimates range from 4800, Ambraseys & Tchalenko 1969b, to 16 000 (sic.), *Aftab-i Sharq*: 1347.6.14. About 10 000 is the generally accepted figure that includes casualties caused by the aftershocks and injuries that became fatal up to six months after the event (Red Lion & Sun Statistics).

396 Houses in the epicentral area are of adobe construction, with domed adobe brick roofs or vaults. For the effects of the earthquake on this type of construction see Hossein-Javaheri 1972; Tchalenko & Ambraseys 1973.

397 The fault break was mapped immediately after the earthquake by Ambraseys & Tchalenko 1969a, 1969b; Brown 1969; Eftakhar-Nezhad et al. 1968; Gansser 1969; Tchalenko & Ambraseys 1970; Tchalenko 1970. Later the eastern part of the fault zone was surveyed by Tchalenko & Berberian 1975, and the western part, northeast of Firdaus, by N.N. Ambraseys.

398 Gansser 1969, considers that the fault break in the Nimbluk valley was most unexpected and not related to clear surface features of recent tectonics, while Niazi 1969, on the evidence of the survival of a number of early monuments in Dasht-i Biyaz, Kakhk and Gunabad, alludes to a seismic quiescence of the region for 800 years, cf. events for 1238, 1549 and 1678, Ambraseys & Melville 1977, which corrects and brings up to date the account of the seismicity of the region in Ambraseys & Tchalenko 1969a, 1969b.

399 It was not possible to establish, even approximately, the age of the major disused *qanats* in the Nimbluk valley. Dasht-i Biyaz as a district is first mentioned in the fourteenth century, Mustaufi, *Nuzhat*: 183 and the nearby Gunabad was already famous in the tenth century for its extensive *qanats*, described as often four leagues in length and as much as 700 *gaz* in depth (nineteen kilometres long and 630 metres deep) (sic.). In 1872, Dasht-i Biyaz had four *qanats*, one of which, allegedly built by the Gabrs (Zoroastrians), never ran dry, Smith 1876: 345. According to local information, until recently the *qanat* system shown in plate 25 was *vaqf* property of the shrine of Imam Riza in Mashhad, and it was built 900 years ago. According to others, the oldest *qanats* in this part of the region always belonged to the Sufi shrine of Bidukht and they should not be more than 300 years old (personal communication Husain Tabandeh, Bidukht).

400 For other cases of almost vertical fault scarps of considerable throw in incompetent superficial deposits, see Slemmons 1957; Steinbrugge & Moran 1957; Florensov & Solonenko 1963; Steinbrugge & Cloud 1962; Ambraseys et al. 1972.

401 Day & Wright 1969; see also Plan Organisation 1968.

402 King et al. 1975; McEvilly & Niazi 1975; Mohajer-Ashjai et al. 1975.

403 Crampin 1969a, 1969b; Sobouti 1972; McEvilly & Niazi 1975.

404 Press Reports; Heuckroth & Karim 1970.

405 Tchalenko & Ambraseys 1973. Subsequent visits to the region confirmed the adverse effects on the social and economic life of the survivors caused by this centralisation and indiscriminate re-grouping of settlements.

406 Most studies of the Dasht-i Biyaz earthquake fail to separate the effects of the two consecutive events and all published isoseismal maps of the event present a rather exaggerated picture of its effects; see Tchalenko & Ambraseys 1973: 371. Subsequent field studies in the Firdaus region carried out in connection with the Dustabad earthquake of 23 September 1947, helped disentangle and clarify the effects of these two 1968 earthquakes.

407 The Dasht-i Biyaz earthquake was strongly felt in the region of Firdaus where it caused some damage, killing twenty-two people in villages to the east of the town. In Firdaus itself damage was small enough to attract evacuees and casualty cases from villages to the east, such as Charmeh, Naudeh and Mus'abi. At the time of the second earthquake, Firdaus was crowded with refugees and had a population in excess of 12 000; see references given for previous event and Daldy 1968, Anonymous 1968.

408 There is no evidence of ground deformations where the alignment of these fractures crosses deeply eroded gullies and the tracks from Fathabad and Fariduni to Baghistan Bala.

409 The destructive earthquake of 14 September 1968 and the unprecedented rains early in 1972 had already weakened many houses in the region.

410 The final official damage and casualty estimates were found to be unreliable. In many instances casualty figures were reckoned by subtracting the number of survivors from the official census of the villages. Later surveys showed that of the total number killed, 15% were men, 25% women and 60% children.

411 The earthquake was investigated in the field by Ambraseys *et al.* 1972; Haghipour *et al.* 1972; Razani & Lee 1973*a*, 1973*b*; Dewey & Grantz 1973; Moinfar 1972; McQuillan 1973; Sobouti *et al.* 1972. The seismological aspects of the event were studied by Dewey & Grantz 1973; Niazi 1973, 1977; North 1972.

412 The quality of building materials used for the construction of new houses was appalling. School buildings and other amenities built on the occasion of the 2500th anniversary of the Persian Empire, a year before the earthquake, collapsed completely. In these structures reinforced concrete consisted only of a mixture of clay mortar and large stones with a few reinforcing bars and stirrups; Moinfar 1972, Ambraseys *et al.* 1972, Razani & Lee 1973*a*, 1973*b*.

413 Ambraseys *et al.* 1972. Haghipour *et al.* 1972 are also inconclusive in defining the primary or secondary nature of these features in that the small surface fractures observed may not reflect the causative fracture. Sobouti *et al.* 1972 mention a fifteen-kilometre long trace of fractures of which, however, we could find no evidence on the ground.

414 McEvilly & Razani 1973; Razani & Lee 1973*b*; Razani 1974.

415 Shortly after the earthquake a strong-motion network of accelerographs (SMA-1) was deployed in the region. Records of aftershocks near Qir and 'Aliabad ('Abbad) show peak ground accelerations between 20% and 40%*g* at frequencies of three to five hertz, Ambraseys *et al.* 1972.

416 Earlier earthquakes in 1440 and possibly around the turn of the twentieth century (1903 Nov. 14?), left no early monuments in the region to be damaged by the 1972 event, see Gropp 1970: 184, 222. The seventeenth—eighteenth-century caravanserais at Baba Najim and Madkhun suffered some damage.

417 Arsovski 1978; Ambraseys *et al.* 1979.

418 Ambraseys *et al.* 1979.

419 Berberian *et al.* 1979*a* report a 19.5-kilometre long rupture of pure right-lateral motion and a significant change of the slip vector since Early Quaternary, from thrust to strike-slip. Our field observations indicate far less spectacular effects: a rupture only half as long with predominantly oblique compression features and in places with strike-slip displacements which are the result of the squeezing out of gouge, or of genuine tectonic origin, not a surprise since the overall strike-slip displacement during Holocene is more than 100 metres. Maximum displacements on individual features measured after the earthquake did not exceed fifteen centimetres; apparent displacements, associated with slumping of the ground did, however, reach thirty-five centimetres.

420 Information on this event is based on published reports by Sharp & Orsini 1978, Haghipour *et al.* 1979, Berberian 1979*a* and press reports. Unpublished information was supplied by the Plan & Budget Organisation and Institute of Geophysics in Tehran.

421 A SM-1 type accelerograph in Tabas, run by the Plan & Budget Organisation, was triggered by the main shock and continued to record until it ran out of film. Within the first fifteen minutes that followed the earthquake, it recorded twenty-five after-shocks. It is interesting that the focal distance of the main shock from Tabas, based on shear-wave arrival minus trigger time $(S - T)$, is half as large as that deduced from the teleseismic epicentral location, i.e. $(S - T) = 4.25$ seconds. Moreover, with the exception of the first aftershock which showed shear-wave arrival minus P-wave time of $(S - P) = 6.5$, all the other after-shocks recorded showed $(S - P) = 1.5(\pm 0.8)$ seconds. Obviously, the epicentre reported by the PDE $(33.386°N-57.434°E)$ is in error.

422 Berberian *et al.* 1979*b*, Berberian 1979*b*. For the geology of the region see Stöcklin *et al.* 1965, Ruttner *et al.* 1968. No proper field mapping of the fault break has so far been published.

423 This earthquake preceded by a few hours the departure of the Shah from Iran and the Islamic revolution of Farvardin-mah.

424 Based on information in *Kaihan*: 1358, nos, 10856−9, Haghi-pour & Amidi 1980 and private communications.

425 Berberian 1976: 350 was told by local people that at the time of the earthquake of 2 July 1972, a four-metre high scarp appeared running from the Chah Anjir region to a location north of Bidkarz, a distance of about ten kilometres. A site visit showed that the fault, which appears to be a normal fault with a throw up to four metres but averaging about two metres, is obviously an old feature, certainly pre-dating the small 1972 earthquake. It seems to control the course of a seasonal stream and various tracks across it were not dislocated during the 1972 earthquake. Moreover, 1955 aerial photographs clearly show the scarp and the local people confirm this earlier existence. A similar conclusion was drawn by K. McCue who visited the site in 1975 (personal communication).

426 It is surprising the effect that the passage of a field party often has on the imagination of local people, not only in Iran, but also in the Balkans and the Middle East. The news of foreigners looking for cracks in the ground produced by earthquakes spreads quickly, quite often amplified, Fraser 1825: 579, Mitford 1884: ii, 44. The passage of more parties asking the same question, however, has astonishing effects. The locals not only lead you right away to the would-be fault break, but also surprisingly enough offer an explanation for the phenomenon, often echoing discussions they overheard or quasi-scientific explanations that the party's driver has given them. In the case of the 1957 and 1958 earthquakes, the local people, having been subjected to the passage of no less than five such parties within a short period of time, were less reliable than before.

427 Hedin 1892*a*: 320; *Istiqlal-i Iran*: 1328, no. 152; Musketoff & Orloff 1893: 411. Richter 1961, considers that the fluctuations of the sea level by more than one metre in the evening of 26 April 1960, which was observed simultaneously by all stations in the southern Caspian area, was connected with the Lar earthquake of 24 April (*sic.*).

428 Kotzebue 1819: 165; Armstrong 1831: 127. The Persian press, after 1857, is full of prediction stories, some of them rather amusing, cf. Wilson 1896: 143, 168, 224.

Chapter 4

1 The sources consulted include those listed in Wood 1942, in the Reports of the British Association for the Advancement of Science (1895—1938), and in the archives of De Bilt, *Die Erdbebenwarte zu Laibach.*

2 Asmara operated only a few months and the seismoscope records of Massawa have not been preserved.

3 The seismographic station in Tehran, the first to be installed in the area, came into operation in 1958 and was followed a year later by the station in Shiraz. In 1962 a secondary station was set up at Safid Rud (SEE, WRI) and soon after in 1964—5, the stations in Tabriz, Mashhad and Kirmanshah came into operation, Shiraz, Tabriz and Mashhad being equipped with instruments of the world-wide standardised seismograph network (WWSSN). In

1971 Mashhad began to expand its regional network with sub-ordinate stations equipped with MEQ-600/800 recorders, five of them today covering the northeast and east-central part of the country. A little later, the Atomic Energy Organisation of Iran began to operate local networks of MEQ-800 recorders. In 1975 six instruments were deployed in the region of Bushire within a radius of 37 kilometres from $29.1°$N-$51.1°$E. In August 1976 another network of seven recorders was installed in the Isfahan region within a radius of 42 kilometres from $32.6°$N-$51.7°$E. Early in 1977 another network of five recorders began to operate around Tehran, deployed 30 kilometres around Amirabad. In the same year a permanent, long-period array was established in the region of Saveh southwest of Tehran, consisting of seven stations deployed around $35.45°$N-$50.73°$E. See *Publications*, Institute of Geophysics, Tehran University, nos. 1–58, 1960–73; *Bul. Seism. Network*, Ferdowsi University, Mashhad, vol. 1–2, 1976–9; *Bulletins Seism. Networks*, Bushire, Isfahan, Tehran, Atomic Energy Organisation of Iran, 1976 onwards; *Bul. d'Etudes*, Station Seism. Chiraz, nos. 1–111, Shiraz 1961–3.

4 The use of high-speed computing facilities for the reduction of data was considered at the IUGG meetings in Rome and Toronto in 1954 and 1957 respectively. However, ISS reckoned that they did not appear to be practical for this purpose, *IUGG Comptes Rendus* no. 11, p. 32, & no. 12, p. 51, Strasbourg 1955, 1958.

5 In a total of 28 710 closely printed pages of the ISS Bulletins.

6 Earthquake Files, Institute of Geological Sciences, Edinburgh. Also appended in the M.Ph. thesis of M.S. Nabavi, entitled the 'Seismicity of Iran', University of London, 1972.

7 Much of the unpublished consular correspondence, both Russian and British, has been published in Ambraseys & Moinfar 1973 and Ambraseys 1974a. See also *IO* 7 228, 229. Similar suspicions were aroused recently by the geophysical expedition to the Karakorum organised by the Royal Geographical Society, which the Russian press considered to be of a 'mysterious' nature, with ulterior motives, see *Pravda*: 1980.9.7.

8 The only damaging shock in the area remembered by local people is that of 23 April 1957 which killed a number of people at Kujur Mithqal ($33.29°$N-$52.46°$E).

9 Q is the distance–depth factor, as defined by Gutenberg & Richter 1956a.

10 This is well below the minimum distance of about $16°$ recog-nised to be the limit, below which m_B is strongly affected by regional inhomogeneities of the crust and upper mantle requiring correction.

11 The differences between published m (USCGS) and m (MOS) values are mainly due to the different response of the WWSSN (short-period) and MOS (medium-period) instruments and also because of USCGS reading amplitudes within the first 5 seconds as compared to within 20 seconds read by MOS. The Q-values of Gutenberg & Richter are based on amplitudes from medium-period instruments, so the USSR procedure is closer to the original M_B scale.

12 Rothé's *The Seismicity of the Earth 1953–1965* intended to be a continuation of Gutenberg & Richter's work, contains magni-tudes of 90 earthquakes in Persia for that period. These were calculated not from amplitude–period data, but from the aver-age of weighted individual station magnitude estimates or from semi-empirical formulae. On examination, most magnitudes shown in Rothé's work as M are in fact broad-band m-estimates, particularly of the smaller events and they must be used with caution. Peronaci's catalogue which is often quoted, contains only six magnitudes calculated from amplitude data of the station in Rome; the remaining 60 magnitude assessments are from Gutenberg & Richter, Strasbourg and Uppsala, see Miya-mura 1976a, 1976b, Rothé 1969, Peronaci 1958, 1959.

13 In the distance range $10°$ to $20°$, small differences in the location of normal shocks may account for rather large differ-ences in the value of the Q-factor.

14 For an example of this kind of confusion, see Fisher & Guidroz 1964, Fisher *et al.* 1964.

15 This is not absolutely true for the whole of the Eastern Mediterranean and Western Asia, for which (as for other regions) Gutenberg & Richter's catalogue is far less complete, with many large magnitude events missing, particularly the addenda to the second edition of the catalogue, Duda 1965, Miyamura 1976a, 1976b.

16 The decision to estimate all magnitudes possible before 1963 was in fact taken too lightly, without realisation at the time of the difficulties that lay ahead. The difficulty of retrieving early station bulletins alone, had it been known in advance, would almost certainly have frustrated the effort.

17 The earliest events in the region for which we have trace ampli-tudes are those of 20 June and 8 July 1895. These were recorded in Strasbourg by a Rebeur–Ehlert seismograph which wrote records identified as those of the shocks in the Zagros and Krasnovodsk respectively.

18 For a larger area that includes Eastern Europe and Western Asia up to $90°$E, that is for a larger number of events of greater mag-nitude than those in table 4.2, (q) was found to be 4.52.

19 Results from equation (4.2) have, however, helped us to assess magnitudes for events in other parts of the Middle East.

20 It is for exactly the same reason that ISS had to reduce the num-ber of its epicentral locations much later in 1953.

21 Before 1963, the only two seismic stations in Persia, in Tehran and Shiraz, which began to operate in 1958 and 1959 respect-ively, did not publish amplitude–period data. Nor did the secondary station at Safid Rud report such date (see above, note 3 (p. 195) and figure 4.2).

22 For distances less than $10°$, regional variations of structure become very important, excluding the use of any general formula.

23 Kárník's station corrections for European stations have been derived from an identical expression but of opposite sign, Kárník 1968.

24 For this reason it is not meaningful to attempt to identify those stations whose corrections are statistically significantly different from zero for the whole period studied.

25 In fact they calculated the magnitude of 63 earthquakes in all, of which 28 have been assigned a magnitude class rather than a specific magnitude value; another 4 events are of intermediate depth.

26 These are kept in the archives of the Millikan Memorial Library of the California Institute of Technology.

27 This observation is based on the comparison of estimates we made for a much larger area that comprises Eastern Europe, Turkey and Western Asia up to $90°$E.

28 For the procedures used by different stations to estimate magni-tudes, see Båth 1969, Lee & Wetmiller 1978 and Miyamura 1978. Tehran and Tabriz began to determine magnitudes of near earthquakes after 1971; for details see Adams 1977.

29 I_J is the Intensity in the Japanese (JMA) scale; $I = 0.5 + 1.5(I_J)$.

30 Equation (4.7) gives results identical to those that can be obtained from a similar expression derived by A. Galanopoulos for Greece, Kárník 1968: 65. For Persia, equation (4.7) is valid up to $M = 7.5$, beyond which no macroseismic data are available. Larger magnitudes derived from (4.7), therefore, are approxi-mate.

31 The elimination of $\log(r')$ from equations (4.6) and (4.7) gives $M = 0.9(I_0 - 1)$ which slightly over-estimates M from (4.8).

32 Note that as defined on p. 32, (i_n) decreases with increasing values of n, i.e. $r_n/r_j < 1$; $i_n - i_j < 0$.

Chapter 5

1 Lee & Brillinger 1979 discuss the quantification of these factors for the Chinese historical record and make an attempt to correct early seismicity, which they express in terms of the number of earthquakes reported, by applying probability functions for population changes and document survival. But seismicity can hardly be expressed solely in terms of the number of earth-quakes per unit time, regardless of their magnitude distribution. Moreover, probability functions for population densities and document survival tell us nothing about the thresholds of per-ception or of reporting, which vary with magnitude for different events. Also, the application of a stationary stochastic point process may not be appropriate to intraplate regions.

The population distribution in China is quite different from that in Persia, where areas of very low density do not exceed more than 350 kilometres in width. Furthermore, because we have used not only Intensity to rate the size of an event, but also radii of perceptibility and other indirect evidence, the chances of missing out large shocks in such areas has been reduced. Another difference between China and Persia is that in the latter printing and the dissemination of information through the press came very late in the last century, almost nine centuries after China.

2 We could find no specific reference to an earlier large earthquake in Tabas, and all the evidence points to a relative local quiescence of at least a millennium. The earliest references to Tabas by Baladhuri: 403, late in the ninth century, and by later Arab and Persian writers, do not allude to any major events. It is not improbable, however, that Tabas was affected earlier, perhaps by the earthquakes of A.D. 763 or July 840 which occurred in Khurasan. The earlier event, at least, seems to have affected a mountainous, country region, possibly Kuhistan in which Tabas is located. Without further details, the account of the ground deformations caused by the earthquake of 763 is insufficient to locate the event.

The earliest man-made structure in Tabas, the Minar-i Kabir or Minarik as the local people used to call it, was a slender, eleventh-century Saljuq tower which was still standing up to a height of forty metres, although deteriorating fast in 1880, Stewart 1911, and was a little damaged and leaning when it was photographed in February 1906 by Hedin 1910, Plates 150, 151 and 154. Berberian 1979*a*: 213 and 1979*b*: 1870 attributes the destruction of this monument to the earthquake of 1978. In fact, the minaret was pulled down in the 1960s (see below, note 17) and its existence for at least 900 years supports the view that during that period there would have been no locally catastrophic earthquake.

3 Goudarzi *et al.* 1972–4; Berberian 1977: 121; Neghabat & Liu 1977; Hattori 1979; Rowshandel *et al.* 1979.

4 It is the smaller earthquakes in the Zagros that have been followed by disproportionally long periods of aftershocks, not noticed elsewhere in Iran. For instance, the earthquakes of December 1864, June 1872, February 1874, 27 July 1946, and 15 July 1917, were followed by sequences that lasted 59, 83, 180, 398 and 132 days respectively. It is also in the Zagros that swarms of small shocks have been noticed, particularly in the regions of Malamir (Izeh), Ardal, Saimareh, and recently in Nurabad, where between May and September 1976 more than twelve hundred shocks caused great concern but no damage, Goudarzi & Mostaanpour 1976. There is no evidence that occasional paroxysms in minor seismicity during the last 150 years in Persia can be considered as forerunners of large shocks.

5 Ibn al-Faqih: 228, 257; al-Muqaddasi: 384; al-Mas'udi, *Tanbih*: 49; al-Ghazzi in Taher 1975: 71.

6 Rich 1836: i, 187; Babin & Houssay 1892; Houssay 1894; Brugsch 1862: ii, 125; Windt 1891: 161; Sawyer 1891: annex A.

7 Tradition has it that the Hilleh-rud received a portion of the waters turned by the earthquake from the Kulil river of Ganaveh, the remainder of that stream having burst northward towards Bihbahan and found exits in the Hindijan-rud (Ab-i Shirin), in an affluent of the river meeting it near Zitun (Zidan), and in the southernmost tributary of the Gharahi, Pelly 1863*b*: 144; 1863*c*.

8 Ouseley in 1811 noticed that vestiges of antiquity spared by time in the region of Bushire had been destroyed by earthquakes, alluding perhaps to the Ganaveh tradition, Ouseley 1819: i, 194. The subsidence of the ground in the vicinity of Bushire *c.* 1780 is mentioned also by Michaux 1911: 380 and it may be due to submarine sliding; we have been unable to consult the original *Journal de Michaux* which until recently was kept in the Bibliothèque de Versailles.

9 Mustaufi, *Nuzhat*: 48; earthquakes felt in Aghda and Yazd usually originate from the region of Shiraz, Petermann 1861: i, 214. Yazd has suffered minor damage from small local shocks, the most recent instance being on 3 May 1931; see also Ambraseys 1979: 67.

10 Strabo: i.3.19; xi.9.1; al-Mas'udi: 49; Chardin 1811: iii, 285; Coon 1951, 1957: 127, 178. Al-Biruni: 22, explains that an earthquake was responsible for the uprooting of a tree in Gurgan. The tree was deposited in a spring and rose up and down as the volume of water fluctuated with the seasons. Also in Muslim legend Saveh was famous for its great lake which had suddenly dried up as a result of an earthquake on the night of the birth of the Prophet.

11 Mustaufi, *Nuzhat*: 75; see tables 5.1 and 5.2 for destructive events in Tabriz. During the first quarter of the last century earthquake shocks were felt every year, Malcolm 1827: ii, 205,

and the inhabitants thought very little of them, Armstrong 1831: 127, usually expecting a large shock every forty years, Kotzebue 1819: 165. See also Melville 1981.

12 See p. 187 (note 259). Smith 1876: 331; Landor 1902: ii, 128; the local people near Nasratabad Sipi attribute the presence of tarantulas to earthquakes which open cracks in the ground through which large numbers of them crawl out, Kennard 1927: 182. Today people in the Neh (Nihbandan) area still attribute many of the features of the country to earthquakes that occurred long before their grandfathers' time. See also Coon 1957: 127, 178.

13 A very well documented event in unpublished and published sources, Egerton 1893; Griesbach 1893; McMahon & McMahon 1897; Davison 1893; McMahon 1897 and more recently Lawrence & Yeats 1979; the associated fault break is perfectly visible on the ground.

14 See for instance, MacGregor 1882: 168; Landor 1902: ii, 317, 411; Zugmayer 1905, 1913.

15 Stories about earthquakes in the Kopet Dagh and in the Atrak do exist, but almost all of them seem to stem from events of the last two centuries. Tsimbalenko 1899, late in the last century, was told that shocks in these regions occur regularly once every twenty years and earlier, engineers building the Transcaspian railway line reported local legends about similar events occurring once every generation.

16 The oldest minaret standing seems to be that of the *masjid-i jami'* at Na'in (tenth century) and the highest the twelfth-century Minar-i Ziyar near Isfahan, which is almost 50 metres high with an aspect ratio (quotient of height to average diameter) of about ten. The tallest in Isfahan is the twelfth-century Minar-i Masjid-i 'Ali, which stands 48 metres high and has an aspect ratio of about eleven. The Minar-i Sariban, also in Isfahan and the Minar-i Barsiyan near the city, come second with heights of 44 and 34 metres respectively but with smaller aspect ratios. The next group of tall minarets is in Herat, in the complex of the *musalla* of Gauhar Shad. In 1885 there were nine minarets, ranging in height between thirty-six and forty-five metres, Yate 1888: 30. Two of them collapsed in the earthquake of 10 September 1931 and another fell in the shock of 24 September 1950, leaving seven minarets leaning in different directions, Byron 1937: 99; Dupree 1971; Blunt 1957; Caspani & Cagnacci 1951; Emanuel 1939: 217; private communication Dr Schlumberger. The rest of the minarets standing are between ten and thirty metres high (see text, p. 31).

Free-standing minarets in Persia were generally constructed of brick masonry with lime mortar, and in contrast with similar structures in Turkey and in the Balkans which are often built of stone, they had much smaller aspect ratios, rarely exceeding ten, being on average about six. The material properties of these polygonal or cylindrical structures, sometimes tapered or step-tapered, vary widely. Nothing is known about their mechanical properties, except that some of these structures were extremely flexible before being strengthened in recent times, oscillating perceptibly under the weight of a person ascending to the top of the structure. Two brick couplets taken after the Buyin Zahra earthquake in 1962 from the *imamzadeh* of Palangabad gave moduli of elasticity of about $E = 30\,000$ kg/ cm^2. Unlike ordinary buildings made of brick, minarets in Persia can be affected by both near and large distant earthquakes because of their comparatively large flexibility and low strength. In the first instance they lose their top part and in the second they can be bodily overturned.

Figure 5.10 shows the natural periods of oscillation of some of the minarets standing in Iran today. These have been calculated from the dimensions of the structures assuming a Young's modulus of 30 000 kg/cm^2, which is eight times smaller than that calculated from ambient vibration tests carried out on minarets built of limestone in Turkey, Aytun 1972; Arioglu & Anadol 1974.

17 Deliberate demolition of minarets for the use of their bricks as building materials was not uncommon. As late as 1915, the Saljuq minaret near the Shahristan bridge in Isfahan was pulled down for the sake of its bricks and the eleventh-century Minar-i Kabir in Tabas (Archaeological Department Registration no. 245) disappeared in the 1960s. Bricks of the monument can be found incorporated in the walls of modern houses near its site.

Minarets have been pulled down in the early 1800s in Kirman-shah, Mitford 1884: ii, 342.

18 Damage to minarets caused by high winds is mentioned by Ravandi: 291–2, who says that on 9 December 1165 (3 Safar 561) the upper parts of the minarets in Saveh were blown off. These structures are still standing with their top part missing. The minaret of the *masjid-i jami'* in Damghan was struck by lightning and its top destroyed before 1815, Morier 1818: 314. For data on high winds, see Djavadi 1966.

19 Price 1832: i, 63; Lumsden 1822: 140; Lycklama 1872: 49; St John 1876: 92.

20 The decreasing number of standing columns in Persepolis since the early seventeenth century has often been attributed to earth-quakes, Francklin 1790: 219; Binning 1857: i, 11; Petermann 1861: ii, 171 *et seq.*; Wilson 1930: 106; Stahl 1962a: 8; Razani & Lee 1973a. Columns in various degrees of preservation on the site and standing up to eighteen metres high with a diameter of about 1.5 metres were located in the Apadana, at the Gate and on the plain to the southeast of the site. Some travellers report only the number of columns standing in one of these locations while others record the total number extant throughout the area. Others would count only complete columns or include incom-plete ones consisting of a pedestal and a standing drum only. Therefore, the actual number of columns standing during a par-ticular period is not always possible to ascertain.

    Referring specifically to the Apadana, Figueroa in 1619 saw twenty columns standing. Two years later della Valle reports twenty-five. In 1628 Herbert found only nineteen and so did Mandelslo in 1638 and Thevenot in May 1665. However, a few months earlier, in March 1665, Tavernier reports only twelve columns erect, although Bruin found it hard to believe that Tavernier had ever been on the spot. In 1672 Struys of whose work Chardin does not seem to think very highly, saw eighteen columns and in February 1674 Chardin*, who surveyed the site carefully, found thirteen complete columns standing and another seven consisting of a pedestal and a drum or part of it still stand-ing, Chardin 1811: viii, 277. However, Chardin's 1674 edition shows nineteen columns standing in the Apadana, two at the Gate and one in the plain (Plate lii). The following plate (Plate liii) shows only eighteen columns in the Apadana and the rest as in Plate lii. Bembo, a few months later in June 1674, reports seventeen standing columns, while Fryer in 1677 says that there were eighteen. Jager* and Kaempfer* c. 1685, Bruin* in 1704 and Niebuhr* in 1765, all found seventeen columns which are shown in the Plates that accompany their description of Persepolis. The number of columns found by Franklin in 1787, Hollingbery in 1800 and Ouseley* in 1811 had decreased to fifteen, while Morier early in 1809 saw sixteen and only thir-teen were reported by Johnson in 1817. This is the number recorded by Flandin & Coste* in 1840 and by subsequent travellers: Texier*, Bode, Binning, Petermann, Ussher, Mounsey, Stolze & Andreas*, Dieulafoy*. For some unknown reason, Porter* in 1818 gives fifteen columns, Lumsden in 1820 gives fourteen, Rich in 1821 gives fifteen and Stack in 1881 reports only twelve. The two columns at the Gate are still standing but the solitary pillar that stood on the plain was pulled down by nomads shortly before 1811, Ouseley 1819: ii, 236. Travellers marked with an asterisk include illustrations of Persepolis in their descriptions; see also Curzon 1966: ii, 150 & 162.

    It seems, therefore, that the first diminution in the num-ber of columns from about twenty to nineteen, would have occurred between 1621 and 1628, and this could have been the result of the earthquake of 1623 in the Marv-dasht. During the following half century two more columns were destroyed but there is no evidence that this was caused by earthquakes. It seems more likely that it was the handiwork of the Governor of Shiraz in the early 1630s, who attempted to obliterate what was

left standing in Persepolis, Chardin 1811: viii, 406. Two more columns were destroyed some time between 1765 and 1787. Should this have been the result of an earthquake, for which there is no evidence, the most likely event would have been the shock of April 1765 which was experienced by Niebuhr in Shiraz shortly after his return from Persepolis. There, he had found seventeen columns standing, one with its upper drum overhang-ing the shaft as shown in drawings made eighty years earlier by Witsen 1694, which suggested to him that earthquakes in Persepolis should not have been excessively severe. As a matter of fact, this upper drum and part of a capital in the southwest corner of the colonnade are still in the same precarious position today. The last two columns fell some time between 1811 and 1817, reducing the number of those standing from fifteen to thirteen. During that period an earthquake in 1812 caused con-siderable damage in Shiraz and it may be associated with the loss of these two columns in Persepolis, but again there is no direct evidence for this. Here, as elsewhere in the Middle East, the collapse of columns is seldom attributed specifically to earth-quakes. Occasionally, it is ascribed to deliberate demolition for use of the fragments as building materials or for making lime. More often much of the damage that has led to their premature overturning by an earthquake has been due to the deliberate hewing of their drums for the lead of their dowels. The robbing of lead dowels was common practice among villagers and nomads for many centuries, and it is one of the prime factors that has hastened the collapse of columns. Cutting off a wedge from a drum creates a metastable shaft, and since it is easier to hew columns at their base from the steps of the crepidoma, their collapse in the event of an earthquake is more likely to be away from the cella.

21 Magnitudes of historical events were calculated directly from equations (4.6) to (4.8). Because they are based on minimum estimates of the extent of damage and felt areas, they are likely to be under-estimated. For events prior to 1840, therefore, cal-culated magnitudes in the range $6.0 \leqslant M < 7.5$, were increased by 0.25 before being used in equation (5.2).

22 Ambraseys 1971; Oliveira 1980; Shih *et al.* 1974.

23 North 1973, in calculating slip rates for the Zagros zone included a number of early, mislocated events, such as the Ravar, Lalehzar and Buhabad earthquakes of 1911, 1923 and 1933, as well as a few intermediate depth shocks that lie outside the zone.

24 A Sassanian canal between Shush and Ahvaz that crosses the Shaur anticline has cut down about 3.7 metres below its original bed, scouring during the past seventeen centuries having offset the slow rise of the anticline. Another canal of the same period, now abandoned, shows a rise of the anticline by 18.3 metres. Lees & Falcon 1952, and Lees 1955, have calculated for the Shaur anticline a rate of uplift as as much as 11.0 mm/year.

25 The term 'rigid' is often misunderstood, even by geologists, Berberian 1977: 203. A piece of crust is rigid when it resists deformation, with the result that it reacts by breaking rather than folding. The Lut, for instance, is a relatively rigid mass if compared with the surrounding mountain structures, and par-ticularly with the flysch belt that frames it in the east. Else-where, Tertiary and, to a lesser extent, Mesozoic beds display a subhorizontal attitude but they are dissected by many fractures and faults in all directions, a fact pointed out by J. Stöcklin.

26 The indiscriminate use of a periodic or purely cyclic activity for prediction purposes, without proper interpretation of its signifi-cance, may lead to results as absurd as the use of the 'lows' in figures 4.3 and 4.7 to predict the next World War!

27 There is no other justification for fitting a linear magnitude–frequency relation to cumulative data than for cosmetic pur-poses, see Båth 1978, 1979b, Lomnitz-Adler & Lomnitz 1978, and Tamaki 1963.

References

Sources referred to in the text and notes are listed by author and date of publication or edition, with the exception that oriental historical sources are given by author alone. Newspapers cited are not listed but they are described in chapter 1. Titles of journals and other series of publications are abbreviated according to the system in the British Standards Institution, *The Abbreviation of Titles of Periodicals*, BS 4148: part 2: 1975.

Collections of sources

Camb. Hist. Iran *Cambridge History of Iran*, Cambridge 1968 onwards; references to various chapters are generally listed independently under author and date.

Camb. Hist. Islam *Cambridge History of Islam*, 2 vols. Cambridge 1970, as above.

EI[1,2] *Encyclopedia of Islam*, 1st ed. 4 vols. (1913–42), 2nd ed. (1954–), Leiden and London.

FO *Foreign Office Records*; reference is made to class number and document number. The main classes used are General Correspondence before 1906 (*FO* 60), Political Correspondence after 1906 (*FO* 371) and Embassy and Consular Archives (*FO* 248). These are in the Public Record Office at Kew.

Hakluyt *Hakluyt Society Publications*, London, Series I (1847–98), series II (1899–). Reference is generally made independently by author and date; note also Hakluyt, *The Principal Navigations*, vols. 1–12, ed. Glasgow 1903–5 and Purchas, *Hakluytus Posthumus*, vols. 14–33, ed. Glasgow 1905–7.

IO *India Office Records*; reference is made to the Political and Secret Correspondence unless otherwise indicated, i.e. L/P & S/7/228 is given as *IO* 7 228. The other main class used is *IO* R/15, Persian Gulf Residency and Agency Records.

Oriental sources

Arabic, Armenian, Greek, Persian, Turkish and Syriac historical sources, and also modern works and secondary sources written in oriental languages (mainly Armenian and Persian). The latter are referred to by author and date. The works of oriental authors writing in European languages are listed under occidental sources.

Abū Dulaf, ed. & trans. V. Minorsky, *Abu Dulaf Mis'ar ibn Muhalhil's travels in Iran (c. A.D. 950)*. Cairo 1955.

Abu 'l-Fidā', *al-Mukhtaṣar fī akhbār al-bashar*. Istambul 1286/1869.

Abū Ḥāmid, *Dhail-i Saljūqnāmeh-yi Ẓahir al-Din Nishāpūri*. Tehran 1332/1953.

Abu 'l-Ḥasan Ghaffārī, *Tārikh-i gulshan-i murād*. British Museum ms. Or. 3592.

Afsari, R. (1977) Zalzaleh-hā-yi ma'rūf-i Kirmān. *Vaḥid*, 219–20, 69–71. Tehran 1356sh.

Agapius, Maḥbūb al-Manbijī, *Kitāb al-'unwān*, ed. & trans. A. Vasiliev, *Patrologia Orientalis*. Paris 1909—15.

Aḥmadī, Shaikh Yaḥyā, *Firmāndihān-i Kirmān*, ed. Bāstānī Pārīzī, Tehran 1354/1975.

Alishan, L.M. (1881) *Shirak* (Shirac, canton d'Ararat, pays de la Grande Arménie). Venice.

Ananun Vanetsi, *Chronicle* in Hakobyan 1951, 1956.

Anastasius, *Historia ecclesiastica . . . etc.*, ed. J.P. Migne, *Patrologia Graeca*, **108**, Paris 1861.

Andreas Evdoketsi, *Chronicle* in Hakobyan 1951.

Anonymous Armenian Chronicle in Hakobyan 1951.

Al-Anṭāki, *Dhail tārīkh Sa'īd b. Biṭrīq*, ed. & trans. Kratchovsky & Vasiliev, *Patrologia Orientalis*. Paris 1924, 1932.

Āqānajafī Qūchānī, *Siyāhat-i sharq*, ed. Shākirī, Mashhad 1351/1972.

Ar'aqel Tavrizhetsi, *Patmut'ivn Hayots*, in Brosset 1874.

'Arīb ibn Sa'd al-Qurṭubī, *Ṣilat tārīkh al-Ṭabarī*, ed. de Goeje, Leiden 1897.

Asoghik, Step'annos Tarontsi, trans. Gelzer & Burckhardt, *Des Stephanos von Taron armenische Geschichte*. Leipzig 1907. See also Macler 1917a.

Bāfqī, Muḥammad Mufīd, *Jāmi'-i Mufīdī*, ed. Īraj Afshār, Tehran 1340—2/1961—3.

Balādhurī, Aḥmad ibn Yaḥyā, *Futūḥ al-buldān*, ed. de Goeje, Leiden 1866.

Bar Hebraeus, Girgis Abu 'l-Faraj, trans. E.A.W. Budge, *Chronography*, London 1932 from *Makhtebhanuth Zabhne* (Bedjan's text).

Bar Hebraeus, Girgis Abu 'l-Faraj, *Tārīkh mukhtaṣar al-duwal*, ed. Ṣāliḥānī, Beirut 1958.

Al-Bīrūnī, Abu 'l-Raiḥān, *Kitāb taḥdīd nihāyat al-amākin li-taṣḥīḥ nasāfat al-masākin*, trans. Jamīl 'Alī, Beirut 1967.

Cedrinos, Georgios, *Synopsis historion*, ed. I. Bekker, *Corpus Scriptorum Historiae Byzantinae*, **24**, Bonn 1838.

Chakhathuno (1842) *Description of Edchmidazin and of the five districts of Ararat*. 2 vols. Edjmiadsin.

Chronicon miscellaneum ad annum 724 pertinens, trans. Chabot, *Corpus Scriptorum Christianorum Orientalium*, Paris 1904.

Codex Sinaiticus 34, in Garitte 1958.

Daulatābādī (1964) *Zalzaleh-hā-yi Tabrīz. Rev. de la Fac. de lettres de l'Université de Tabriz*, **2**, 1—26. Summer 1343sh.

Daulatshāh Samarqandī, *Tadhkirat al-shu'arā'*, ed. E.G. Browne. London 1901.

Al-Dhahabī, Muḥammad ibn Aḥmad, *Kitāb duwal al-Islām*, with *Dhail* by al-Sakhawī. Hyderabad 1337/1919.

Dīvānbegī, Riżā 'Alī (1969) Tanazzul. *Sālnāmeh-yi dunyā*, **25**, 182—91, Tehran 1348sh.

Edjmiadsin Manuscript no. 977/970, in Garegin 1951: i, 89.

Elias Bār Sinaia, Nisibinus, trans. L.J. Delaporte, *Elie Bar-Sinaya Metropolitain de Nisibe chronographie* (from British Museum ms. Add. 7197), in *Bibliothèque de l'École des Hautes Études*, fasc. 181, Paris 1910.

Ep'rikean, H.S. (1903) *Patkerazard bnashkharhik bar'aran*. Venice.

Evliyā Chelebī, *Seyāhatnāmesi*, ed. Jūdat & 'Āṣim, Istanbul 1314/1896.

Fāmī, Thiqat al-Dīn, *Tārīkh-i Harāt*, in Barbier de Meynard 1860.

Fānī, Ḥājjī Muḥammad, *Tārīkh-i zalzaleh-yi Qūchān*. Āstāneh Quds library, Mashhad.

Fasā'ī, Ḥasan, *Fārs nāmeh-yi Nāṣirī*. 2 vols. in 1, lith. Tehran 1313/1895.

Faṣīḥ Khwāfī, *Mujmil-i Faṣīḥī*, ed. Maḥmūd Farrukh, 3 vols. Mashhad 1340/1961.

Gardīzī, 'Abd al-Hayy, *Kitāb zain al-akhbār*, ed. Muḥammad Nāẓim, Berlin 1928.

Garegin, Kat'oghikos I (1951) *Yishatakaranq jer'agrads* (Colophons 5th century A.D. to 1250). Antilias, Lebanon.

Ghaffārī, Aḥmad ibn Muḥammad, *Nigāristān*, quoted by Khanikoff, Excursion à Ani en 1848, in Brosset 1849: 121—52.

Ghaffārī, Aḥmad ibn Muḥammad, *Tārīkh-i jahān-ārā*. Tehran 1343/1964.

Al-Ghazzī, Kamāl al-Dīn, *al-Tadhkirat al-Kamāliyya*, in Taher 1975.

Grigor Kemakhetsi, *Chronicle* in Hakobyan 1956.

Grigor Urakh Karintsi, *Chronicles* in Hakobyan 1951.

Grigor Varagetsi, *Chronicle* in Hakobyan 1956.

Gulrīz, Muḥammad 'Alī (1961) *Mīnūdar yā bāb al-jinnāt-i Qazvīn*, Tehran 1340sh.

Ḥāfiẓ-i Abrū, Nūr al-Dīn, *Jughrāfiyā: qismat-i rub'-i Khurāsān: Harāt*, ed. Mā'il Naravī, Tehran 1349/1970.

Ḥāfiẓ-i Abrū, Nūr al-Dīn, *Majmū' al-tawārīkh*, in F. Tauer, *Cinq opuscules de Hafiz-i Abru concernant l'histoire de l'Iran au temps du Tamerlan*. Prague 1959.

Ḥājjī Khalīfeh, Muṣṭafā, *Taqwīm al-tawārīkh*, with continuation by Ibrāhīm Muteferriqa, Istambul 1146/1733.

Hakob Tivriktsi, *Chronicle* in Hakobyan 1956.

Hakobyan, V.A. (1951—6) *Manr Zhamanakagrut'yunner XIII—XVIII DD.* (Armenian short chronicles, 13th—18th centuries). 2 vols. Erivan.

Ḥamza ibn Ḥasan al-Iṣfahānī, *Kitāb sīni mulūk al-arḍ wa 'l-anbiyā'*, ed. J.M.E. Gottwaldt, Leipzig 1844—8.

Hāshimī, Muḥammad Ṣadr (1948) *Tārīkh al-jarā'id wa 'l-majallāt-i Īrān*. 4 vols. Tehran 1327—31sh.

Hātif Iṣfahānī, Aḥmad, *Dīvān*, ed. Vaḥīd Dastagirdī, Tehran 1332/1953.

Ḥazīn Lāhijī, *Tadhkirat al-aḥwāl*, ed. & trans. F.C. Belfour, *The life of Sheikh Mohammed Ali Hazin*, London 1830.

Hovhannes Avagerets, *Chronicle* in Hakobyan 1956.

Ḥudūd al-'ālam, trans. & comment. V. Minorsky, 2nd ed. Bosworth, London 1970.

Ḥusainī, Majd al-Dīn, *Kitāb zinat al-majālis*. Lith. Tehran 1359/1940.

Husameddin, H., *Amasya Tarikhi* in Rifat Arinci, *Chorumlu*, Chorum 1939—46.

Ibn al-Athīr, 'Izz al-Dīn, *al-Kāmil fī 'l-tārīkh*, ed. C.J. Tornberg, Leiden 1851—76.

Ibn al-Balkhī, *Fārs nāmeh*, ed. Le Strange & Nicholson, London 1921.

Ibn Baṭṭūṭa, ed. & trans. Defrémy & Sanguinetti, *Voyages d'Ibn Batoutah*. 4 vols. Paris 1914.

Ibn Bazzāz al-Ardabīlī, *Ṣafwat al-ṣafā'*. Lith. Bombay 1329/1911.

Ibn al-Faqīh, Aḥmad ibn Muḥammad, *Kitāb al-buldān*, ed. de Goeje, Leiden 1885.

Ibn al-Fautī, 'Abd al-Razzāq, *al-Ḥawādith al-jāmi'a wa 'l-tajārib al-nāfi'a*, ed. Muṣṭafā Jawād, Baghdad 1351/1932.

Ibn Funduq, 'Alī ibn Zaid, *Tārīkh-i Baihaq*, ed. A. Bahmanyār, Tehran 1317/1938.

Ibn al-'Imād, 'Abd al-Hayy, *Shadharāt al-dhahab fī akhbār man dhahaba*. Cairo 1350—1/1931—2.

Ibn Isfandiyār, Muḥammad ibn al-Ḥasan, *Tārīkh-i Ṭabaristān*, abridged trans. E.G. Browne, London 1905.

Ibn Iyās, Muḥammad ibn Aḥmad, *Badā'i' al-ẓuhūr fī waqā'i' al-duhūr*. 2 vols. Cairo 1311—14/1893—6.

Ibn al-Jauzī, 'Abd al-Raḥmān, *Kitāb al-muntaẓam fī tārīkh al-mulūk wa 'l-umam*. Hyderabad 1938—41.

Ibn al-Jauzī, 'Abd al-Raḥmān, *Mukhtaṣar al-muntazam (Shudhūr al-'uqūd)*. Cairo National Library, ms. *tārīkh* 95 (*M*).

Ibn Kathīr, Ismā'īl ibn 'Umar, *al-Bidāya wa 'l-nihāya*. 14 vols. Cairo 1351—8/1932—9.

Ibn Khallikān, Shams al-Dīn, *Wafāyat al-a'yān*. Cairo 1882.

Ibn Miskawaih, *Kitāb tajārib al-umam*, ed. & trans. Amedroz & Margoliouth, Oxford 1921.

Ibn Rustah, Abū 'Alī, *Kitāb al-a'lāq al-nafīsa*, ed. de Goeje, Leiden 1891.

Ibn al-Shiḥna, *Rauḍat al-munāẓir fī akhbār al-awa'il wa 'l-awākhir*, on the margins of the Bulāq (Cairo 1874) ed. of Ibn al-Athīr's *al-Kāmil*, vols. 7—9.

Idrisī, *Kitāb nuzhat al-mushtāq*, trans. Jaubert, *Géographie d'Edrisi*, 2 vols. Paris 1836—40.

Imdād Hasan (1960) *Shirāz dar gudhashteh va ḥāl*. Shiraz 1339sh.

Isahak Vardapet, *Chronicle* in Hakobyan 1951.

Isfizārī, Mu'in al-Dīn, *Rauḍāt al-jannāt fī auṣāf madīnat Harāt*, ed. M. Kāzim Imām, 2 vols. Tehran 1338/1959.

Ittiḥādiyyeh, M. (date?) *Safar nāmeh-yi Buntān* (Travels of Bontemps). Tehran.

Īzad Pināh, Ḥamīd (1971) *Āthār-i bāstānī va tārīkhī-yi Luristān*. Tehran 1350sh.

Jābirī, Ḥasan (1943) *Tārīkh-i Iṣfahān va Ray*. Isfahan 1322sh.

Al-Jazzār, Abu 'l-Ḥasan, *Taḥsīn al-manāzil min haul al-zalāzil*, ed. M.A. Ṭāhir, Traité de la fortification des demeures contre l'horreur des séismes. *Annales Islamologiques*, **12**, 131—59 (Cairo 1974).

Juvainī, 'Alā' al-Dīn, *Tārīkh-i Jahān-gushāy*, ed. M. Qazvīnī, 3 vols. London 1913, 1916, 1937.

Kabābī, Muḥammad 'Alī (1963) *Bandar 'Abbās va khalīj-i Fārs*. Tehran 1342sh.

Kaihān, M. (1932) *Jughrāfiyā-yi mufaṣṣal-i Īrān*. 3 vols. Tehran 1311sh.

Kārang, A.A. (1972) *Āthār-i bāstānī-yi Āzarbāyjān, āthār va ibniyeh-yi tārīkhī-yi shahristān-i Tabrīz*. vol. 1, Tabriz 1351sh.

Karbalā'ī, Ḥāfiẓ Ḥusain, *Rauḍat al-jinnān wa jannāt al-janān*, ed. al-Qurrā'ī, vol. 1, Tehran 1344/1965.

Karīmān, Ḥusain (1970) *Ray-i bāstān*. 2 vols. Tehran 1349sh.

Karīmī, Bahman (1965) *Rāhnumā-yi āthār-i tārīkhī-yi Shīrāz*. Tehran 1344sh.

Kāshānī, Abu 'l-Qāsim, *Tārīkh-i Ūljā'ītū*, ed. M. Hambly, Tehran 1348/1969.

Kasravī, Aḥmad (1929) *Shahryārān-i gumnām, II, Ravvādiyān*. Tehran 1308sh.

Kasravī, Aḥmad (1956) *Muqālāt-i Kasravī*, ed. Yaḥyā Zukā', Tehran 1335sh.

Khalīfeh Nīshāpūrī, *Tārīkh-i Nīshāpūr*, ed. Bahman Karīmī, Tehran 1339/1960.

Khatjikyan, L.S. (1950) *XIV dari hayeren dzer'agreri hishatakaranner*. Erivan.

Khatjikyan, L.S. (1955) *XV dari hayeren dzer'agreri hishatakaranner*. Erivan.

Kirakos Gandzaktsi, *Patmut'ivn Hayots, 300–1265*. Venice 1865.

Maftūn, 'Abd al-Razzāq Dunbulī, *Tajribat al-aḥrāt wa tasliyat al-abrār*, ed. Ṭabāṭabā'ī, 2 vols. Tabriz 1349–50/1970–1.

Mahjūrī, Isma'īl (1966) *Tārīkh-i Māzandarān*. 2 vols. Sari 1345sh.

Majd al-Dīn Khwāfī, *Raudeh-yi khuld*, ed. M. Farrukh, Tehran 1346/1968.

Al-Maqrīzī, Taqī al-Dīn, *Kitāb al-sulūk li-ma'rifat duwal al-mulūk*, ed. Sa'īd 'Āshūr, Cairo 1970.

Mar'ashī, Ẓahīr al-Dīn, *Tārīkh-i Gīlan va Dailamistān*, ed. M. Sutūdeh, Tehran 1347/1968.

Martiros Khalifa, *Chronicle* in Hakobyan 1956.

Maskhūr, M.J. (1973) *Tārīkh-i Tabrīz tā pāyān-i qarn-i nuhum hijrī*. Tehran 1352sh.

Al-Mas'ūdī, 'Alī ibn al-Ḥusain, *Kitāb al-tanbīh wa 'l-ishrāf*, ed. de Goeje, Leiden 1894.

Al-Mas'udī, 'Alī ibn al Ḥusain, *Murūj al-dhahab*, ed. & trans. Barbier de Meynard & Pavet de Courteille, Paris 1861–77.

Michel le Syrien, *Chronique*, trans. J.B. Chabot, Paris 1899–1910, reprint Brussels 1963.

Michel le Syrien, (L): *Chronique de Michel le Grand*, trans. Langlois, Venice 1868, from the Armenian version of Prêtre Ischok.

Mihdī, Muḥammad, *Niṣf-i jahān fī ta'rīf-i Iṣfahān*, ed. M. Sutūdeh, Tehran 1340/1961.

Mīrkhwānd, *Tārīkh raudat al-safā'*, ed. 'Abbās Parvīz, Tehran 1338–9/1959–60.

Moses Kaghankatuatsi, *History of Aghvank'*. Tiflis 1912.

Muḥammad Ṭāhir ibn 'Abd al-Qāsim, *'Ajā'ib al-ṭabaqāt*. Royal Asiatic Society ms. P. 179.

Munajjim Yazdī, *Tārīkh-i Shāh 'Abbās*. British Museum ms. Or. 6263.

Munshī, Iskandar Beg, *Tārīkh-i 'ālam-ārā-yi 'Abbāsī*, ed. Iraj Afshār, Tehran 1350/1971.

Al-Muqaddasī, Muḥammad ibn Aḥmad, *Ahsan at-taqāsim fī ma' rifat al-aqālim*, ed. de Goeje, Leiden 1877.

Muṣṭafavī, M.T. (1963) *Iqlīm-i Pārs*. Tehran 1342sh.

Mustaufī, Ḥamdallāh, *Nuzhat al-Qulūb*, ed. Le Strange, London 1915.

Mustaufī, Ḥamdallāh, *Tārīkh-i guzīdeh*, facs. ed. E.G. Browne, London 1910.

Muvāḥid, J. (1970) *Bastak va Khalīj-i Fārs*. Tehran 1349sh.

Nādir Mīrzā, *Tārīkh va jughrāfī-yi dār al-sulṭaneh-yi Tabrīz*. Lith. ed., reprint Tehran 1351/1972.

Na'īma, Muṣṭafā, *Tārīkh-i Na'īma*. 2 vols. Istambul 1147/1734.

Nakhjuvānī, H.H. (1964) *Mawādd al-tawārīkh*. Tehran 1343sh.

Narāqī, Ḥasan (1966) *Tārīkh-i ijtimā'ī-yi Kāshān*. Tehran 1345sh.

Narāqī, Ḥasan (1969) *Āthār-i tārīkhī-yi shahristānhā-yi Kāshān va Naṭanz*. Tehran 1348sh.

Narshakhī, Abū Bakr, *Tārīkh-i Bukhārā*, ed. Mudarris Riżāvi, Tehran 1317/1939.

Nāṣir-i Khusrau, *Safar nāmeh*, ed. & trans. C. Schefer, Paris 1881.

Naṭanzī, Mahmūd ibn Hidāyatallāh, *Naqāvat al-āthār fī dhikr al-akhyār*, ed. Iḥsān Ishrāqī, Tehran 1350/1971.

Nīmdihī, 'Abd al-Karīm, *Kanz al-mā'āni*. Esad Efendi Istambul ms. 884, quoted by Aubin 1973.

Nīmdihī, 'Abd al-Karīm, *Ṭabaqāt-i Maḥmūd Shāhī*. Cambridge University Library, ms. N.271.17.

Orbelean, Step'annos Siunetsi, *Patmut'ivn nahangin Sisakan*. Paris 1859.

Pūr-i Bahā Jāmī, *Majmū'a-yi dīvānhā-yi qadīm*. Cambridge University Library, ms. E.G. Browne Collection, V: 65(7), fols. 224r–225v.

Qal'eh Bandī (1969) *Tārīkh va jughrāfīyā-yi shahristān-i Bihshahr*. Tehran 1348sh.

Qatrān, *Dīvān-i Ḥākim Qatrān-i Tabrīzī*, ed. Nakhjuvānī, Tabriz 1333/1954.

Qazvīnī, Abu 'l-Ḥasan, *Fawā'id al-Ṣafawiyya*. Cambridge University Library, ms. Oo.6.41.

Qūchānī, Afshār (1929) article in *Iṭṭilā'āt* no. 791, Tīr 1308.

Qudāma, Abu 'l-Faraj, *Kitāb al-kharāj*, ed. de Goeje, Leiden 1889.

Qummī, Qāḍī Aḥmad, *Khulāṣat al-tawārīkh*, ed. Iḥsān Ishrāqī, Tehran (in preparation).

Al-Qūṣī, Aḥmad, *al-Barākin wa 'l-zalāzil*. Cairo 1907 (Cairo National Library, ms. *ṭabi'īyāt* 114 (Taimūr).

Rá'īn, Ismā'īl (1970) *Īrāniyān-i Armani*. Tehran 1349sh.

Rashīd al-Dīn, Fażlallāh, *Jāmi' al-tawārīkh*, ed. Bahman Karīmī, Tehran 1338/1959.

Rashīd al-Dīn, Fażlallāh, (II): *Tārīkh-i fīrqeh-yi rafīqan va ṭāyifeh-yi dä'iyān-i Ismā'īliyyeh*, ed. Dabīr Siyyāqī, Tehran 1337/1958.

Rashīd, Mehmet, *Tārīkh-i Rashīd*. 6 vols. Constantinople 1282/1865.

Rāvandī, Abū Bakr, *Rāḥat al-ṣudūr wa āyat al-surūr*, ed. Muḥammad Iqbāl, London 1921.

Riżā Qulī Khān Hidāyat, *Raudat al-safā' Nāṣirī*, a continuation of Mīrkhwānd's *Raudat al-safā'*, ed. Parvīz, vol. 8 onwards, Tehran 1339/1960.

Rūmlū, Ḥasan, *Ahsan al-tawārīkh*, ed. C.N. Seddon, Baroda 1931.

Ṣadīqī, M. (1972) *Jughrāfīyā-yi Gīfān*. Doctoral thesis, Faculty of Letters, Mashhad University, 1351sh.

Saifī Haravī, *Tārīkh nāmeh-yi Harāt*, ed. al-Ṣiddīqī, Calcutta 1944, reprint Tehran 1973.

Ṣā'igh, Sulaimān, *Tārīkh al-Mauṣil*. Cairo 1342/1923.

Sāmī, A. (1968) *Shiraz*. Shiraz 1347sh.

Samuel Anetsi, *Samuel Qahanayi Anedsvoy Havaqmunq i Grots Patmagrats*, ed. Arshak Ter-Miqelean, Vagharshapat 1893.

Samuel Anetsi, (C): Continuator (1168–1427), Matenadaran ms. 5120, 5619 in Abich 1882.

Ṣanī' al-Dauleh, Muḥammad Ḥasan Khān, *Mir'āt al-buldān-i Nāṣirī*. 4 vols. lith. Tehran 1294–7/1877–80.

Ṣanī' al-Dauleh, Muḥammad Ḥasan Khān, *Muntaẓam-i Nāṣirī*. 3 vols. lith. Tehran 1298–1300/1880–2.

Ṣanī' al-Dauleh, Muḥammad Ḥasan Khān, *Maṭla' al-shams*. 3 vols. lith. Tehran 1301–3/1883–6.

Ṣanī' al-Dauleh, Muḥammad Ḥasan Khān, *Rūznāmeh-yi khāṭirat-i I'timād al-Sulṭaneh*, ed. Iraj Afshār, Tehran 1345/1966.

Shākirī, R.A. (1897) *Zalzaleh-yi Qūchānī-yi 1311*. Library of the Bank Melli, Tehran 1315 H.

Shākirī, R.A. (1967) *Jughrāfīyā-yi tārīkhī-yi Qūchān*. Mashhad 1346sh.

Shūshtarī, Imām (1952) *Tārīkh va jughrāfīyā-yi Khūzistān*. Tehran 1331sh.

Sibṭ ibn al-'Ajamī, *Kunūz al-dhahab fī tārīkh al-Ḥalab*, in Sauvaget, *Materiaux pour servir à l'histoire de la ville d'Alep*, ii. Beirut 1950.

Sibṭ ibn al-Jauzī, Shams al-Dīn, *Mir'āt al-zamān*, facs. ed. Jewett, Chicago 1907.

Sīdī 'Alī Re'īs, trans. H. Moris, *Relation des voyages de Sidi-Aly*. Paris 1827.

Siunetsi, Step'annos (1885) *Patmut'ivn tohn Sisakanatsn*. Tiflis.

Strabo, ed. & trans. H.L. Jones, *The Geography of Strabo*, Loeb Classical library, London 1969.

Sutūdeh, M. (1970) *Az Āstārā tā Istārbād*. 2 vols. Tehran 1349–51sh.

Al-Suyūṭī, Jalāl al-Dīn, *Kashf al-ṣalṣala 'an waṣf al-zalzala*, ed. 'Abd al-Laṭīf Sa'dānī, Fez 1971.

Al-Suyūṭī, *Tārīkh al-khulafā'*, ed. 'Abd al-Hamīd, Cairo 1964.

Tābandeh, Sulṭānhusain (1969) *Tārīkh va jughrāfīyā-yi Gunābād*. Tehran 1348sh.

Al-Ṭabarī, Muḥammad ibn Jarīr, *Tārīkh al-rusul wa 'l-mulūk*, ed. de Goeje, Leiden 1883–4.

Ṭabāṭabā'ī, Muḥammad Riżā, *Tārīkh-i aulād al-aṭhār*, in Daulatābādī 1964.

Taher, M.A. (1974) see al-Jazzār.

Taher, M.A. (1975) Textes d'historiens Damascènes sur les tremblements de terre du XIIᵉ siècle de l'Hégire (XVIIᵉ–XVIIIᵉ s.). *Bull. d'Etudes Orientales*, 27, 51–108. Damascus.

Ṭāhiriyā (1968) *Dāmghān-i shish hazār sāleh*. Tehran 1347sh.

Tarbiyat Muḥammad 'Alī, *Dānishmandān-i Āzarbaijān*. Tehran 1314/1935.

Tārīkh-i Sīstān, ed. Bahār, Tehran 1314/1935.

Tchamtchean, Miqayel, *Patmut'ivn Hayots*, 3 vols. Venice 1781–6.

Ter Hovhanianc, H. (1880) *History of New Julfa and Isfahan*. 2 vols. New Julfa.

Theophanis, *Chronographia*, ed. J.P. Migne, *Patrologia Graeca*, 108, Paris 1861.

Thomas Artsruni, *Histoire des Ardzrouni*, in Brosset 1874.
Al-'Umarī, Yāsīn al-Khaṭīb, *al-Āthār al-jāliya fī 'l-ḥawādith al-arḍiya*, Iraq Academy Baghdad, ms.
Vaḥid Māzandarāni, G. (1965) *Māzandarān va Astarābād*. Tehran 1344sh. (Farsi trans. with notes of Rabino 1928.)
Yaghmā'ī, Iqbāl (1947) *Jughrāfiyā-yi tārīkhī-yi Dāmghān*. Tehran(?) 1326sh.
Ya'qūbī, Aḥmad, *Kitāb al-buldān*, ed. de Goeje, Leiden 1892.
Ya'qūbī, Aḥmad, *Tārīkh*, ed. Th. Houtsma, Leiden 1883.
Yāqūt ibn 'Abdallāh al-Ḥamawi, *Mu'jam al-buldān*, ed. Wüstenfeld, Leipzig 1866—73.
Yohannes Mamigonian, *Patmut'ivn Taron*. Venice 1832. See also J.-R. Emine, Jean Mamigonien, continuation de l'histoire de Daron, in V. Langlois, *Collection des historiens anciens et modernes de l'Arménie, etc.*, vol. 1, Paris 1867.
Yovhannes Kat'oghikos, *Patmut'ivn Yovhannu Kat'oghikosi*. Jerusalem 1867.
Zain al-'Ābidīn Shirvāni, *Bustān al-siyāḥa*, lith. Tehran 1310/1892.
Zakariyā Qazvīni, *Kitāb āthār al-bilād*, ed. Wüstenfeld, Göttingen 1848.
Zarkūb, Abu'l-'Abbās, *Shīrāz nāmeh*, ed. Bahman Karīmī, Tehran 1931.
Żarrābī, 'Abd al-Raḥīm, *Tārīkh-i Kāshān*, ed. Iraj Afshār, Tehran 1341/1962.
Zunūzī, Muḥammad Ḥusain, *Baḥr al-'ulūm*, quoted by Mashkūr 1973.

Occidental sources

All works in European languages, including Russian, cited in the text. They are listed by author and date of edition. Reference is made only to the first page of articles in periodicals, or exceptionally to a page of specific interest.

Abbott, K. (1855) Geographical notes taken during a journey in Persia in 1849 and 1850. *J. R. Geogr. Soc.*, 25: 1.
Abdalian, S. (1935a) La sismicité de l'Arménie dans le passé. *Rev. Geogr. & Geol. Dyn.*, 8: 96.
Abdalian, S. (1935b) Etude sur le violent tremblement de Zanguezour en Armenie Sovietique. *Rev. Geogr. & Geol. Dyn.*, 8: 51.
Abdalian, S. (1953) Le tremblement de terre de Toroude en Iran. *La Nature*, no. 2333: 314 Paris.
Abdalian, S. (1963) Le seisme du 1er Septembre 1962. *Publ. Inst. Geophys. Univ. Tehran*, no. 15: 43.
Abdalian, S. (1964) Seismo-tectonique de l'Iran. *Publ. Inst. Geophys. Univ. Tehran*, no. 16.
Abich, H. (1847) Geognostische Reise zum Ararat und Verschüttung des Thales von Arguri 1840. *Monatsber. Ges. Erdkd. Berlin*, NS 4: 28.
Abich, H. (1857) Sur les derniers tremblements de terre dans la Perse septentrionale et dans le Caucase. *Bul. Cl. Phys.-Math. Acad. Imp. Sci. St. Petersbourg*, 14 (4): 49.
Abich, H. (1858) Tremblement de terre observé à Tabriz en Septembre 1856, etc. *Bul. Cl. Phys.-Math. Acad. Imp. Sci. St. Petersbourg*, 16 (22): 337.
Abich, H. (1882) *Geologische Forschungen in den Kaukasischen Ländern*. Ed. A. Hölder, vol. 2, Vienna.
Abich, H. (1896) *Aus Kaukasischen Ländern*. 2 vols., Vienna.
Academia Sinica (1956) *Chronological tabulation of Chinese Earthquake Records*. 2 vols., Peking (in Chinese).
Academia Sinica (1977) *Catalogue of Strong Earthquakes in China*. Institute of Geophysics, Peking (in Chinese).
Adams, R. (1977) Survey of practice in determining magnitudes: Europe, Asia, Africa, Australasia and the Pacific. *World Data Centre A*, Report SE-8, Washington.
Adle, Ch. (1971) Contribution à la géographie historique du Damghan. *Le monde Iranien et l'Islam*, 1: 69.
Afshar, K. (1960) The Lar earthquake of 24th April 1960. *Publ. Inst. Geophys. Univ. Tehran*, no. 4.
Agamennone, G. (1895) Tremblement de terre de la Mer Caspienne de la nuit 8—9 juillet 1895. *Bul. Met. Sis. Obs. de Constantinople*.
Agamennone, G. (1896) Liste des tremblements de terre en Orient 1894—1896. *Bul. Met. Sis. Obs. de Constantinople*.
Agamennone, G. (1897) Il terremoto di Kischm. *Bul. Soc. Sism. Ital.*, 3: 49.
Agamennone, G. (1900) Liste des tremblements de terre observés en Orient 1896. *Beitr. zur Geophys.*, 4: 118.
Ainsworth, W. (1841) An account of a visit to the Chaldeans, inhabiting Central Kurdistan; and of an ascent of the peak of Rowandiz-Tur Sheikhiwa — in the summer of 1840. *J. R. Geogr. Soc.*, 11: 21.

Alexander, J.E. (1827) *Travels from India to England and a Journey through Persia, Asia Minor etc. in the Years 1825—26*. London.
Alishan, L.M. (1882) Nachrichten über Erdbeben und vulkanologische Phänomene, in H. Abich, *Geologische Forschungen in den Kaukasischen Ländern*, vol. 2. Vienna.
Allemagne, H.-R. d' (1911) *Du Khorassan au pays des Bakhtiaris*, 4 vols, Paris.
Allenbach, P. (1966) Geologie und Petrographie des Damavand und seiner Umgebung. *Mitth. Geol. Inst. Eidg. Tech. Hochsch. Zurich*, N.F. 63.
Alsan, E., Tezucan, L. & Båth, M. (1975) An earthquake catalogue for Turkey for the interval 1913—1970. *Jt. Publ. Kandilli-Uppsala Seismol. Inst.*, no. 7—75, Uppsala.
Alsinawi, S. & Ghalib, H. (1975) Historical seismicity of Iraq. *Bul. Seismol. Soc. Am.*, 65: 541.
Ambraseys, N. (1961) On the seismicity of South-West Asia; data from a XV-century Arabic manuscript. *Rev. Etud. Calamités*, 37: 18.
Ambraseys, N. (1962a) *On the Regional Conditions and Damage during the Buyin-Zara (Iran) Earthquake of 1st September 1962*. UNESCO, 2 vols, Paris.
Ambraseys, N. (1962b) A note on the chronology of Willis' list of earthquakes in Palestine and Syria. *Bul. Seismol. Soc. Am.*, 52: 77.
Ambraseys, N. (1963) The Buyin-Zara, Iran, earthquake of September 1962; a field report. *Bul. Seismol. Soc. Am.*, 53: 705.
Ambraseys, N. (1965) An earthquake engineering study of the Buyin-Zara earthquake of September 1st 1962. *Proc. 3rd World Conf. Earthq. Eng.*, Auckland, 3, paper V-7.
Ambraseys, N. (1968) Early earthquakes in North-Central Iran. *Bul. Seismol. Soc. Am.*, 58: 485.
Ambraseys, N. (1971) Value of historical records of earthquakes. *Nat.*, 232: 375.
Ambraseys, N. (1973) Dynamics and response of foundation materials in epicentral regions of strong earthquakes. *Invited Paper, 4th World Conf. Earthq. Eng.*, Rome, 1: CXXVI.
Ambraseys, N. (1974a) The seismicity of Iran; the Silakhor, Lurestan, earthquake of 23rd January 1909; part II. *Annal. di Geofis.*, 27: 399.
Ambraseys, N. (1974b) The historical seismicity of North-Central Iran. *Geol. Surv. Iran*, Rep. no. 29: 47. Tehran.
Ambraseys, N. (1975a) Seismicity and earthquake risk at Gol-e Gohar, Kerman. *Rep. Natl. Iranian Steel Ind.*, Tehran: 19.
Ambraseys, N. (1975b) Studies in historical seismicity and tectonics, in *Geodynamics Today*, 1: 7, Royal Society, London.
Ambraseys, N. (1976) Earthquake epicentres in Iran. *Proc. CENTO Semin. Earthq. Hazard Minimization*, 1: 70. Tehran.
Ambraseys, N. (1977a) The Chahar Mahal earthquake of 6th April 1977. *CENTO Reconnaissance Mission, Plan & Budget Organ. Publ.* Tehran.
Ambraseys, N. (1977b) Earthquake stability of columns. *Proc. 1st Symp. ICOMOS on Prot. of Monuments*, UNESCO, Paris.
Ambraseys, N. (1978) The relocation of epicentres in Iran. *Geophys. J. R. Astron. Soc.*, 53: 117.
Ambraseys, N. (1979) A test case of historical seismicity: Isfahan and Chahar Mahal, Iran. *Geogr. J.*, 145: 56.
Ambraseys, N. & Melville, C. (1977) The seismicity of Kuhistan, Iran. *Geogr. J.*, 143: 179.
Ambraseys, N. & Moinfar, A. (1973) The seismicity of Iran; the Silakhor, Lurestan, earthquake of 23rd January 1909. *Annal. di Geofis.*, 26: 659.
Ambraseys, N. & Moinfar, A. (1974a) The seismicity of Iran; the Firuzabad, Nehavend, earthquake of 16th August 1958. *Annal. di Geofis.*, 27: 1.
Ambraseys, N. & Moinfar, A. (1974b) The seismicity of Iran; the Karkhaneh, Kangavar, earthquake of 24th March 1963. *Annal. di Geofis.*, 27: 23.
Ambraseys, N. & Moinfar, A. (1975) The seismicity of Iran; the Turshiz, Kashmar—Khorasan earthquake of 25th September 1903. *Annal. di Geofis.*, 28: 253.
Ambraseys, N. & Moinfar, A. (1976) Iran earthquakes 1969. *Tech. Res. & Stand. Bur. Plan & Budget Org.*, Publ. no. 63. Tehran.
Ambraseys, N. & Moinfar, A. (1977a) The Kaj Darakht, Torbat-i Haidariyyeh earthquake in Iran of 25th May 1923. *Annal. di Geofis.*, 30: 3.
Ambraseys, N. & Moinfar, A. (1977b) Iran earthquakes 1968. *Tech. Res. & Stand. Bur. Plan & Budget Org.*, Publ. no. 69. Tehran.

Ambraseys, N. & Moinfar, A. (1977c) The Torud earthquake of 12th February 1953. *Annal. di Geofis.*, **30**: 185.

Ambraseys, N. & Tchalenko, J. (1968) Dasht-e Biaz, Iran, earthquake of August 1968. *Nat.*, **220**: 903.

Ambraseys, N. & Tchalenko, J. (1969a) The Dasht-e Bayaz earthquake of August 31, 1968 in Iran. *Bul. Seismol. Soc. Am.*, **59**: 1751.

Ambraseys, N. & Tchalenko, J. (1969b) Dasht-e Bayaz earthquake of 31 August 1968. *UNESCO* Publ. no. 1214/BMS, Paris.

Ambraseys, N., Arsovski, M. & Moinfar, A. (1979) The Gisk earthquake of 19 December 1977 and the seismicity of the Kuhbanan fault-zone. *UNESCO* Publ. no. FMR/SC/GEO/79/192, Paris.

Ambraseys, N., Moinfar, A. & Amin, M. (1975) Iran earthquakes 1970. *Tech. Res. & Stand. Bur. Plan & Budget Org.*, Publ. no. 53. Tehran.

Ambraseys, N., Moinfar, A. & Peronaci, F. (1973) The seismicity of Iran; The Farsinaj, Kermanshah, earthquake of 13th December 1957. *Annal. di Geofis.*, **26**: 679.

Ambraseys, N., Moinfar, A. & Tchalenko, J. (1971) The Karnaveh, Northeast Iran, earthquake of 30th July 1970. *Annal. di Geofis.*, **24**: 475.

Ambraseys, N., Moinfar, A. & Tchalenko, J. (1972) Ghir earthquake of 10th April 1972. *UNESCO* Publ. no. 2789/RMP.RD/SCE, Paris.

Andrews, P. (1973) The white house of Khurasan; the felt tents of the Iranian Yomut and Göklen. *Iran*, **11**: 93.

Anonymous (1783) *J. de Paris*, no. 196: 814.

Anonymous (1825) Tremblement de terre en Perse. *J. des Voyages, Devouvertes et Navigations Modernes, ou Archive Geogr. du XIX siecle*, **25**, Paris.

Anonymous (1839) Earthquakes at Tabriz. *Silliman's Journal*, now = *Am. J. Sci.*, **37**: 351.

Anonymous (1840) in *Zhurnal Ministerstvo Vnutrennosti Diel*, **8**, nos. 37, 38. St Petersburg.

Anonymous (1845) Partial fall of Ararat. Appendix to F. Parrot's *Journey to Ararat*, London.

Anonymous (1872a) in *Izv. Kavkaz. Otd. Russ. Geogr. O-Va.*, **1** (2). Tiflis.

Anonymous (1872b) *Nat.*, **6**: 39.

Anonymous (1872c) *Proc. Geol. Soc. London*, **29**: 271.

Anonymous (1894) The city of Kushan. *Sci. Am. Supl.*, no. 943: 15072.

Anonymous (1895) The great earthquake in Persia. *The Chilean Times de Valpariso*, 4 May 1895.

Anonymous (1899) Materiali dlia izuchenie zemletriasenii Rosii. *Prilozh. Izv. Russ. Geogr. O-Va.*, **35**. St Petersburg.

Anonymous (1900) Stat'i i soobschcheniia. *Izv. Kavkaz. Otd. Russ. Geogr. O-Va.*, **13** (2): 64. Tiflis.

Anonymous (1935) The Trans-Iranian section, Iranian State Railways. *The Railw. Gazz.*, **63**: 145, & **69**: 165.

Anonymous (1936) Trans-Iran railway. *J. R. Cent. Asian Soc.*, **23**: 506.

Anonymous (1937) The Trans-Iranian railway. *The Railw. Gazz.*, **66**: 1112.

Anonymous (1945) Earthquake in the Arabian Sea. *Nat.*, **156**: 712.

Anonymous (1960) Earthquakes in Mazanderan. Append. J, *Water Resour. of Mazanderan*: 154; Sir Alexander Gibb, London.

Anonymous (1968) Kakhk earthquake. *Nat.*, **219**: 996.

Anonymous Merchant (1873) The travels of a merchant in Persia. Tr. Ch. Grey, *Hakluyt Soc. Publ.* no. 49.

Apalloff, B. (1956) Kolebanya urovnya Kaspiiskogo more. *Tr. Inst. Okeanol. Akad. Nauk.*, **15**.

Arago, F. (1859) *Oeuvres completes de Francois Arago*, **12**, Paris.

Arioglu, E. & Anadol, K. (1974) On the earthquake resistance of the Süleymaniye mosque, Istanbul. *Proc. 5th World Conf. Earthq. Eng.*, Rome, **2**: 2309.

Armstrong, T.B. (1831) *Journal of Travels in the Seat of War*, London.

Arnold, A. (1877) *Through Persia by Caravan*. 2 vols., London.

Arpat, E., Saroglu, F. & Iz, H. (1977) Caldiran depremi. *Yeryuvari ve Insan*, **2**: 29. Ankara.

Arsovski, M. (1978) The Iranian earthquake of December 20, 1977; seismotectonic characteristics. Part I: *Partial Rep. Earthq. Rec. Mission*, UNESCO, April 1978; Part II: *Publ. Iziis*, Skopje.

Aslam, M., Godden, W. & Scalise, Th. (1980) Earthquake rocking response of rigid bodies. *J. Struct. Div. Am. Soc. Civ. Eng.*, **106** (2).

Assereto, R. (1966) Geological map of Upper Djadjerud and Lar Valleys, Central Elburz, Iran. 1:50,000 map with explanatory notes. *Ist. di Geol. dell'Univ. di Milano*, Ser. G. Publ. no. 232.

Atlas = Kondorskaia, N. & Shebalin, N. (1977) *Novii katalog silniikh zemletriasenii na territorii SSSR*. Izd. Akad. Nauk. SSSR. Moscow. See also 1962 edition (*Atlas* 1962).

Aubin, J. (1959) La ruine de Siraf et les routes du Golfe Persique aux XIe et XIIe siècles. *Cah. de Civilis. Medievale*, **2**: 295.

Aubin, J. (1967) Un santon Quhistani de l'époque timuride. *Rev. Etud. Islamique*, **35**: 185.

Aubin, J. (1969) La survie de Shilau et la route du Khunj-o-Fal. *Iran*, **7**: 21.

Aubin, J. (1971) Réseau pastorale et réseau caravanier. Les grand' routes du Khurassan à l'époque mongole. *Le monde Iranien et l'Islam*, **1**: 105.

Aubin, J. (1973) Le Royaume d'Ormuz au début du XVI siècle. *Mare Luso-Indicum*, **2**: 77.

Aubin, J. (1977a) Le propriété foncière en Azarbaydjan sous les Mongols. *Le monde Iranien et l'Islam*, **4**: 79.

Aubin, J. (1977b) La question de Sirğan au XIIIe siècle. *Stud. Iran.*, **6**: 285.

Aubin, J. (1979) La guerre au Kirman et l'an mil. *Stud. Iran.*, **8**: 213.

Aytun, A. (1972) Yapilarin doğal titresim periyodlarinin deneysel yolla ölçümü. *Türk. Deprem Sorunu Sempozyumu*: 1. Ankara.

Babin, C. & Houssay, F. (1892) A travers la Perse meridionale. *Le Tour du Monde*, no. 1647–60: 65. Paris.

Baker, V. (1876) *Clouds in the East*, London.

Ballantine, H. (1879) *Midnight Marches through Persia*, Boston.

Ballore, M. de (1906) *La Géographie Seismologique*, Paris.

Ballore, M. de (1924) *La Géologie Sismologique*, Paris.

Baratta, M. (1901) *I Terremoti d'Italia*, Turin.

Barazangi, M. & Dorman, J. (1969) World seismicity maps compiled from ESSA, Coast and Geodetic Survey, epicentral data 1961–1967. *Bul. Seismol. Soc. Am.*, **59**: 369.

Barbier de Meynard, C. (1857) Description historique de la ville de Kazvin. *J. Asiat.*, **10**: 257.

Barbier de Meynard, C. (1860) Extraits de la chronique Persane de Herat. *J. Asiat.*, **16**: 461.

Barthold, V. (1930) *Historico-Geographical Survey of Iran*. Trans. H. Sardavar, Tehran (in Persian).

Båth, M. (1958) The energies of seismic body waves and surface waves. in H. Benioff *et al.*, *Contributions in Geophysics: In Honour of Beno Gutenberg*, London.

Båth, M. (1969) Handbook on earthquake magnitude determinations. *Publ. Seismol. Inst.*, Uppsala.

Båth, M. (1978) A note on recurrence relations for earthquakes. *Tectonophysics*, **51**: T23.

Båth, M. (1979a) Teleseismic magnitude relations. *Rep. Seismol. Inst., Uppsala*, no. 2–79.

Båth, M. (1979b) A note on frequency and energy relations for earthquakes. *Tectonophysics*, **56**: T27.

Båth, M. & Duda, S. (1979) Some aspects of global seismicity. *Rep. Seismol. Inst., Uppsala*, no. 1–79.

Baumgarten, V. (1896) Poyezdka po vostochnoi Persii L. Gv. Volynskavo polka poruchika Baumgartena v 1894 g. (Geografichesko-torgovoye isledovabie), *Sbornik Mater. po Azii*, **63**: 1.

Bayer, K., Heuckroth, L. & Karim, R. (1969) An investigation of the Dasht-e Bayaz, Iran earthquake of August 31, 1968. *Bul. Seismol. Soc. Am.*, **59**: 1793.

Beckett, Ph. (1953) Qanats around Kerman. *J. R. Cent. Asian Soc.*, **40**: 47.

Beer, A. & Stagg, J.N. (1946) Seismic sea-wave of November 27 1945. *Nat.*, **158**: 63.

Bell, C.M. (1840) Geological notes on part of Mazunderan. *Tr. Geol. Soc. London*, **5**: 577.

Bell, J. of Antermony (1788) *Travels from St. Petersburgh in Russia to Various Parts of Asia*. 2 vols. Edinburgh.

Bell, M.S. (1889a) A visit to the Karun river and Kum. *Blackwood's Edinburgh Magazine*, **145**, April: 453.

Bell, M.S. (1889b) Isfahan to Bushire; roads and resources of southern Persia. *Blackwood's Edinburgh Magazine*, **145**, July: 96.

Bellew, H.W. (1874) *From the Indus to the Tigris; a Narrative of a Journey through the Countries of Balochistan, Afghanistan, Khorassan and Iran, in 1872*. London.

Bembo, A. (1803) Viaggio e giornale per parte dell'Asia di quattro anni incirca fatto de ma Ambrosio Bembo. In I. Morelli's *Disertazione intorno ad alcuni viaggiatori eruditi Veneziani poco noti*, Venice.

Berberian, M. (1976) Contribution to the seismotectonics of Iran: Part II. *Geol. Surv. Iran*, Rep. no. 39, Tehran.

Berberian, M. (1977) Contribution to the seismotectonics of Iran: Part III. *Geol. Surv. Iran*, Rep. no. 40, Tehran.

Berberian, M. (1979*a*) Tabas-e-Golshan, Iran, catastrophic earthquake of 16th September 1978. *Disasters*, 2: 207.

Berberian, M. (1979*b*) Earthquake faulting and bedding thrust associated with the Tabas-e-Golshan, Iran, earthquake of September 16th 1978. *Bul. Seismol. Soc. Am.*, 69: 1861.

Berberian, M. & Papastamatiou, D. (1978) Khurgu, North Bandar Abbas, Iran, earthquake of March 21st 1977. *Bul. Seismol. Soc. Am.*, 68: 411.

Berberian, M., Asudeh, I. & Arshadi, S. (1979*a*) Surface rupture and mechanism of the Bob-Tangol, southeastern Iran, earthquake of 19th December 1977. *Earth Planet. Sci. Lett.*, 42: 456.

Berberian, M., Asudeh, I., Bilham, R., Scholtz, C. & Soufleris, C. (1979*b*) Mechanism of the main shock and the aftershock study of the Tabas-e-Golshan, Iran, earthquake of September 16, 1978. *Bul. Seismol. Soc. Am.*, 69: 1851.

Berlage, H. (1932) Seismometer. In B. Gutenberg, *Handbuch der Geophysik*, 4: 299, Berlin.

Bernay (1863) Tremblement de terre à Chiraz. *Petit J. Quot.* 23 April, Paris.

Berninghausen, W. (1966) Tsunamis and seismic seiches reported from regions adjacent to the Indian Ocean. *Bul. Seismol. Soc. Am.*, 56: 69.

Berryat, J. (1761) Liste chronologique des éruptions de volcans, des tremblements de terre etc. *Collect. Acad., Partie Etrang.*, 6, Paris.

Berthier, F., Billiault, J.P., Halbronn, B. & Maurizot, P. (1974) Etude stratigraphique, pétrologique et structural de la région de Khorramabad, Zagros, Iran. Theses, Doctorat de 3e Cycle, Université de Grenoble.

Binning, R.B.M. (1857) *A Journal of Two Years' Travel in Persia, Ceylon etc.* 2 vols, London.

Bishop-Bird, I.L. (1891) *Journeys in Persia and Kurdistan.* 2 vols, London.

Bivar, A.D. (1972) The tomb at Resget; its architecture and inscriptions. *Proc. 5th Int. Congr. Iran. Art and Archeol.*, Tehran, 2: 15.

Bjeshdsian, Father Minas (1830) *Voyage en Pologne.* Venice.

Blair-Fish, P. & Bransby, P. (1972) Flow patterns. *Am. Soc. Mech. Eng., Trans., J. of Eng. for Ind.*, 72-MH-6.

Blaramberg, I.F. (1850) Topograficheskoe i statisticheskoe opisanie vostochnago berega Kaspiiskago-moria ot Astrabadskago-zaliva do misa Tiuk-Karagana. *Zap. Imp. Russ. Geogr. O-Va.*, 4: 49.

Blau, O. (1863) Vom Urmia-See nach dem Van-See. *Petermann's Mitt.*, 6: 201.

Blunt, W. (1957) *A Persian Spring.* London.

Bode, de C.A. (1845) *Travels in Luristan and Arabistan.* 2 vols., London.

Bontemps, A. *Relation d'un voyage en Turquie et en Perse pendant l'année 1807.* See Ittihādiyyeh.

Bosworth, C.E. (1968) *Sistan under the Arabs.* Rome.

Bosworth, C.E. (1973) *The Ghaznavids* 2nd ed. Beirut.

Bosworth, C.E. (1977*a*) *The Later Ghaznavids: Splendour and Decay.* Edinburgh.

Bosworth, C.E. (1977*b*) *The Medieval History of Iran, Afghanistan and Central Asia.* Variorum Reprints, London.

Boullaye-le-Gouz, La (1657) *Les Voyages et observations du Sieur de la Boullaye-le-Gouz.* Paris.

Bout, R., Derruau, M., Dresch, J. & Peguy, Ch. (1961) Observations de géographie physique en Iran septentrional. *Mem. et Doc. CNRS*, 8, Paris.

Bowen, H. (1931) The *sar-gudhasht-i sayyidnā*, the 'Tale of the three schoolfellows' and the *wasaya* of the Nizam al-Mulk. *J. R. Asiat. Soc.*: 771.

Brant, J. (1841) Notes of a journey through a part of Kurdistan, in the summer of 1838. *J. R. Geogr. Soc.*, 10: 341.

Brayley-Hodgetts, E.A. (1916) *Round about Armenia.* London.

Brittlebank, W. (1873) *Persia during the Famine.* London.

Brosset, M.F. (1841) Note sur le village Arménien d'Agorhi et sur le couvent de St. Jacques. *Bul. Sci. Acad. Sci.*, 8: 41, St Petersburg.

Brosset, M.F. (1849) *Rapports sur un voyage archéologique dans la*

Géorgie et dans l'Arménie en 1847–1848. 1, report no. 3, St Petersburg.

Brosset, M.F. (1860) *Les ruines d'Ani.* St Petersburg.

Brosset, M.F. (1861) *Les ruines d'Ani capitale de l'Arménie sous les rois Bagratides, aux X et XI siècles.* St Petersburg.

Brosset, M.F. (1874–6) *Collection d'historiens Arméniens.* 2 vols., St Petersburg.

Brown, J.A. (1963) The earthquake disaster in Western Iran September 1962. *Geography*, 48: 184, Sheffield.

Brown, R.D. (1969) Geological appraisal of the 1968 earthquakes in Khorassan Province, Iran. *US Dep. Inter. Geol. Surv.*, Proj. Rep. IR-3, Washington.

Browne, E.G. (1910) *The Persian Revolution, 1905–1909.* Cambridge.

Browne, E.G. (1914) *The Press and Poetry of Modern Persia.* Cambridge.

Brugsch, H. (1862) *Reise der preussischen Gesandtschaft nach Persien 1860 und 1861.* 2 vol., Leipzig.

Bruin, Cornelis de (1737) *Travels into Muscovy, Persia and Parts of the East-Indies etc.* 2 vol. London.

Bryce, J. (1877) *Transcaucasia and Ararat.* London.

Brydges, H. Jones (1834) *An Account of the Transactions of His Majesty's Mission to the Court of Persia in the Years 1807–1811.* Vol. 1. London.

Bubnov, S. (1968) Potres v Khorassanu. *Gradbeni Vestn.*, 11: 201, Ljubljana.

Buhse, F.A. (1855) *Eine Reise durch Transkaukasien und Persien in den Jahren 1857–1859.* Moscau Kaiserlichen Univ., Moscow.

Buist, G. (1852) The volcanoes of India. *Trans. Bombay Geogr. Soc.*, 10: 139.

Bullen, K. (1933) The constants of seismological observatories. *Publ. Br. Assoc. Adv. Sci.*, Newport, Isle of Wight.

Bulliet, R. (1976) Medieval Nishapur: a topographic and demographic reconstruction. *Stud. Iran.*, 5 (1): 67.

Bune, V. & Vvedenskaya, N. (1970) Correlation of M_{LH} and m_{pv} by data of the network of seismic stations of the USSR. *Geophys. J. R. Astron. Soc.*, 19: 533.

Bunge, A. (1860) Die Russische Expedition nach Chorassan in den Jahren 1858 und 1859. *Petermann's Mittheil*, 6: 205.

Buniyatov, Z.M. (1977) Sevedeniia o zemletriasenilakh v nekotorikh srednevekovikh Arabskikh istochnikakh. *Izv. Akad. Nauk. Azerbaidzhansk. SSR.*, no. 5: 93.

Burgess, Ch. & E. (1942) *Letters from Persia Written by Charles and Edward Burgess 1828–1855.* Ed. B. Schwartz, New York.

Burnes, A. (1834) *Travels into Bokhara.* 2 vols. London.

Burney, C. (1973) Excavations at Haftavan Tepe 1971. *Iran*, 11: 158.

Busk. H.G. (1926) The Shimbar Valley landslip dam, Bakhtiari country, South Persia. *Geol. Mag.*, 63: 355.

Busse, H. (1975) Iran under the Buyids. *Camb. Hist. Iran*, 4: 250.

Butterfield, R. & Andrawes, K. (1972) An investigation of a plane strain continuous penetration problem. *Geotech.*, 22: 597.

Byron, R. (1937) *The Road to Oxiana.* London.

Byus, E.I. (1948) Seismicheskie usloviya zakavkaz'ya. *Izd. Akad. Nauk Gruz. SSR*, Tiflis.

Byus, E.I. (1952) Seismicheskie osnov'y seismogeografii zakavkaz'ya. *Izd. Akad. Nauk Gruz. SSR*, Tiflis.

Byus, E.I. (1955) Kvoprosu o khode seismicheskoy aktivnosti v zakavkazye. *Izd. Akad. Nauk Gruz. SSR*, Tiflis.

Cahen, C. (1978) Faḍluwayh le Shāvankāreh. *Stud. Iran.*, 7: 111.

Canard, M. (1965) La campagne Arménienne du Sultan Salguqide Alp Arslan et la prise d'Ani en 1064. *Rev. Etud. Arméniennes*: 239.

Carli, G.R. (1697) *Cronologia historica . . . da Hazi Halife Mustafa, etc.* Venice.

Caspani, E. & Cagnacci, E. (1951) *Afghanistan crocevia dell'Asia.* Milan.

Central Meteorological Observatory (1952)*Magnitude Catalogue of Major Earthquakes in the Vicinity of Japan 1885–1950.* Tokyo.

Chardin, J. (1811) *Les voyages du Chevalier Jean Chardin en Perse et autres lieux de l'Orient.* Ed. Langlès, Paris.

Chaybany (1971) *Les voyages en Perse et la pensée française au XVIIIe siècle.* Paris.

Chirikoff, E.I. (1875) Putevoi zhurnal. *Zap. Imp. Russ. Geogr. O-Va.*, 9: 1. St Petersburg.

Clavijo, Ruy Gonzalez de (1928) *Historia del Gran Tamerlan*, tr. Le Strange, *Embassy to Tamerlane, 1403–1406.* London.

Conolly, A. (1838) *Journey to the North of India overland from England through Russia, Persia and Afghanistan.* 2 vols. London. Cf. *J. R. Geogr. Soc.*, 4 (1834): 279.

Coon, C.S. (1951) *Cave Explorations in Iran 1949.* Museum Monographs, The Univ. Museum Univ. Pennsylvania, Philadelphia.

Coon, C.S. (1957) *Seven Caves.* London.

Coronelli, P. (1693) De' tremuoti accaduti dal diluvio universale fin all' anno 1693. In *Epitome Cosmografica*: 286.

Crampin, S. (1969a) Aftershocks of the Dasht-e Bayaz, Iran, earthquake of August 31st, 1968. *UNESCO* Publ. no. 1214/BMS, Paris.

Crampin, S. (1969b) Aftershocks of the Dasht-e Bayaz, Iran, earthquake of August 1968. *Bul. Seismol. Soc. Am.*, 59: 1823.

Cuinet, V. (1890–1895) *La Turquie d'Asie.* 4 vols. Paris. Vol. 2, 1892.

Curzon, G. (1966) *Persia and the Persian Question.* 2nd ed. 2 vols., London.

Daldy, A.F. (1968) Visit to Iran in December 1968; Ferdous. *Internal Report*: Build. Res. Stn., Garston.

Dalyell, R.A. (1862) Earthquake of Erzerum, June 1859. *Proc. R. Geogr. Soc.*, 6: 62.

Daulier-Deslandes, Vendomois, A. (1926) The beauties of Persia, trans. A.T. Wilson, *Persia Society*, London.

Davison, C. (1893) Note on the Quetta earthquake of December 20th 1892. *Geol. Mag.* 3rd decade, 10: 356.

Davison, C. (1924) *A History of British Earthquakes.* Cambridge.

Day, J.B.W. & Wright, E.P. (1969) Hydrogeological studies in the Nimbaluk and Gonabad areas of Khorassan, Iran. Unpublished Report, Inst. Geol. Sci. London.

Dellenbach, J. (1964) *Contribution à l'étude géologique de la région située à l'est de Tehran, Iran.* Thèse Univ. Strasbourg.

Despeyroux, J. & Lescuyer, M. (1963) Problèmes de la protection parasismique en Iran; Buyin Zahra. *SOCOTEC.* Rep. no. SO/JD 63–57, Paris.

Dewey, J.W. & Grantz, A. (1973) The Ghir earthquake of April 10, 1972 in the Zagros mountains of southern Iran; seismotectonic aspects and some results of a field reconnaissance. *Bul. Seismol. Soc. Am.*, 63: 2071.

Dieulafoy, J. Mme (1887) *La Perse, la Chaldée et la Susiane*, Paris.

Dilley, A. & Dimand, N. (1931) *Oriental Rugs and Carpets.* New York.

Djavadi, Sh. (1966) *Climats de l'Iran.* Monogr. de Metéorol., Paris.

Dmitriev-Mamonov, A.N. (1903) *Putevoditel' po Turkestanu*, St Petersburg.

Douglas, D., Young, J. & Lilwall, R. (1974) Computer programs for epicentre determination. *AWRE* Rep. no. 28/74.

Dressdnische Gelehrte Anzeigen (1756) Nachrichten von Erdbeben. No. 2–40, Dresden.

Drouville, G. (1825) *Voyage en Perse fait en 1812 et 1813.* 2 vols. Paris.

Du Cerceau (1728) *The History of the Revolution of Persia, taken from the Memoirs of Father Krusinski, Procurator of the Jesuits at Ispahan.* 2 vols. London.

Duda, S. (1965) Secular seismic energy release in the circum-Pacific belt. *Tectonophysics*, 2: 409.

Dupré, A. (1819) *Voyage en Perse fait dans les années 1807, 1808 et 1809.* 2 vols. Paris.

Dupree, N.H. (1971) *A Historical Guide to Afghanistan.* Afghan Air Authority, Kabul.

Durand, E.R. (1902) *An Autumn Tour in Western Persia.* London.

Durri Efendi (1810) *Relation de Dourry Efendy, ambassadeur de la Porte Othomane auprès du roi de Perse.* Paris.

Dyer, R.E.H. (1921) *The Raiders of the Sarhad.* London.

Dzhanashvili, M. (1902) Zemletriacenii v proshlom. *Izv. Kavkaz. Otd. Russ. Geogr. O-Va.*, 15: 319, Tiflis.

Eastwick, E.B. (1864) *Journal of a Diplomat's Three Years Residence in Persia.* 2 vols. London.

Eftakhar-Nezhad, J., Haghipour, A. & Davoudzadeh, M. (1968) Earthquake Disaster in Khurasan. *Publ. Inst. Geophys. Univ. Tehran*, no. 46.

Egerton, R.W. (1893) Effects of earthquakes on North-Western railway, India. *Engineering*, May 19: 698.

Elwell-Sutton, L.P. (1968) The Iranian Press, 1941–1947. *Iran*, 6: 65.

Emanuel, W.V. (1939) *The Wild Asses*, London.

Ergin, K., Guclu, U. & Uz, Z. (1967) A catalogue of earthquakes for Turkey and surrounding area; A.D. 11–1964. *Maden Fak., Arz Fizigi Enst.*, 24: 169, Istanbul.

Ethé, H. (1903) *Catalogue of Persian Manuscripts in the Library of the England–Indian Office.* Oxford.

Fagergren (1853) Tremblement de terre en Perse; destruction de Chiraz dans la nuit du 21 au 22 avril 1853. *Nouv. Ann. Voyages*, 5e Ser., 35: 187.

Falcon, N. (1974) Southern Iran: Zagros mountains. Mesozoic–Cenozoic orogenic belts. Ed. A.M. Spencer, *Geol. Soc. London*, 4: 199.

Fernberger, G.C. (1898) *Die Reise des Hans Christoph Freiherrn von Teufel etc.* Ed. G.E. Friess, *xxxii Program des k.k. Ober-Gymnasiums Seitenstetten*, Linz.

Feuvrier, Dr (1899) *Trois ans à la cour de Perse.* Paris.

Figueroa, don Garcia de Silva (1667) *L'ambassade de Don Garcia de Silva Figueroa en Perse.* Paris.

Filadelfin, A. (1860) Zemletransenii v Shemakha i Erzerum v mai 1859 goda. *Sh. Sotr. R. G. O. Akad. Stampa*, Abuha.

Fisher, R. & Guidroz, R. (1964) World collection and evaluation of earthquake data; 1963 seismicity. *Air Force Cambridge Res. Lab.* Rep. 64-964, Cambridge, Mass.

Fisher, R., Baker, R. & Guidroz, R. (1964) Worldwide collection and evaluation of earthquake data; 1960 seismicity. *Special Rep. Terr. Sci. Lab.*, Proj. 8652, no. 3, Cambridge, Mass.

Flandin, E. (1851) *Voyage en Perse de MM. Eugene Flandin, peintre, et Pascal Coste, architecte pendant les années 1840 et 1841.* 2 vols. Paris.

Flandin, E. & Coste, P. (1843–54) *Voyage en Perse.* Pahlavi Commem. Repr. Ser., Tehran 1976.

Florensov, N. & Solonenko, V. (1963) Gobi-altaiskoe zemletryasenie. *Akad. Nauk. USSR Publ.*. Moscow.

Fookes, P. & Knill, J. (1969) The application of engineering geology in the regional development of Northern and Central Iran. *Eng. Geol.*, 3: 90.

Forbes, R. (1931) *Conflict, Angora to Afghanistan.* London.

Forster, G. (1798) *A Journey from Bengal to England.* 2 vols. London.

Fort, Charles (1973) *The Book of the Damned.* London.

Franken, H.J. (1964) The stratigraphic context of the clay tablets found at Deir 'Alla. *Palest. Explor. Q.*: 73.

Francklin, W. (1790) *Observations Made on a Tour from Bengal to Persia in the Years 1786–1787.* London.

Fraser, J.B. (1825) *Narrative of a Journey into Khorasan in the Years 1821–2.* London.

Fraser, J.B. (1838) Notes on the country lying between the meridians of 55° and 64°, and embracing a section of the Elburz mountains in Northern Khorasan. *J. R. Geogr. Soc.* 8: 308.

Freedman, H. (1967) Estimating earthquake magnitude. *Bul. Seismol. Soc. Am.*, 57: 747.

Freville, N. (1964) *The Bridge of the Maiden.* London.

Freygang, W. and F. (1816) *Relation d'un voyage à Tauris en 1812.* Hamburg.

Fryer, J. (1698) *A New Account of East India and Persia in Eight Letters.* London.

Fuchs, C.W. (1886) Statistik der Erdbeben 1865–1885. *Sitzungsber. Kais. Akad. Wiss. Math.-Naturwiss. Cl.*, 92 (3).

Gabriel, A. (1929) *Im Weltfernen Orient.* Munich.

Gabriel, A. (1935) *Durch Persiens Wüsten*, Stuttgart.

Gabriel, A. (1952) *Die Erforschung Persiens.* Vienna.

Gansser, A. (1969) The large earthquakes of Iran and their geological frame. *Eclogae. Geol. Helv.*, 62 (2): 443.

Gardane, A. de (1809) *Journal d'un voyage dans la Turquie d'Asie et la Perse fait en 1807 et 1808.* Paris.

Garitte, G. (1958) Le calendrier Palestino–georgien du Sinaiticus 34. *Subsidia Hagiographica* no. 30. Brussels.

Garthwaite, G. (in preparation) *Khan and Shahs; the Bakhtiyaris in Iran.* (personal communication)

Garza, T. & Lomnitz, C. (1979) The Oaxaca gap: a case history. *Pure & Appl. Geophys.*, 117: 1187.

Gaspar de San Bernardino, in H. Muaray (1820) *Historical Account of Discoveries and Travels in Asia.* 1: 382. Edinburgh.

Gates, G., Page, W., Savage, W. & Zuberi, S. (1977) The Makran earthquake November 28, 1945; a 1975 field appraisal of geologic and cultural effects. *Proc. 6th World Conf. Earthq. Eng.* New Delhi, 2: 529.

Gazetteer of Persia (1885–1918) 4 vols. India Office, London.

Gerasimov, I. (1978) The past and future of the Aral and the Caspian

Seas in W. Brice, *The Environmental History of the Near and Middle East*, New York and London.

Gobineau, A. de (1859) *Trois ans en Asie, de 1855 à 1858*. Paris.

Godard, A. (1934) Les monuments de Maragha. *Soc. Etud. Iran. l'art Persan*, Publ. no. 9. Paris.

Godard, A. (1936) Les tours de Ladjim et de Resget. *Athar-i Iran*, 1: 109.

Godard, A. (1937) Isfahan. *Athar-i Iran*, 2: 115.

Godard, A. (1941) The architecture of the Islamic period; a survey of Persian art. *Ars Islamica*, 8: 11.

Gojkovic, S. (1973) The Qal'eh Hasan Ali cryptoexplosion structures. *Geol. Surv. Iran*, Rep. no. Yu/52.

Gollancz, H. (1927) *The Settlement of the Order of Carmelites in Mesopotamia*. Oxford.

Gorshkov, G.P. (1947) Zemletriasennia Turkmenii. *Trudi Seismol. Inst. Akad. Nauk*, no. 122, Moscow.

Gorshkov, G., Spesivtsev, V. & Popov, V. (1941) Katalog zemletriaseni na territorii SSSR. *Trudi Seismol. Inst. Akad. Nauk*, no. 95, Moscow.

Goudarzi, K.M. (1975) On the seismicity of Iran during the year 1971. *J. Earth & Sp. Phys.*, 4: 21, Tehran.

Goudarzi, K.M. & Ghaderi-Tafreshi, M. (1976) The Qaen, Khorassan earthquake of November 7, 1976. *Proc. CENTO Semin. Earthq. Hazard Minimization*, Tehran.

Goudarzi, K.M. & Mostaanpour, M. (1976) A preliminary report on seismic swarm in Nourabad. *Proc. CENTO Semin. Earthq. Hazard Minimization*, Tehran. *Tech. Res. & Stand. Bur. Plan & Budget Org.*, Publ. no. 70.

Goudarzi, K.M., Hossein-Javaheri, J., *et al.* (1972–4) Seismic zoning of the Iranian Plateau. Various papers in *J. Earth & Sp. Phys.*, 1–3, Tehran.

Gouin, P. (1979) *Earthquake History of Ethiopia and the Horn of Africa*. Int. Dev. Res. Cent., Ottawa.

Gouvea, A. de (1646) *Relation des grandes guerres et victoires obtenues par le Roy de Perse etc.* Rouen.

Greenwood, P. (1976) A new and eyeless cobitid fish from the Zagros mountains, Iran. *J. Zool. London*, 180: 129.

Griesbach, C.L. (1893) Notes on the earthquake of Beluchistan on the 20th December 1892. *Rec. Geol. Surv. India*, Pt. 1: 27, Pt. 2: 57.

Grodenkoff, N.I. (1883) *Voyna v Turkmenii; pokhod Skobeleva v 1880–1881 gg.* vol. 1, St Petersburg.

Gropp, G. (1970) Bericht über eine Reise in West- und Sudiran. *Archeol. Mitt. aus Iran*, 3.

Grumel, V. (1958) *Traite d'études Byzantines; la chronologie*. Bibl. Byzantine, Paris.

Gubbins, R.E. (1944) Incidence of earthquakes in Iraq. Unpublished Rep. No. 14/3226, Director General of Irrigation, Baghdad.

Gülkan, P., Gürpinar, A., Celebi, M., Arpat, E. & Gencoglu, S. (1978) *Engineering report on the Muradiye-Caldiran, Turkey, earthquake of 24 November 1976*. Natl. Acad. Sci., Washington.

Gutenberg, B. (1956) Great earthquakes 1896 to 1903. *Trans. Am. Geophys. Union*, 37 (5): 608.

Gutenberg, B. & Richter, C. (1942) Earthquake magnitude, intensity, energy and acceleration. *Bul. Seismol. Soc. Am.*, 32: 163.

Gutenberg, B. & Richter, C. (1956a) Magnitude and energy of earthquakes. *Ann. di Geofis.*, 9: 1.

Gutenberg, B. & Richter, C. (1956b) Earthquake magnitude, intensity, energy and acceleration. *Bul. Seismol. Soc. Am.*, 46: 105.

Gutenberg, B. & Richter, C. (1965) *Seismicity of the Earth and Associated Phenomena*. 3rd edition, Hafner.

Haghipour, A. & Amidi, M. (1980) Geotectonics of the Ghaenat earthquakes of northeast Iran of November 14 to December 9, 1979. *Geol. Surv. Iran*. Intern. Rep.

Haghipour, A., Amidi, M. & Aghanabati, A. (1979) Geotectonics of the Tabas earthquake, East Iran. *Bul. Iran. Pet. Inst.*, no. 76: 11, Tehran.

Haghipour, A., Iranmanesh, M. & Takin, M. (1972) The Ghir earthquake in Southern Persia; a field report and geological discussion. *Geol. Surv. Iran*, Intern. Rep. no. 52.

Hagiwara, T. & Naito, T. (1959) *A Report of the Japanese Mission to Iran for Investigating the Problem Related to Disastrous Earthquakes in Iran*. Tokyo.

Hale, F. (1920) *From Persian Uplands*. London.

Hambly, G. (1964) An introduction to the economic organisation of early Qajar Iran. *Iran*, 2: 69.

Hammer-Purgstall, J. von (1827–35) *Geschichte des Osmanischen Reiches*. 10 vols., Pest.

Hammer-Purgstall, J. von (1844) *Histoire de l'Empire Ottoman, depuis son origine jusqu'à nos jours*. French tr. M. Dochez, 3 vols. Paris.

Hansman, J. (1968) The problems of Qūmis. *J. R. Asiat. Soc.*: 111.

Hanway, J. (1753) *An historical account of the British Trade over the Caspian Sea, with . . . the Revolutions of Persia*. 4 vols., London.

Harris, C.P. (1969) The Persian Gulf submarine telegraph of 1864. *Geogr. J.*, 135 (2): 169.

Harrison, J.V. (1936) Kuh-Galu, south-west Iran. *Geogr. J.*, 88: 20.

Harrison, J.V. (1946) Southwest Persia, a survey of the Pish-i-kuh in Luristan. *Geogr. J.*, 108: 56.

Harrison, J.V. & Falcon, J. (1937) The Saimarreh landslip, southwest Iran. *Geogr. J.*, 80: 42.

Hattori, S. (1979) Seismic risk maps in the World. *Bul. Int. Inst. Seismol. & Earthq. Eng.*, 17: 74, Tokyo.

Hay, N.R. (1921) *Two years in Kurdistan; Experiences of a Political Officer, 1918–1920*. London.

Hedin, S. (1892a) Der Demavend nach eigener Beobachtung. *Verhandlung der Ges. für Erdkd. zu Berlin*, 19: 304.

Hedin, S. (1892b) Genom Khorasan och Turkestan. Minnen frän en resa i Centralasien 1890 och 1891. Stockholm.

Hedin, S. (1910) *Overland to India*. 2 vols. London.

Hedin, S. (1918) *Eine Routenaufnahme durch Ostpersien*. 2 vols. Stockholm.

Hell, Xavier Hommaire de (1854–60) *Voyage en Turquie et en Perse*. 4 vols. Paris.

Hendricks, H.E. (1962) Report on Iran's earthquake disaster of September 1962. Unpubl. Rep., US Aid Mission to Iran, Task Force on Earthquake Rehabilitation.

Herbert, Sir Thomas (1634) *A Relation of Some Yeares Travaile, Begunne Anno 1626 etc.* London, and 2nd ed. *Some Yeares Travels into Africa and Asia the Great etc.*, London 1638.

Heuckroth, L.E. & Karim, R.A. (1970) *Earthquake History, Seismicity and Tectonics of the Regions of Afghanistan*. Fac. of Eng., Kabul Univ.

Hieronimo di Santo Stefano (1857) Account of the journey of Hieronimo di Santo Stefano. *Hakluyt Soc. Publ.*, no. 22.

Hitchen, C.S. (1946) Earthquake messages from the Mutasarrifs of Sulaimaniyah and Mosul. Unpubl. Rep., Adm. Off., Baghdad.

Hoff, K. von (1840) Chronik der Erdbeben und Vulkan-Ausbrüche, mit vorausgehender Abhandlung über die Natur diesser Erscheinungen. In *Gesch. Ueberlieferung nachgew. naturl. Veränder. Erdoberfläche*, vols. 2, 3 and 4, Gotha.

Hoffman, G. (1880) Auszüge aus Syrischen Akten persischer Märtyrer. *Abh. für d. Kunde des Morg.*, 7 (3). Leipzig.

Hollingbery, W. (1814) *A Journal of Observations made during the British Embassy to the Court of Persia in the Years 1799, 1800 and 1801*. Calcutta.

Holmes, W.R. (1845) *Sketches on the Shores of the Caspian*. London.

Hossein-Javaheri (1972) Sun-dried building in Khorassan and their resistance to earthquakes. *J. Earth & Sp. Phys.*, 1: 7. Tehran.

Housner, G. (1963) The behaviour of inverted pendulum structures during earthquakes. *Bul. Seismol. Soc. Am.*, 53.

Houssay, F. (1894) La structure du sol et son influence sur la vie des habitants; études sur la Perse méridionale. *Ann. de Geogr.*, 3: 278.

Houtum Schindler, A. (1877a) Notes on some antiquities found in a mound near Damghan. *J. R. Asiat. Soc.*, 9: 425.

Houtum Schindler, A. (1877b) The Persian Government telegraphs. *J. Soc. Telegr. Eng.*, 4 (13): 262.

Houtum Schindler, A. (1879) Reisen in südwestlichen Persien. *Z. der Ges. für Erdkd. zu Berlin.*, 14: 38, 81.

Houtum Schindler, A. (1881a) Neue Angaben über die Mineralreichthümer Persiens und Notizen über die Gegend westlich von Zendjan. *Jahrb. der Kais. Kön. Geol. Reichsanst.*, 31 (2): 169. Vienna.

Houtum Schindler, A. (1881b) Reisen im südlichen Persien 1879. *Z. der Ges. für Erdkd. zu Berlin.*, 16: 307.

Houtum Schindler, A. (1883) Reisen im nordwestlichen Persien 1880–1882. *Z. der Ges. für Erdkd. zu Berlin*, 18: 320.

Houtum Schindler, A. (1896) *Eastern Persian Iraq*. London.

Howard, H.H. (1975) Gilbert's cow story. *Earthq. Eng. Res. Inst., Newsletter*, 9 (1): 103. Los Angeles.

Howell, B. & Schultz, T. (1975) Attenuation of Modified Mercalli intensity with distance from the epicentre. *Bul. Seismol. Soc. Am.*, 65: 650.

Hübschmann, H. (1904) *Die altarmenischen Ortsnamen.* Strasbourg.

Huckriede, R., Kürsten, M. & Venzlaff, H. (1962) Zur Geologie des Gebietes zwischen Kerman und Sagand (Iran). *Geol. Jb.*, 51: 197.

Huntington, E. (1905) The basin of eastern Persia and Sistan, in R. Pumpelly, *Explorations in Eastern Turkestan.* Carnegie Inst., Washington.

Huot, J.J.N. (1837) Nouveau cours élémentaire de géologie. *Libr. Encyclop. Roret*, Paris, 1: 108.

Jackson, A.V.W. (1911) *From Constantinople to the Home of Omar Khayyam.* New York.

Jackson, J. (1980) Reactivation of basement faults and crustal shortening in orogenic belts. *Nat.*, 283: 343.

Jager, Herbert de (1911) in F. Valentijn, *Oud en nieuw Oost Indien.* vol. 5; also *Oud-Holland*, vol. 29, Amsterdam.

James, G.A. & Ghashghaie, M. (1960) The Lar earthquake of April 24, 1960. Unpubl. Rep., Iran. Oil Explor. and Prosp. Co., Masjid-i Sulaiman.

Jaubert, P.A. (1821) *Voyage en Arménie et en Perse fait dans les années 1805 et 1806.* Paris.

Johnson, J. (1818) *A Journey from India to England through Persia, Georgia, Russia etc. in the Year 1817.* London.

Kaempfer, E. (1712) *Amoenitatum exoticarum etc. rerum Persicarum.* Pahlavi Commem. Repr. Ser., Tehran 1976.

Kagan, Y. & Knopoff, L. (1979) Spatial distribution of earthquakes; the two-point correlation function. *Geophys. J. R. Astron. Soc.*, 56: 67.

Kanamori, H. & Abe, K. (1979) Re-evaluation of the turn-of-the-century seismic peak. *Publ. Seismol. Lab. California Inst. Technol.*, Pasadena.

Kárník, V. (1968–71) *Seismicity of the European Area*, 2 vols. Prague and Dordrecht.

Kawasumi, H. (1951) Measure of earthquake danger and expectancy of maximum intensity throughout Japan as inferred from the seismic activity in historical times. *Bul. Earthq. Res. Inst.*, 29: 469, Tokyo.

Kawasumi, H. (1954) Intensity and magnitude of shallow earthquakes. *Publ. Bur. Cent. Seismol. Int.*, Ser. A, Trav. Sci. fac. 19: 99, Strasbourg.

Kazemzadeh, F. (1968) *Russia and Britain in Persia, 1864–1914.* New Haven and London.

Kennard, C. (1927) *Suhail.* London.

Khanikoff, N. (1861) *Memoire sur la partie méridionale de l'Asie Centrale.* Paris.

Khodzko, A.I. (1853) Opisanie dorogi, vedushchei iz Nishapur vdol' iuzhnikh predelov Khorasanskago Kurdistana do Sharuda, 1834. *Zap. Imp. Russ. Geogr. O-Va.*, 7 (2): 237. St Petersburg.

Khoshbakht-Marvi, A. (1977) A brief report of the Qaen earthquake of November 7, 1976, in Khorassan, Iran. *Proc. CENTO Semin. Earthq. Hazard Minimization*, Tehran. *Tech. Res. & Stand. Bur. Plan & Budget Org.*, Publ. no. 70: 280.

King, G., Bilham, R., Campbell, J., McKenzie, D. & Niazi, M. (1975) Detection of elastic strainfields caused by fault creep events in Iran. *Nat.*, 253: 430.

Kirnos, D., Kharin, D. & Shebalin, N. (1961) Istoriya razvitiya instrumentalnikh seismicheskikh nablyudeny v SSR. In *Atlas*, ed. Moscow 1961: 9.

Kleiss, W. (1968) Das Kloster des Heiligen Thaddäus – Kara Kilise – in Iranisch-Azerbaidjan. *Istanbuler Mitt.*, 18: 304.

Kleiss, W. (1969) Bericht über zwei Erkundungsfahrten in Nordwest-Iran. *Archaeol. Mitt. aus Iran*, N.F., 2.

Knopoff, L. & Kagan, Y. (1977) Analysis of the theory of extremes as applied to earthquake problems. *J. Geophys. Res.*, 82: 5647.

Kobayashi, H. (1963) *Buyin Earthquake of Sept. 1, 1962, Iran.* Tokyo Institute of Technology.

Kortazzi, J. (1900) Les perturbations du pendule horizontal à Nicolajew en 1897, 1898 et 1899. *Beitr. zur Geophys.*, 4: 383, Leipzig.

Kotzebue, Moritz von (1819) *Narrative of a Journey to Persia in the Suite of the Russian Embassy, 1817.* London.

Kremer, A. von (1880) Uber die grossen Seuchen des Orientz nach arabischen Quellen. *Zitzungsberichte der Philosoph-Historichen Classe der Kaiserl. Akad. der Wiss. in Wien*, 96: 69.

Krinsley, D.B. (1972) A geomorphological and paleoclimatological study of the playas of Iran. 2 vols. *Air Force Cambridge Res. Lab.*, Rep. 70.0503.

Kroell, A. (1977) Louis XIV, la Perse et Muscate. *Le Monde Iranien et l'Islam*, 4: 1.

Krumbach, G. (1949) Turkmenische Erdbeben von 5 Oktober 1948. *Urania*, 12 (2): 70, Jena.

Lambton, A.K.S. (1970) Persia: the breakdown of society. In *Camb. Hist. Islam*, 1: 430.

Landor, A.H.S. (1902) *Across Coveted Lands.* 2 vols. London.

Lawrence, R. & Yeats, R. (1979) Geological reconnaissance of the Chaman fault in Pakistan. Geol. Surv. Pakistan, *Geodynamics of Pakistan*: 351. Quetta.

Layard, A.H. (1846) A description of the province of Khuzistan. *Geogr. J.*, 16: 1.

Layard, A.H. (1887) *Early Adventures in Persia, Susiana and Babylonia.* 2 vols. London.

Lee, S.P. (1958) A practical magnitude scale. *Acta Geophys. Sin.*, 7: 98, Peking.

Lee, W. & Brillinger, D. (1979) On Chinese earthquake history – an attempt to model an incomplete data set by point process analysis. *Pure & Appl. Geophys.*, 117 (6): 1229.

Lee, W. & Wetmiller, R. (1978) Survey of practice in determining magnitudes: North, Central and South America. *World Data Cent. A*, Rep. SE-9, Washington.

Lee, W., Wu, F. & Jacobsen, C. (1976) A catalogue of historical earthquakes in China compiled from recent Chinese publications. *Bul. Seismol. Soc. Am.*, 66: 2003.

Lees, G.M. (1955) Recent earth movements in the Middle East. *Geol. Rundsch.*, 43: 221.

Lees, G. & Falcon, N. (1952) The geographical history of the Mesopotamian plains. *Geogr. J.*, 118: 24.

Le Strange, G. (1905) *The Lands of the Eastern Caliphate.* Cambridge.

Lewis, B. (1966) *The Arabs in History.* 4th ed. London.

Lockhart, L. (1960) *Persian Cities.* London.

Loftus, W.K. (1855) On the geology of portions of the Turko-Persian frontier. *Q. J. Geol. Soc. London*, 11: 247.

Lomnitz, C. (1970) Casualties and behaviour of populations during earthquakes. *Bul. Seismol. Soc. Am.*, 60: 1309.

Lomnitz-Adler, J. & Lomnitz, C. (1978) A new magnitude–frequency relation. *Tectonophysics*, 49: 237.

Longrigg, S.M. (1925) *Four Centuries of Modern Iraq.* Oxford.

Lorimer, J.G. (1915) *Gazetteer of the Persian Gulf, Oman and Central Arabia.* 2 vols. Calcutta.

Lotichius, J.P. (1646–50) *J.P. Lotichii rerum Germanicarum sub Matthia, etc.*, Frankfurt am Main.

Lucas, P. (1705) *Voyage du Sieur Paul Lucas au Levant.* 2 vols. The Hague.

Lumsden, T. (1822) *A Journey from Merut in India to London during the Years 1819 and 1820.* London.

Lycklama a Nyeholt, T.M. (1872) *Voyage en Russie, au Caucase et en Perse . . . exécuté pendant les années 1866, 1867 et 1868.* Paris.

Lynch, H.F.B. (1901) *Armenia. Travels and Studies.* London.

Lynch, T.K. (1869) On Consul Taylor's journey to the sources of the Euphrates. *Proc. R. Geogr. Soc.* 13: 243.

Lysakowski, C. (1906) Tremblements de terre arrives dans la Russie d'Asie et en Perse depuis 10 ans. *Bul. Soc. Astron. Franc.*, 20: 45.

Lysakowski, C. (1910) Tremblements de terre de la Perse du 23 janvier (Burujird); de la Bulgarie 15 febrier 1909. *Bul. Soc. Astron. Franc.*, 24: 45.

MacGregor, C.M. (1879) *Narrative of a Journey through the Province of Khorasan in 1875.* 2 vols. London.

MacGregor, C.M. (1882) *Wanderings in Balochistan.* London.

Macler, E. (1917a) *Histoire universelle par Etienne Asolik de Taron.* Paris.

Macler, E. (1917b) Erzurum ou topographie de la haute Arménie. *J. Asiat.*, 13.

Maevski, F. (1899) Kuchanskoe zemletriasenie 5 Ianvaria 1895 g. *Mater. dlia Izuch. zemletriasenii Rosii, Prilozenie II, Izv. Gos. Geogr. O-Va.*, St Petersburg.

Malcolm, J. (1827) *Sketches of Persia.* 2 vols. London.

Malinovski, N. (1935) Katalog zemletryasenii v AzSSR. *Tr. Azarb. Itdelen. Zak. FAN*, 10, Baku.

Mallet, R. (1850–8) Report on the facts of earthquake phenomena.

Rep. BAAS, London. (1850, 1852, 1853, 1854, 1858) References in the text are to the BAAS pagination of the combined edition.

Mallowan, M.E. (1966) *Nimrud and its Remains*. London.

Manandian, H.A. (1965) *The Trade and Cities of Armenia in relation to Ancient World Trade*. Lisbon.

Mandelslo, J.A. (1727) *Voyages célèbres et remarquables faits de Perse aux Indes Orientales par Jean Albert de Mandelslo*. 2 vols. Amsterdam.

Manestey, S. (1812) Itinerary from Koom to Sultanieh, in J. Morier, *Journey through Persia*: 411, London.

Marco Polo (1903) *The Book of Ser Marco Polo*. Tr. H. Yule, 3rd ed. revised H. Cordier, 2 vols. London.

Mariastein Manuscript (1721) Cont. Anni. Frat. Minor. Conv., Thann, Abbaye Mariastein.

Matheson, S. (1976) *Persia: an Archaeological Guide*. London.

Maunsell, F.R. (1890) *Reconnaissance in Mesopotamia, Kurdistan, North-West Persia and Luristan from April to October 1888*. 2 vols. Simla.

McEvilly, T.V. & Niazi, M. (1975) Post-earthquake observations at Dasht-e Bayaz, Iran. *Tectonophysics*, 26: 267.

McEvilly, T.V. & Razani, R. (1973) The Qir, Iran earthquake of April 10, 1972. *Bul. Seismol. Soc. Am.*, 63: 339.

McKenzie, D. (1972) Active tectonics of the Mediterranean region. *Geophys. J. R. Astron. Soc.*, 30: 109.

McMahon, H. (1897) The southern borderlands of Afghanistan. *Geogr. J.* 9: 393.

McMahon, C. & McMahon, H. (1897) Notes on some volcanic and other rocks which occur near the Baluchistan-Afghan frontier, between Chaman and Persia. *Q. J. Geol. Soc.*, 53 (211).

McQuillan, H. (1973) A geological note on the Qir earthquake S.W. Iran, April 1972. *Geol. Mag.*, 110 (3): 243.

Melville, C. (1978) Arabic and Persian source material on the historical seismicity of Iran from the 7th to the 17th centuries A.D. Ph.D. Thesis, Cambridge.

Melville, C. (1980) Earthquakes in the history of Nishapur. *Iran*, 18: 103.

Melville, C. (1981) Historical monuments and earthquakes in Tabriz. *Iran*, 19: 159.

Merewether, W. (1852) A report of the disastrous consequences of the severe earthquake etc. *Bombay Geogr. Soc.*, 10: 284.

Meshkati, N. (1974) *List of the Historical Sites and Ancient Monuments of Iran*. Natl. Organ. Prot. Hist. Monuments, Publ. no. 5, Tehran.

Michaux, A. (1911) *Voyage en Syrie et en Perse 1782–5*. Ed. E.T. Hamy, Geneva.

Milne, J. (1911) Catalogue of Destructive Earthquakes. *BAAS*: 650.

Minorsky, V. (1930) Transcaucasica. *J. Asiat.*, 217: 41.

Minorsky, V. (1953) *Studies in Caucasian History*. London.

Minorsky, V. (1964) *Iranica*. Tehran.

Minster, J. & Jordan, T. (1978) Present-day plate motions. *J. Geophys. Res.*, 83: 5331.

Mirza Shirazi (1864) A brief account of the province of Fars. *Trans. Bombay Geogr. Soc.*, 17: 175.

Mitchell, R. (1958) Instability of the Mesopotamian Plains. *Bul. Soc. Geogr. Egypte*, 31: 127.

Mitford, E.L. (1884) *A Land-March from England to Ceylon Forty Years ago*. 2 vols. London.

Miyamura, S. (1976a) Historical development of global seismological observations. *Bul. Int. Inst. Seismol.*, 14: 21. Tokyo.

Miyamura, S. (1976b) Provisional magnitudes of middle American earthquakes not listed in the magnitude catalogue of Gutenberg–Richter. *Bul. Int. Inst. Seismol.*, 14: 41, Tokyo.

Miyamura, S. (1978) Magnitude of earthquakes. *Int. Inst. Seismol.*, Publ. no. 11. Tokyo.

Mohajer, G.A. (1964) The Qazvin earthquake. *Bul. Iran. Pet. Inst.*, 15: 428, Tehran (in Farsi).

Mohajer, G.A. & Pierce, G.R. (1963) Qazvin, Iran earthquake. *Bul. Am. Pet. Geol.*, 47 (10): 1878.

Mohajer-Ashjai, A. (1974) Strain and slip measurements along active faults in the Tehran region. *Geol. Surv. Iran*, Rep. no. 29: 139. Tehran.

Mohajer-Ashjai, A., Behzadi, H. & Berberian, M. (1975) Reflections on the rigidity of the Lut block and recent crustal deformation in eastern Iran. *Tectonophysics*, 25: 281.

Moinfar, A.A. (1969) Dasht-e-Bayaz and Ferdows earthquakes; an engineering study of the Khorassan–Iran earthquakes of August

31 and September 1st 1968. *Tech. Res. & Stand. Bur. Plan & Budget Org.*, Publ. no. 21, Tehran.

Moinfar, A.A. (1972) Preliminary study of Ghir–Fars–Iran earthquake of 10th April 1972. *Tech. Res. & Stand. Bur. Plan. & Budget Org.*, Publ. no. 10, Tehran (in Farsi).

Moinfar, A.A. (1975) The earthquake of Sarkhun, Bandar Abbas, Iran. *Tech. Res. & Stand. Bur. Plan & Budget Org.*, Publ. no. 46, Tehran

Moinfar, A.A., Banisadr, M. & Tabarsi, M. (1973) The Bandar Abbas earthquake of November 8 1971. *Tech. Res. & Stand. Bur. Plan & Budget Org.*, Publ. no. 13, Tehran (in Farsi with English summary).

Molnar, P. & Tapponnier, P. (1977) Relation of the tectonics of eastern China. *Geol.*, 5: 212.

Monier, Père (1723) Mémoire de la mission d'Erivan. In *Nouveaux Mémoires des Missions de la Compagnie de Jésus dans le Levant*, 3: 228, Paris.

Monteith, W. (1852) Notes sur la position de plusieurs anciennes villes situées dans les plaines d'Ararat et de Nakhtchevan et sur les bords de l'Araxe. *Nouv. Ann. Voyage*, 5e ser., 32: 129.

Monteith, W. (1857) Notes on the routes from Bushire to Shiraz. *J. R. Geogr. Soc.*, 27: 108.

Monti, V. (1907) Notizie sui terremoti osservati in Italia durante l'anno 1905. *Bul. Soc. Sism. Ital.*, 12: 211.

Morgan, J. de (1894) *Mission scientifique en Perse; I–II: études géographiques*. Paris.

Morier, J. (1812) *A Journey through Persia, Armenia, and Asia Minor to Constantinople in the Years 1808 and 1809*. London.

Morier, J. (1818) *A Second Journey through Persia, Armenia and Asia Minor . . . etc. between 1810 and 1816*. London.

Mounsey, A.H. (1872) *A Journey through the Caucasus and the Interior of Persia*. London.

Muraviev, N.N. (1871) *Journey to Khiva through Turcoman Country in 1819–20*. Calcutta.

Murray, C.A. (1859) On some mineral springs near Tehran. *Q. J. Geol. Soc. London*, 15: 198.

Musketoff, I.V. (1891) Materialy dlya izucheniya zemletriasenii Rossii. *Prilozh. Izv. Russ. Geogr. O-Va.*, 27, St Petersburg.

Musketoff, I.V. (1899) Materialy dlya izucheniya zemletriasenii Rossii. *Prilozh. Izv. Russ. Geogr. O-Va.*, 35, St Petersburg.

Musketoff, I.V. & Orloff, A. (1893) *Katalog zemletriesenii Rossiskoi Imperii*. St Petersburg.

Nabavi, M. (1970) Le seisme de Macou. *Publ. Inst. Geophys. Univ. Tehran*, no. 51.

Napier, G.C. (1876) *Collection of Journals and Reports on Special Duty in Persia*. London: HMSO.

Nau, F. (1911) Notices des manuscrits Syriaque, Ethiopens et Mandeens, entrés à la Bibliothèque Nationale de Paris depuis l'édition des catalogues. *Rev. de l'Orient Chretien*, 2nd Series, 6.

Negahban, E. (1971) article in *Kayhan International*, 13th March.

Neghabat, F. & Liu, S. (1977) Earthquake regionalisation of Iran. *Proc. 6th World Conf. Earthq. Eng.*, New Delhi, 2: 531.

Nejjar, S. (1974) *Kashf ac-calcala 'an wacf az-zalzala*. Cahiers Centre Univ. de la Rech. Sci., 3. Rabat.

Newberie, J. (1905) Two voyages of master John Newberie, in *Hakluytus Posthumus or Purchas his Pilgrimes*, 8: 452, Glasgow.

Newmark, N. & Rosenblueth, E. (1971) *Fundamentals of Earthquake Engineering*. New Jersey.

Niazi, M. (1968) Fault rupture in the Iranian Dasht-e-bayaz earthquake of August 1968. *Nat.*, 220: 569.

Niazi, M. (1969) Source dynamics of the Dasht-e Bayaz earthquake of August 31, 1968. *Bul. Seismol. Soc. Am.*, 59: 1843.

Niazi, M. (1973) Focal mechanism of Ghir earthquake of April 10, 1972, and tectonics of southern Iran. *68th Annu. Natl. Meet. Seismol. Soc. Am.*, Golden, Colorado.

Niazi, M. (1977) Premonitory potential of short period surface waves. *Proc. CENTO Semin. Earthq. Hazard Minimization*, Tehran. *Tech. Res. & Stand. Bur. Plan & Budget Org.*, Publ. no. 70: 499.

Niazi, M. & Taheri, S. (1981) Seismicity of the Iranian Plateau and bordering regions. *Bul. Seismol. Soc. Am.* (forthcoming).

Niazi, M., Asudeh, I., Ballard, G., Jackson, J., King, G. & McKenzie, D. (1978) The depth of seismicity in the Kermanshah region of the Zagros Mountains in Iran. *Earth Planet. Sci. Lett.*, 40: 270.

Niebuhr, C. (1776–80) *Voyage en Arabie et en d'autres pays circonvoisins*. 2 vols. Amsterdam.

Nikitin, A. (1857) The travels of Athanasius Nikitin of Twer, ed. R.M. Major, *Hakluyt Soc. Publ.*, no. 22.

North, R. (1972) A thrust event in southern Iran. *Lincoln Lab. Semi-Ann. Tech. Summ.* ESD-TR-72-354, Dec. 1972: 60.

North, R. (1973) Seismic source parameters. Ph.D. Thesis, Cambridge. *Nat.*, 252: 560.

Nowroozi, A.A. (1971) Seismo-tectonics of the Persian Plateau, Eastern Turkey, Caucasus and Hindu-Kush regions. *Bul. Seismol. Soc. Am.*, 61: 317.

Nowroozi, A.A. (1976) Seismotectonic provinces of Iran. *Bul. Seismol. Soc. Am.*, 66: 1249.

Nowroozi, A.A. (1979) Reply to M. Berberian on the comparison between instrumental and macroseismic epicentres. *Bul. Seismol. Soc. Am.*, 69: 641.

Oddone, E. (1907) Les tremblements de terre ressentis pendant l'annee 1904. *Publ. Bur. Cent. Seismol. Int.*, Ser. B, Strasbourg.

Oldham, Th. (1883) A catalogue of Indian earthquakes. *Mem. Geol. Surv. India*, 19: 163, Calcutta.

Oliveira, C. (1980) Seismic hazard analysis for zones of intermediate seismicity. *Proc. 7th World Conf. Earthq. Eng.*, Istanbul, 1: 269.

Olivier, G.A. (1807) *Voyage dans l'empire Othoman, l'Egypte et la Perse.* 3 vols., Paris.

Olson, R.W. (1975) *The Siege of Mosul and Ottoman–Persian relations, 1718–1743.* Indiana.

Omote, S., Kobayshi, H., Nakagawa, K., Kawabata, S. & Nakaoka, E. (1962) A preliminary report on the Qazvin earthquake of September 1, 1962. Unpubl. Rep., Jpn. Mission for Tech. Assist., Tehran.

Orsolle, E. (1885) *Le Caucase et la Perse.* Paris.

Otter, J. (1743) *Voyage en Turquie et en Perse.* 2 vols. Paris.

Ouseley, W. (1819) *Travels in Various Countries of the East more particularly Persia in 1810, 1811 and 1812.* 3 vols. (vol. 2: 1821, vol. 3: 1823) London.

Page, W., Alt, J., Cluff, Ll. & Plafker, G. (1979) Evidence for the recurrence of large-magnitude earthquakes along the Makran coast of Iran and Pakistan. *Tectonophysics*, 52: 536.

Pagirev, D. (1909) Gibel' Akhori 20 iyun 1840 g. *Izv. Kavkaz. Otd. Russ. Geogr. O-Va.*, 19.

Pakdaman, K. (1968) Technical information on the earthquakes of August 31st and September 1st 1968 in the East Iran–South Khorassan. Unpubl. Rep. Minist. Housing, Tehran.

Parrot, F. (1845) *Journey to Ararat.* London.

Patterson, J. (1908) Seismic disturbances, Simla Observatory. *Mon. Weather Rev.*, July, Calcutta.

Pelly, L. (1863a) Report of the acting political resident, Persian Gulf. *Trans. Bombay Geogr. Soc.*, 17: 20.

Pelly, L. (1863b) Remarks on the tribes, trade and resources around the shore line of the Persian Gulf. *Trans. Bombay Geogr. Soc.*, 17: 32.

Pelly, L. (1863c) Recent tour round the northern portion of the Persian Gulf. *Trans. Bombay Geogr. Soc.*, 17: 113.

Pelly, L. (1864) Remarks on a recent journey from Bushire to Shiraz. *Trans. Bombay Geogr. Soc.*, 17: 141.

Pendse, C.G. (1946) A short note on the Makran earthquake of 28th November 1945. *J. Sci. & Ind. Res. India*, 5 (3): 106.

Penton, E. (1902) A journey from Quetta to Meshed via the Nushki-Sistan trade route. *Geogr. J.*, 20: 80.

Peronaci, F. (1958) Sismicita dell'Iran. *Ann. di Geofis.*, 11: 55.

Peronaci, F. (1959) Contributo alla conoscenza delle caratteristiche sismiche dell'Iran settentrionale. *Ann. di Geofis.*, 12: 523.

Perrey, A. (1845) Liste de tremblements de terre ressentis en Europe et dans les parties adjacentes de l'Afrique et de l'Asie pendant les années 1843 et 1844. *C. R. Acad. Sci.* 20: 1444.

Perrey, A. (1850) Sur les tremblements de terre ressentis dans la pénisule Turco-Hellenique et en Syrie. *Mem. Acad. R. Sci. Belg.*, 23, Brussells.

Perrey, A. (1854) Note sur les tremblements de terre en 1853. *Bul. Acad. R. Sci. Bruxelles*, 21 (1): 457.

Perrey, A. (1860) Note sur les tremblements de terre en 1857 avec suppléments pour les années antérieurs 1842– . *Mem. Cour. et autres Mem. Acad. R. Belg.*, 10 (4): 3.

Perrey, A. (1862) Note sur les tremblements de terre en 1858, avec suppléments pour les années antérieures 1843– . *Mem. Cour. Bruxelles*, 12 (4): 3.

Perrey, A. (1864) Note sur les tremblements de terre en 1862 avec supplément pour les années antérieurs. *Mem. Cour. Bruxelles*, 16 (5–6): 1.

Perrey, A. (1865) Note sur les tremblements de terre en 1863 avec

suppléments pour les années 1843–1862. *Mem. Cour. Bruxelles*, 17 (5): 1.

Perrey, A. (1866) Note sur les tremblements de terre en 1864 avec suppléments pour les années 1843–1863. *Mem. Cour. Bruxelles*, 18 (4): 1.

Perrey, A. (1867) Note sur les tremblements de terre en 1865 avec suppléments pour les années antérieurs de 1843–1861. *Mem. Cour. Bruxelles*, 19 (3): 1.

Perrey, A. (1872a) Note sur les tremblements de terre en 1868 avec suppléments pour les années antérieurs de 1843–1867. *Mem. Cour. Bruxelles*, 22 (3): 1.

Perrey, A. (1872b) Note sur les tremblements de terre en 1869 avec suppléments pour les années 1843–1868. *Mem. Cour. Bruxelles*, 22 (4): 1.

Perrey, A. (1875) Note sur les tremblements de terre en 1871 avec suppléments pour les années antérieurs de 1843–1870. *Mem. Cour. Bruxelles*, 24 (4): 1.

Perry, J.R. (1979) *Karim Khan Zand; a History of Iran 1747–1779.* Chicago.

Petermann, H. (1861) *Reisen in Orient.* 2 vols. Leipzig.

Petrescu, G. & Purcaru, G. (1964) The mechanism and stress pattern at the focus of the September 1, 1962, Buyin-Zara (Iran) earthquake. *Ann. de Geophys.*, 20 (3): 1.

Petroff, M.P. (1955) *Bibliografiya po geografii Irana.* Izd. Akad. Nauk Turk. SSR, Ashkhabad.

Petrusevich, N.G. (1880) Iugo-vostochnoi pribrezhie Kaspiiskago moria i dorogi ot' nego v' merv'. *Izv. Kavkaz. Otd. Russ. Geogr. O-Va.*, 11: 124, Tiflis.

Philippe de la très Sainte Trinité (1669) *Voyage d'Orient.* Lyons.

Pinar, N. & Lahn, E. (1952) Turiye depremleri izahli katalogu. *Bayindirlik Bakanligi, Yapi ve Imar Isleri Reisligi*, 36 (6): 153, Ankara.

Plan Organisation (1968–9) *Proposals for the Re-development of the Khurasan Earthquake Regions.* Study Rep. Publ., 7 vols., Tehran 1347–8 (in Farsi).

Planhol, X. de (1968) Geography of settlement, in *Camb. Hist. Iran*, 1: 409.

Pontevès de Sabran, J. de (1890) *Notes de Voyage d'un hussard, un raid en Asie.* Paris.

Pope, A.U. & Ackerman, P. (1939) *A Survey of Persian Art.* Oxford.

Poppe, B., Naab, D. & Derr, J. (1978) Seismograph station codes and characteristics. *US Geol. Surv.*, Circ. No. 791, Arlington, Va.

Porter, R. Ker (1821) *Travels in Georgia, Persia, Armenia, Ancient Babylonia during the Years 1817 to 1820.* 2 vols. London.

Poser, Heinrich von (1675) *Tagebuch; Reise von Konstantinopel aus durch die Bulgaren, Armenien, Persien und Indien.* Jena.

Poullet, Sieur du (1668) *Nouvelles relations du Levant.* 2 vols. Paris.

Preece, J.R. (1879) Telegraphs in Persia. *J. Soc. Telegr. Eng.*, 8 (29): 403.

Price, W. (1832) *Journal of the British Embassy to Persia.* 2 vols. London.

Quittmeyer, R.C. & Jacob, K.H. (1979) Historical and modern seismicity of Pakistan, Afghanistan, Northwestern India, and Southeast Iran. *Bul. Seismol. Soc. Am.*, 69: 773.

Rabino, H.L. (1913) Liste des journaux de Perse et les journaux publiés hors de Perse en langue Persanes. *Rev. du Monde Musulman*, 22: 289.

Rabino, H.L. (1917) Les provinces Caspiennes de la Perse; le Guilan. *Rev. du Monde Musulman*, 32.

Rabino, H.L. (1928) *Mazanderan and Asterabad.* London.

Rabino, H.L. (1936) Les dynasties du Mazandaran. *J. Asiat.*, 228: 397.

Rabino, H.L. (1946) *Great Britain and Iran: Diplomatic and Consular Officers.* London.

Radde, G. (1898) Transkaspien und Nord-Chorassan. *Petermanns Mitt.*, Ergänzungsheft no. 126: 194.

Rawlinson, H.C. (1840) Notes on a journey from Tabriz, through Persian Kurdistan to the ruins of Takhti-Soleiman, etc. *J. R. Geogr. Soc.*, 10: 1.

Razani, R. (1974) An engineering investigation of the Qir earthquake of 10 April 1972 in Southern Iran. *Iran. J. Sci. Tech.*, 3 (1): 1.

Razani, R. & Lee, K.L. (1973a) *The Engineering Aspects of the Qir Earthquake of 10 April 1972 in Southern Iran.* Nat. Acad. Sci., Washington.

Razani, R. & Lee, K. (1973b) An engineering study of the Qir earthquake in Southern Iran, April 10, 1972. *Proc. 5th World Conf. Earthq. Eng.*, Rome, Paper no. 46.

Rebeur-Paschwitz, E. (1895) Horizontalpendel-Beobachtungen auf der

Kaiserlichen Universitäts-Sternwarte zu Strassburg 1892–1894. *Neitr. zur Geophys.*, 2: 211.

Reinemund, J.A. (1968) Preliminary appraisal of the need for geologic investigations of the Khorassan earthquake, Iran. UN Dep. Inter., *Geol. Surv. Proj. Rep. IR-2.*

Reinegg, M. (1796) *Beschreibung Kaukasus.* 2 vols., St Petersburg.

Reitlinger, G. (1932) *A Tower of Skulls.* London.

Rich, C.J. (1836) *Narrative of a Residence in Koordistan and on the Site of Ancient Nineveh; with Journal of a Voyage down the Tigris to Baghdad; and an Account of a Visit to Shirauz and Persepolis.* 2 vols. London. See summary in *J. R. Geogr. Soc.*, 6: 351.

Richard, Abbé (1771) *Histoire naturelle de l'air et des météores.* Paris.

Richards, F. (1931) *A Persian Journey.* London.

Richter, C. (1935) An instrumental magnitude scale. *Bul. Seismol. Soc. Am.*, 25: 1.

Richter, C. (1958) *Elementary Seismology.* Freeman.

Richter, V.G. (1961) Vertical movements of the earth's crust and fluctuation in the level of the Caspian. *Geogr. v Shkole*, 2: 16. Moscow.

Ritter, K. (1840) *Die Erdkunde von Asien.* 12 vols. Berlin (vols. 8 and 9).

Rivandeneyra, D.A. (1880) *Viaje al interior de Persia.* 3 vols. Madrid.

Rochechouart, Comte J. de (1867) *Souvenirs d'un voyage en Perse.* Paris.

Rodler, A. (1888) Geologische Expedition in das Bachtyaren-Gebirge im westlichen Persien. *Anz. d. k. Akad. Wiss. Math. Naturwiss. Cl.*, 25: 199. Vienna.

Rodler, A. (1890) Bericht über eine geologische Reise im westlichen Persien. *Sitzungsber. Kais. Akad. Wiss. Math. Naturwiss. Cl.*, 98: 28. Vienna.

Ronaldshay, Earl of (1902) A journey from Quetta to Meshed via the Nushki-Sistan trade route. *Geogr. J.*, 20: 70.

Roscher, W.H. (1909) Die Tessarakontaden und Tessarakontadenlehren der Griechen und anderer Völker. *Ber. der Philologisch-historischen Kl. der Königlich Sächsischen Ges. der Wiss. zu Leipzig*, 61 (2).

Rothé, J.P. (1946) Le séisme du 27 novembre 1945, et l'hypothèse de Suess sur la cause du déluge. *C. R. Acad. Sci.*, 222: 301.

Rothé, J.P. (1949) Chronique seismologique. *Rev. Etud. Calamités*, 11: 11.

Rothé, J.P. (1959) Chronique seismologique. *Rev. Etud. Calamités*, 36: 16.

Rothé, J.P. (1969) *The Seismicity of the Earth 1953–1965.* UNESCO Earth Sci. Ser., Paris.

Rowshandel, B., Nemat-Nasser, S. & Adeli, H. (1979) A tentative study of seismic risk in Iran. *Iran. J. Sci. Technol.*, 7: 211.

Rudolph, E. (1903) Die Fernbeben des Jahres 1897. *Beitr. zur Geophys.*, 5: 1.

Rustanovich, D.N. (1967) Seismichnost' territorii Turkmenskoy CCP i Ashkhabadskoe zemletriasenie 1948 g. *Vaproc. Inz. Seismol. Byul.*, 12, Akad. Nauk. Moscow.

Ruttner, A., Nabavi, M. & Haijan, J. (1968) Geology of the Shirgesht area – Tabas area – East Iran. *Geol. Surv. Iran*, Rep. no. 4.

Rycaut, P. (1680) *The History of the Turkish Empire from the Year 1623 to the Year 1677.* London.

Saint John, O.B. (1876) On the physical geography of Persia and narrative of a journey through Baluchistan and southern Persia, in F.J. Goldsmid *Eastern Persia*, vol. 1, London.

Saint-Martin, J. (1818) *Mémoires historiques et géographiques sur l'Arménie.* vol. 1. Paris.

Sanjian, A.K. (1969) *Colophons of Armenian manuscripts, 1301–1480.* Harvard Univ. Press.

Saraby, F. & Foroughi, B. (1962) A survey of the Qazvin earthquake of September 1962. *Geol. Surv. Iran*, Intern Rep.

Sarre, F. (1899) Reise von Ardebil nach Zendschan im nord-westlichen Persien. *Petermanns Mitt.*, 45: 215.

Savage, C.D.W. (1957a) Earthquake of 2nd July 1957. Unpubl. Rep. B. 45 Ref. 33/4/86/252. Sir Alexander Gibb, London.

Savage, C.D.W. (1957b) Notes on a fast safari up the Heraz River, August 9th–10th 1957. Unpubl. Rep. B. 58 Ref. 3374/81/272. Sir Alexander Gibb, London.

Savage, C.D.W. (1957c) Notes on a visit to the upper parts of the Kaselyan Basins. Unpubl. Rep. B. 59, Ref. 3374/81/273. Sir Alexander Gibb, London.

Sawyer, H.A. (1891) *Report of a Reconnaissance in the Bakhtiari country, South-west Persia.* Simla.

Schefer, C. (1881) *Sefer nameh; relation du voyage de Nassiri Khosrau.* Translation and Appendices. Paris.

Schmidt, J.J.F. (1879) *Studien über die Erdbeben.* Leipzig.

Schwarz, P. (1910–36) *Iran im Mittelalter nach den arabischen Geographen.* 9 vols. Leipzig, Stuttgart, Berlin.

Sercey, Cte de (1928) *Une ambassade extraordinaire. La Perse en 1839–40.* Paris.

Serena, Carla (1883) *Hommes et choses en Perse.* Paris.

Seyfart, J.F. (1756) *Allgemeine Geschichte der Erdbeben.* Frankfurt and Leipzig.

Seymour, J. (1951) *The Hard Way to India.* London.

Sharp, R. & Orsini, N. (1978) Preliminary Report of the Tabas, Iran, earthquake of September 16, 1978. *Earthq. Eng. Res. Inst., Newsletter*, 12 (6).

Shebalin, N.V. (1974) Ochagi silnjikh zemletriasenii na territorii CCCP, *Izd. Nauk.* Moscow.

Shih, C., Haun, W., Tsao, H., Wu, H., Liu, Y. & Huang, W. (1974) Some features of Chinese seismic activity. *Acta Geophys. Sin.*, 17: 1 (in Chinese).

Sieberg, A. (1917) Catalogue régional des tremblements de terre ressentis pendant l'année 1908. *Publ. Bur. Cent. Seismol. Int.* Ser. B, Strasbourg.

Sieberg, A. (1932) Erdbebengeographie. In B. Gutenberg, *Handbuch der Geophysik*, 4, Berlin.

Siroux, M. (1949) Caravanserails d'Iran et petites constructions routières. *Mem. Inst. Fr. Archeol. Orient. du Caire*, 81: 1.

Slemmons, D. (1957) Geological effects of the Dixie Valley Nevada, earthquakes of December 16, 1954. *Bul. Seismol. Soc. Am.*, 47: 353.

Smith, A. (1953) *Blind White Fish in Persia.* London.

Smith, E. (1876) The Perso-Baluch frontier mission 1870, 1871, and the Perso-Afghan mission 1871, 1872. In F. Goldsmid, *Eastern Persia*, vol. 1, London.

Smith, E. & Dwight, H. (1834) *Missionary Researches in Armenia.* London.

Smith, M.B. (1935) Masdjid-i Djum'a, Demawend. *Ars Islamica*, 2: 153.

Smith, N. (1971) *A History of Dams.* London.

Sobouti, M. (1969) Le séisme de Dasht-e Bayaz dans la province de Khorassan, Iran. *Ann. di Geofis.*, 22 (3): 230.

Sobouti, M. (1972) A statistical study of the amplitude distribution of aftershocks of two great earthquakes of Iran; Buyin Zara–Dasht-e Bayaz. *J. Earth Space Phys. Inst. Geophys. Tehran Univ.*, 1 (1): 34.

Sobouti, M. & Eshghi, I. (1970) Le séisme de Moraveh-Tappeh dans le Turqumin-Sahara de l'Iran. *Publ. Inst. Geophys. Univ. Tehran*, no. 52.

Sobouti, M., Eshghi, I. & Hossein-Javaheri, J. (1972) The Qir earthquake of 10th May 1972. *J. Earth Space Phys. Inst. Geophys. Tehran Univ.*, 1 (2): 17.

Soder, P.A. (1959) Geology of the area north of Saveh. Nat. Iran. Co. Unpubl. Rep. G.R. no. 174.

Soloviev, S. (1955) Klassifikatsi zemletryasenii po velichne ikh energii. *Tr. Geofiz. Inst. Akad. Nauk*, 30, Moscow.

Sondhi, V.P. (1947) The Makran earthquake, 28th November 1945, the birth of new islands. *Indian Miner.*, 1 (3): 146.

Southgate, H. (1840) *Narrative of a Tour through Armenia, Kurdistan, Persia and Mesopotamia.* 2 vols. London.

Sponheuer, W. (1960) Methoden zur Herdtiefenbestimmung in der Makroseismik. *Freiberger Forschungsh. C.* 88, Berlin.

Stack, E. (1882) *Six Months in Persia.* 2 vols. London.

Stahl, A.F. (1911) *Handbuch der regionalen Geologie; Persien.* vol. 5, part v.6; Heidelberg.

Stahl, P. (1960) Enquête macroséismique sur les tremblements de terre à Lar des 24 avril et 3 mai 1960. Unpubl. Rep.

Stahl, P. (1962a) Sur la séismicité de l'Iran. *Bul. Etud.* no. 3, Stn. Seismol. de Shiraz.

Stahl, P. (1962b) Séisme destructeur au sud de Kazvin. *Bul. Seismol. Provis.* Annex Sept–Oct., Shiraz.

Steensgaard, N. (1974) *The Asian Trade Revolution of the 17th Century.* Chicago.

Stein, A. (1937) *Archaeological Reconnaissances in North-Western India and South-Eastern Iran.* London.

Stein, A. (1940) *Old Routes of Western Iran.* London.

Steinbrugge, K. & Cloud, W. (1962) Epicentral intensities and damage in the Hebgen Lake, Montana, earthquake of August 17, 1959. *Bul. Seismol. Soc. Am.*, 52: 181.

Steinbrugge, K. & Moran, D. (1957) Engineering aspects of the Dixie Valley—Fairview Peak earthquakes. *Bul. Seismol. Soc. Am.*, 47: 335.

Stepanian, V.A. (1942) Istoricheskii obzor o zemletriasenikh v Armenii i v prilegaiuschikh rayonakh. *Izd. Arm. FAN, Akad. Nauk*, Erivan.

Stevens, J. (1715) *The History of Persia*. London.

Stewart, C.E. (1881) The country of the Tekke Turkomans and the Tejend and Murghab Rivers. *Proc. R. Geogr. Soc.*, 3: 513.

Stewart, C.E. (1911) *Through Persia in Disguise*. London.

Stöcklin, J. & Nabavi, M. (1973) Tectonic map of Iran (1:2,500,000). *Geol. Surv. Iran*.

Stöcklin, J., Eftekhar-nezhad, J. & Hushmand-zadeh, A. (1965) Geology of the Shotori Range. *Geol. Surv. Iran*, Rep. no. 3.

Stöcklin, J., Eftekhar-nezhad, J. & Hushmand-zadeh, A. (1972) Central Lut reconnaissance; East Iran (& map 1:500,000). *Geol. Surv. Iran*, Rep. no. 22.

Stolze, F. & Andreas, F. (1882) *Persepolis*. Berlin.

Storey, C.A. (1970—2) *Persian Literature*. Reprinted in 2 vols. London.

Stratil-Sauer, G. (1937) Eine Route im Gebiet des Kuh-o-Hazar (Südiran). *Petermann's Mitt.*, 83: 310.

Stratil-Sauer, G. (1950) Birdjand, eine ostpersische Stadt. *Mitt. Geogr. Ges. Wien*, 92 (4—6): 106.

Strauss, Th. (1911) Eine Reise im westlichen Persien. *Petermann's Mitt.*, 57 (1): 65.

Struys, J. (1724) *Les voyages de Jean Struys en Moscovie, Tartarie, en Perse, aux Indes etc.* 3 vols. Rouen.

Stupin (1864) Soobshchenie o zemletryaseni v Ardabilskoe stranakh. *Zap. Kavkaz*, Feb. 1864: 23, Tiflis.

Stürken, A. (1906) Reisebriefe aus dem Persischen Golf und Persien. *Mitt. Geogr. Ges. Hamburg*, 21: 69.

Sykes, P.M. (1897) *Recent Journeys in Persia*. *Geogrl. J.*, 10 (6): 568.

Sykes, P.M. (1902a) *Ten Thousand Miles in Persia*. London.

Sykes, P.M. (1902b) A fourth journey in Persia, 1897—1901. *Geogrl. J.*, 19 (2): 121.

Sykes, P.M. (1910) Historical notes on Khurasan. *J. R. Asiat. Soc.*: 1113.

Sykes, P.M. (1911) A sixth journey in Persia. *Geogrl. J.*, 37: 1, 149.

Taher, M.A. (1979) Corpus des textes arabes relatifs aux tremblements de terre et autres catastrophes naturelles de la conquête Arabe au xii/xviii J.C. Thèse de Doctorat d'Etat, Univ. Paris. 2 vols. in 1 (Fr. and Arabic).

Tamaki, I. (1963) A warning against the use of formula $\log N = a + b$ $(8-M)$, in Geophys. Inst. Kyoto Univ., *Geophys. Papers dedicated to K. Sassa*: 555.

Tams, E. (1908a) Geographische Verbreitung und erdwissenschaftliche Bedeutung der aus den Erdbebenbeobachtungen des Jahres 1903 sich ergebenden Epizentren. *Beitr. zur Geophys.*, 9: 237.

Tams, E. (1908b) Die mikroseismischen Registrierungen einiger Beben des Jahres 1903. *Beitr. zur Geophys.*, 9: 509.

Tancoigne, J.M. (1820) *A Narrative of a Journey into Persia and Residence at Tehran*. London.

Tasios, Th. (1969) An engineering report of the UNESCO reconnaissance mission on Khorassan earthquake of 1968. Unpubl. Rep., Nat. Tech. Univ. Athens.

Tate, G.P. (1910—12) *Seistan; a Memoir on the History Topography, Ruins and People of the Country*. 2 vols., Calcutta.

Tavernier, J.B. (1681) *Les six voyages de Jean-Baptiste Tavernier, qu'il a fait en Turquie, en Perse, et aux Indes*. 3 vols. Paris.

Tchalenko, J.S. (1970) Similarities between shear zones of different magnitudes. *Bul. Seismol. Soc. Am.*, 81: 1625.

Tchalenko, J.S. (1973) Recent destructive earthquakes in the Central Alborz. *Ann. di Geofis.*, 26: 303.

Tchalenko, J.S. (1974) Recent destructive earthquakes in the Central Alborz. *Geol. Surv. Iran*, Rep. no. 29: 97.

Tchalenko, J.S. (1975) Seismicity and structure of the Kopet Dagh (Iran, USSR). *Philos. Trans. R. Soc. London*, 278: 1.

Tchalenko, J.S. (1977) A reconnaissance of the seismicity and tectonics at the northern border of the Arabian plate; Lake Van region. *Rev. de Geogr. Phys. Geol. Dyn.*, 19, fasc. 2: 189, Paris.

Tchalenko, J.S. & Ambraseys, N. (1970) Structural analysis of the Dasht-e Bayaz (Iran) earthquake fractures. *Bul. Geol. Soc. Am.*, 81: 41.

Tchalenko, J.S. & Ambraseys, N. (1973) Earthquake destruction of adobe villages in Iran. *Ann. di Geofis.*, 26: 357.

Tchalenko, J.S. & Berberian, M. (1974) The Salmas (Iran) earthquake of May 6th, 1930. *Ann. di Geofis.*, 27: 151.

Tchalenko, J.S. & Berberian, M. (1975) Dasht-e Bayaz fault, Iran: earthquake and earlier related structures in bedrock. *Bul. Geol. Soc. Am.*, 86: 703.

Tchalenko, J.S. & Braud, J. (1974) Seismicity and structure of the Zagros (Iran): The Main Recent Fault between 33° and 35°N. *Philos. Trans. R. Soc. London*, 277: 1.

Tchalenko, J.S., Berberian, M. & Behzadi, H. (1973) Geomorphic and seismic evidence for recent activity on the Doruneh fault (Iran). *Tectonophysics*, 19: 333.

Tchalenko, J.S., Braud, J. & Berberian, M. (1974a) Discovery of three earthquake faults in Iran. *Nat.*, 248: 661.

Tchalenko, J.S., Berberian, M., Iranmanesh, H., Bailly, M. & Arsovski, M. (1974b) Tectonic framework of the Tehran region. *Geol. Surv. Iran*, Rep. no. 29: 7.

Teixeira, P. (1902) *The Travels of Pedro Teixeira*, tr. W.F. Sinclair, *Hakluyt Soc. Publ.*, NS. no. 9.

Teufel, H.Ch. (1898) *Die Reise des Hans Christoph Freiherrn von Teufel in das Morgenland 1588—1590*. Linz.

Texier, C.F. (1839—52) *Déscription de l'Arménie, la Perse et la Mésopotamie*. 2 vols. Paris.

Theatrum Europaeum (1617—1721), ed. J.B. Abelin, 21 vols. Frankfurt.

Thevenot, J. (1687) *The Travels of M. Thevenot*. Eng. trans. L. Lovell, London.

Thielmann, M. von (1875) *Journey in the Caucasus, Persia, and Turkey in Asia*. 2 vols. London and Leipzig.

Tholozan, J. (1879) Sur les tremblements de terre qui on eu lieu en Orient du 7e au 17e siecle. *C. R. Acad. Sci.*, 88: 1063.

Thompson, R.C. (1937) A new record of an Assyrian earthquake. *Iraq*, 4: 186.

Tipper, G.H. (1921) The geology and mineral resources of Eastern Persia. *Rec. Geol. Surv. India*, 53 (1): 51.

Toksöz, N., Arpat, E., & Saroglu, F. (1977) East Anatolian earthquake of 24 November 1976. *Nat.*, 270: 423.

Tolstov, S. (1960) (ed.) Nizoviya Amu-Daryi, Sarikamish, Uzboi. *Mater. Khorezmskoi Eksped.*, no. 3: 267. Izd. vo Akad. Nauk, Moscow.

Toppozada, T. (1975) Earthquake magnitude as a function of intensity data in California and Western Nevada. *Bul. Seismol. Soc. Am.*, 65: 1231.

Trezel Col. (1821) Notice sur le Gilan et le Mazanderan, in Jaubert, *Voyage en Arménie*, Paris.

Truilhier, M. (1838) Mémoire descriptif de la route de Tehran à Meched et de Meched à Jezd reconnue en 1807. *Bul. Soc. Geogr. Paris.*, 9: 109, 249, 313; 10: 5.

Tsimbalenko, L.I. (1893) Uzasnoie zemletriasenie v gorod Kochane. *Novoie Vremia* no. 6398, December 1893. St Petersburg. See also Persian trans. in *Vahid*, 12 (March 1975): 680.

Tsimbalenko, L.I. (1899) Kuchanskoie zemletriasenie 5-vo noiabria 1893 goda. *Prilozh. Izv. Russ. Geogr. O-Va.*, 35: 11, St Petersburg.

Tuson, P. (1979) *The Records of the British Residency and Agencies: the Persian Gulf*. India Office Lib., London.

Ussher, J. (1865) *A Journey from London to Persepolis*. London.

Valle, Pietro della (1677) *Viaggi di Pietro della Valle il pellegrino*. Bologna.

Vambery, A. (1973) *Arminius Vambery, His Life and Adventures*. New York.

Varthema, Ludovico di (1863) *The Travels of Ludovico di Varthema*, ed. G. Badger, *Hakluyt Soc. Publ.*, no. 32: 94.

Vaughan, H.B. (1890) *Report of a Journey through Persia*. Calcutta.

Vere-Jones, D. (1978) Earthquake prediction — a statistician's view. *J. Phys. Earth*, 26: 129, Tokyo.

Villotte, Père J. (1730) *Voyages d'un missionaire de la Compagnie de Jesus en Turquie, en Perse etc.* Paris.

Vladimirov, L. (1972) Glavit. *Index*, 1: 31, London.

Vogt, J. (1979) Les tremblements de terre en France. *Mem. Bur. Rech. Geol. Minières*, no. 96, Orléans.

Voskoboinikoff (1841) Mount Ararat. *The Athenaeum*, no. 695: 157.

Voskoboinikoff (1847) Eine Reise durch das nördliche Persien. *Arch. Wiss. Kd. Russl.*, 5: 674, Berlin.

Vrolyk, F. (1957a) Mission de protection civile en Iran. Unpubl. Rep., Gouv. Gen. de l'Algérie.

Vrolyk, F. (1957*b*) Le tremblement de terre de 1957 en Iran. Letter to *Inst. Phys. du Globe.*

Wadati, K. (1931) Shallow and deep earthquakes. *Geophys. Mag.*, 4: 231.

Wagner, M. (1848) Reise nach dem Ararat und dem Hochland Armenien. In *Reisen und Landesbeschreibungen* no. 35, Stuttgart.

Wagner, M. (1856) *Travels in Persia, Georgia and Koordistan.* 3 vols. London.

Walther, B.S. (1805) *Die Erdbeben und Vulkane, physisch und historisch betrachtet.* Leipzig.

Walton, H.I. (1865) Correspondence. *Trans. Bombay Geogr. Soc.*, 17: cxxv.

Waterman, L. (1929–31) *Royal Correspondence of the Assyrian Empire.* 4 vols. Univ. Michigan.

Watson, R.G. (1866) *A History of Persia from the Beginning of the 19th Century to the Year 1858.* London.

Weidenbaum, E.G. (1884) Bolshoi Ararat i popitki voskhozhdenia na ego bershinu (Great Ararat and attempts to reach its summit). *Zap. Imp. Russ. Geogr. O-Va.*, 13: 103, Tiflis.

Whitehouse, D. (1968) Excavations at Siraf. *Iran*, 6: 1.

Whitelock, Lt (1838) Descriptive sketches of the islands and coast situated at the entrance of the Persian Gulf. *J. R. Geogr. Soc.*, 8: 170.

Wilbraham, R. (1839) *Travels in the Trans-Caucasian Provinces of Russia and along the Southern Shore of the Lakes Van and Urumiah.* London.

Wilkinson, C.K. (1937) The Iranian expedition, 1936. *Bul. Metrop. Mus. Art*, 32, Oct. 1937 (2): 3.

Wilkinson, C.K. (1975) *Nishapur: Pottery of the Early Islamic Period.* New York.

Wilkinson, J.C. (1964) A sketch of the historical geography of the Trucial Oman. *Geogr. J.*, 130: 337.

Wilkinson, J.C. (1969) Arab settlement in Oman. Oxford D.Phil. Thesis.

Willey, P. (1963) *The Castles of the Assassins.* London.

Willis, B. (1928) Earthquakes in the Holy Land. *Bul. Seismol. Soc. Am.*, 18: 77.

Willmore, P. & Kárník, V. (1971) Manuel pratique des observatoires séismologiques. *Publ. Int. Seismol. Centre*, Edinburgh.

Wills, C.J. (1894) *Behind an Eastern Veil.* London.

Wilson, A.T. (1930) Earthquakes in Persia. *Bul. SOAS*, 6 (1): 103.

Wilson, S.G. (1896) *Persian Life and Customs.* Edinburgh.

Windt, H. de (1891) *A Ride to India across Persia and Baluchistan.* London.

Witsen, N. (1694) Two draughts of the famous Persepolis. *Philos. Trans. R. Soc. London*, no. 210.

Wolff, J. (1845) *Narrative of a Mission to Bokheara.* 2 vols. London.

Wood, H. (1942) A chronological conspectus of seismological stations. *Bul. Seismol. Soc. Am.*, 32: 97.

Wright, D. (1977) *The English amongst the Persians during the Qajar period 1787–1921.* London.

Wu, F. & Ben-Menahem, A. (1965) Surface wave radiation pattern and source mechanism of the September 1, 1962, Iran earthquake. *J. Geophy. Res.*, 70 (16): 3943.

Wulff, H. (1966) *The Traditional Crafts of Persia.* Cambridge, Mass.

Yarham, E.R. (1942) The Trans-Persian railway. *Can. Geogr. J.*, 25: 148. (Also in *J. United Serv. Inst.*, 87: 46, London.)

Yate, C.E. (1888) *Northern Afghanistan or Letters from the Afghan Boundary Commission.* London.

Yate, C.E. (1900) *Khurasan and Sistan.* London and Edinburgh.

York, J., Cardwell, R. & Ni, J. (1976) Seismicity and quaternary faulting in China. *Bul. Seismol. Soc. Am.*, 66: 1983.

Young, T.C. (1968) Godin Tepe. *Iran*, 6: 160.

Zander, G. (1972) La réstauration de quelques monuments historiques d'Isfahan. *5th Int. Congr. Iran. Art and Archaeol.*, Tehran, 2: 246.

Zareh, K. (1963) The reconstruction of the village of Rudak. *Edareh Sakhteman Publ.*, University of Tehran.

Zarudnoi, N. (1901) Ekskursia po Vostochnoi Persii 1898. *Zap. Imp. Russ. Geogr. O-Va.*, 36 (1): 1. St Petersburg.

Zarudnoi, N. (1916) Treti ekskursiz po Vostochnoi Persii, Khorasan, Seistan i Persidski Beludzhistan 1900–1901. *Zap. Imp. Russ. Geogr. O-Va.*, 50: 1, St Petersburg.

Zugmayer, E. (1905) *Eine Reise durch Vorderasien im Jahre 1904.* Berlin.

Zugmayer, E. (1913) Balutschistan; vorläufige Ergebnisse einer Reise in Jahre 1911. *Mitt. Geogr. Ges. München*, 8: 40.

213

INDEX